REALISM AND ANTI-REALISM IN THE PHILOSOPHY OF SCIENCE

BOSTON STUDIES IN THE PHILOSOPHY OF SCIENCE

VOLUME 169

Tscha Hung with C. G. Hempel, Vienna 1982.

REALISM AND ANTI-REALISM IN THE PHILOSOPHY OF SCIENCE

Beijing International Conference, 1992

Edited by

ROBERT S. COHEN
Boston University

RISTO HILPINEN
University of Turku and University of Miami

and

QIU RENZONG
Institute of Philosophy, Chinese Academy of Social Sciences

KLUWER ACADEMIC PUBLISHERS
DORDRECHT / BOSTON / LONDON

Library of Congress Cataloging-in-Publication Data

Realism and anti-realism in the philosophy of science : Beijing International Conference, 1992 / edited by Robert S. Cohen, Risto Hilpinen, Qiu Renzong.
p. cm. -- (Boston studies in the philosophy of science ; v. 169)
"Beijing International Conference on Philosophy of Science"--Pref. p. 10.
Includes index.
ISBN 0-7923-3233-4
1. Realism--Congresses. 2. Science--Philosophy--Congresses. I. Cohen, R. S. (Robert Sonné) II. Hilpinen, Risto. III. Ch'iu, Jen-tsung. IV. Beijing International Conference on Philosophy of Science (1992) V. Series.
Q175.32.R42R43 1995
501--dc20 94-39752

ISBN 0-7923-3233-4

Published by Kluwer Academic Publishers,
P.O. Box 17, 3300 AA Dordrecht, The Netherlands.

Kluwer Academic Publishers incorporates
the publishing programmes of
D. Reidel, Martinus Nijhoff, Dr W. Junk and MTP Press.

Sold and distributed in the U.S.A. and Canada
by Kluwer Academic Publishers,
101 Philip Drive, Norwell, MA 02061, U.S.A.

In all other countries, sold and distributed
by Kluwer Academic Publishers Group,
P.O. Box 322, 3300 AH Dordrecht, The Netherlands.

Printed on acid-free paper

Printed in the Netherlands

TABLE OF CONTENTS

PHILOSOPHICAL ARGUMENTS

REALISM AND GENERAL METHODOLOGY

PHYSICAL SCIENCES

SOCIAL AND BIOLOGICAL SCIENCES

PREFACE

This book contains the main papers from the first Beijing International Conference on Philosophy of Science, held in 1992 and dedicated to issues of 'Realism and Anti-Realism in Science'. The Conference was organized by the Chinese Society for Dialectics of Nature (Philosophy of Nature, Science and Technology), and sponsored by the China Association for Science and Technology (CAST), the China International Conference Center for Science and Technology (ICCST), the Boston University Center for Philosophy and History of Science, the International Union of History and Philosophy of Science through its Division of Logic, Methodology and Philosophy of Science (IUHPS/DLMPS), and by Mr. Joseph C. T. Lee of Hong Kong. The Organizing Committee was chaired by Qiu Liang-hui, the Programme Committee by Robert S. Cohen, but we are especially mindful of the responsible role of our real 'Secretary-General' Qiu Renzong and his endlessly helpful colleague Ju Zhang.

The Conference was blessed by the late Professor Tscha Hung who had hoped to join with us, with Chinese and non-Chinese alike. He had for many years been Director of the Institute for Foreign Philosophy at Beijing University, and indeed for decades of thinking and teaching Tscha Hung was a link between the 'scientific philosophy' of his logical empiricist Vienna and the new time of China after the Second World War. Both the teacher and also the critic of his own mentors and colleagues, he was a scholar who quietly carried the roles of being the Ayer and Hempel of China.

Where might philosophy of science within China be located today? As in the West, modern Chinese philosophers deal with issues in addition to those of science, technology, medicine and the natural world, and they do so within various frameworks, some from traditional Chinese civilization, some from the currents of classical, modern and post-modern Western sources, others within a Chinese Marxist outlook. Philosophy of science however is especially needful of, and open to, international collaboration and mutual learning. Like the sciences, like mathematics, physics, biology, chemistry, computer science, modern logic, nuclear

R. S. Cohen, R. Hilpinen and Qiu Renzong (eds.), Realism and Anti-Realism in the Philosophy of Science, xi–xii.

engineering, like Western medicine and surgery, so philosophy of the sciences should be a field of international cooperation. This Conference was a contribution to the extension of such cooperation in philosophical clarification of the currently living concerns about science, among scientists as well as philosophers.

The renaissance of this endeavor in the People's Republic was evident at this conference, among younger and elder scholars from every part of China. The restrictions imposed by dogmatic habits or doctrines, so hurtful to the honest progress of science and certainly of its philosophical understanding, are hopefully behind our times (however difficult it may be to be freed from habit and its rituals). Even that contentious phrase from old Engels, adapted from Hegel, the 'dialectics of Nature', may be seen to be a suggestive problematic for scientific theorists, or perhaps as heuristic for philosophy of nature and of science-in-society. And within science, 'materialism' may be realist, objectivist, but not thereby opposed to theoretically conceived entities or forces or whatever the scientific imagination may conjure forth. So, 'realism', whether another 'neo-materialism' or not, is an epistemological issue, an ontological puzzle, and even a target for the pragmatic 'realist' of a different sort. Is there a new 'idealism' within the social constructivist interpretations of today, even within the cultural anti-science which has arisen? Our Conference dealt with such matters, perhaps without sufficiently vigorous exposition of the case for anti-realism but the reader will judge.

I am reminded of a contrast drawn by the British philosopher Winston Barnes many years ago. Barnes said that the idealist holds that "There is a Mind with a world in it", while the materialist holds that "there is a world with minds in it". Now we have many-world theories, and non-reductive emergent realisms, and non-ontological non-idealist anti-realisms. The issues were before this gathering of speakers and participants, and revised for this book. What would our friend Tscha Hung have thought?

Robert S. Cohen
(for the editors)

ROBERT S. COHEN

RECOLLECTIONS OF TSCHA HUNG

I first learned of Tscha Hung in 1955 during conversation with Marie Neurath. Ten years earlier, on the last day of Otto Neurath's life, Tscha Hung had come to Otto at his home in Oxford, to talk once more about protocol sentences and the Schlick–Carnap–Neurath debates. Who was this Chinese philosopher, I wondered, and why was he in Oxford, concerned about clarifying logical empiricism, in 1945? But Marie had no notes of Otto's conversations with Tscha Hung, only a pleasant memory of the visitor's sweet and inquisitive intelligence, his modesty, and his ease with both German and English.

During the first of my visits to China, in 1985, I was introduced to Tscha Hung by Fan Dainian; we had in fact exchanged letters before then, mainly about Vienna Circle studies. We talked at some length about developments in the philosophy of science since the Second World War, and to my delight his interests were very far-reaching. We met again in China in 1988, and also during the Schlick–Neurath centenary symposium in Vienna. Our talks were not systematic and the topics tumbled over each other. He knew of my attempt to bring Neurath into greater appreciations, both as critical philosopher and as practical social thinker. This led us not only to Neurath's confrontation with Marxist theory and Communist practice but to Tscha Hung's too. Then we discussed Neurath's optimistic feeling, in his last years, about the future influence of a rational and humane social liberalism in the British mode.

Quickly we turned to issues of science, and the roles of scientists in society. There was no doubt of Tscha Hung's concern about misuse of science and technology, but equally no doubt for him that scientific reasoning is the guide for human life. This was, he agreed, the realistic policy of political and economic forces in the modernization of feudal China, first in the Three People's Principles of the democratic revolution led by Sun Yat-sen before the First World War, and then in the developing Marxist socialisms after that war. Democracy and science were joined in social/perspective, often as a slogan along with other slogans for popular support.

To Tscha Hung, there was an evident task for the philosopher at this

R. S. Cohen, R. Hilpinen and Qiu Renzong (eds.), Realism and Anti-Realism in the Philosophy of Science, xiii–xv.

point, which was to offer clarity about science, and about what is not scientific, and to investigate central concepts of the sciences. For himself this task took two forms: his research, as in his dissertation on the concept of cause in contemporary physics, and in his careful attention to probability; and his teaching, together with his writing of textbooks and editing of translations from Western philosophers. He seemed pleased to be compared in his Chinese university environment with A.J. Ayer in Ayer's own 'Return from Vienna' to write *Language, Truth and Logic*.

Certainly his situation had been difficult, and his Stoic equanimity striking when I saw him. We talked sometimes about Needham's researches in the history of the sciences in ancient and classical China, for which Tscha Hung had great respect, and this led to his questions about Vienna Circle investigations of the social conditions of knowledge in general, and natural sciences and mathematics in particular. He had known Edgar Zilsel in Vienna days, and he was much intrigued by my attempt to sketch Zilsel's works of the 1940s on the historical sociology of science in the period of the scientific revolution and its precursors. All the more, he was eager to read Needham's use of Zilsel's historical theories for the Chinese cases.

Tscha Hung was much taken with all the work on external conditions of science, on what Carnap had called the pragmatics, on the several accounts of the 'historical turn' by Kuhn and others, by the analytic/synthetic debate around Quine, and by both the older and newer apparent relativisms typified by the image of 'Neurath's boat'. After his own adherence to Schlick's teaching in early studies, Hung continued to worry over the coherence/correspondence/formal conceptions of truth, and of knowledge generally. The issue of realism vs. anti-realism, in our international conference, was I suppose the vital issue for him. To his pleasure, this continued to be vital for others as well. With this went his happy recognition that a new generation of philosophers of science were investigating and interpreting the history of the Vienna Circle. He was pleased to accept membership on the Editorial Committee of the Vienna Circle Collection.

What then, I asked, was his own relation to Chinese thought, and to Chinese colleagues, whether scientists or philosophers? He had been a young student of Liang Qi-chao but left China at age 18 for university studies in Germany. His own philosophical formation was deeply within the anti-metaphysical logical empiricism. So, in China, he would be apt to try 'rational reconstruction' of traditional, or 'received', Chinese science and philosophy, to establish criteria for meaning, to identify what is testable, verifiable, probable, certain. But, like Schlick, he saw a role for the metaphysical, the non-scientific, in a different linguistic type. He liked my vague suggestion that perhaps metaphysical phrases and stories, although not strictly cognitive, may convey 'how the world feels'. He had spoken of metaphysics, at its best, as a 'conceptual poem'.

How, I wondered, did he see the other philosopher of his own time in China, one who also had completed a western doctorate in philosophy (Columbia University) well before the Second World War, and one whom I had an opportunity to meet in China. This was, of course, the innovative neo-Confucian Fung Yu-lan, also a surviver through the stages of 20th century China, also at Beijing University. In the early years, just before and after the founding of the People's Republic, they had debated, and engaged in written controversy. Fung brought his own new rational, even formal, reconstruction of the immense cumulative traditional thought, and cultural weight, of China. There was, for Tscha Hung, little understanding of the world-significant achievement of modern science in the ethical focus of Fung's work, and yet he urged me to respect Fung's historical and critical expositions of Chinese classical philosophers.

Moreover, when we talked about the dreams of Marxist humanists for a society within which people might be happy, and about Neurath's praise for Epicurus, Tscha Hung admitted, and lamented, the relative weakness of the Vienna Circle, indeed of the logical empiricists at large, in dealing with problems of ethics, of practical life. When I remonstrated, speaking of 'practical' writings of his beloved Schlick, of his admired Reichenbach, of his friend Feigl, of his hero Russell, he countered with the great example of Kant including moral issues within theoretical philosophy. Even Fung's neo-Confucian doctrine seemed to Hung in danger of losing the intuitive feeling of practical life within a new formalism. I wondered where Tscha Hung might have proceeded in emulating Kant's problematic, but he shook his head, saying he had lost those three decades in isolation, and it was too late now. He surely was no longer isolated in his last years, neither from Western friends nor from Chinese colleagues. He was, throughout his time, a teacher of integrity and a witness to a stage of the history of philosophy.

FAN DAINIAN

HONG QIAN AND THE VIENNA CIRCLE

Hong Qian (Tscha Hung, 1909–1992), a member of the Program Committee of this Conference, should have been here with us if the angel of death had not taken life away from him three and half months ago.

Hong Qian was born on October 12, 1909 in Fujian province but his native town was She county, Anhui province. During his youth, he was a student of the great Chinese scholar Liang Qi-chao. In 1927, encouraged by Liang, Hong went to Jena, Germany in order to study spiritual philosophy under the supervision of Rudolf Eucken. Since Eucken died before Hong's arrival, he began to study physics, mathematics and philosophy at the University of Jena. Deeply attracted by Hans Reichenbach's *Von Kopernikus bis Einstein* (1927) and *Relativitätstheorie und Erkenntnis apriori* (1920), Hong moved to Berlin to listen to Reichenbach's lectures. Knowing that Hong was interested in philosophy, Reichenbach suggested his going to Vienna to be the student of Moritz Schlick.

Hong Qian studied at the University of Vienna after 1928. At Schlick's suggestion, Hong attended Rudolf Carnap's course in mathematical logic, Hans Hahn's course in mathematics, Felix Ehrenhaft's course in physics, and Schlick's, Friedrich Waismann's and Victor Kraft's courses in philosophy. From 1930, at the invitation of Schlick, Hong Qian attended the meetings of the Schlick group held every Thursday evening, and became the unique member of the Vienna Circle from the Far East. He was acquainted with Otto Neurath, Philipp Frank, Herbert Feigl, Kurt Gödel, Theodor Radakovic, Edgar Zilsel, Alfred Tarski, Walter Hollitscher, etc. there. Under Schlick's supervision, Hong Qian worked on his doctorate. He was awarded his doctorate in 1934 for a thesis entitled '*Die Frage der Kausalität in der neuen Physik*'. After that, Hong stayed at the University of Vienna to continue philosophical research. On June 22nd, 1936, Hong's beloved teacher Schlick was assassinated. Thereafter, Hong went back to China early in 1937.

Hong Qian became lecturer of philosophy at Qinghua University when he went back to Peiping in 1937. After the outbreak of the Sino-Japanese War, Hong fled to the Southwest of China, and after passing through

R. S. Cohen, R. Hilpinen and Qiu Renzong (eds.), Realism and Anti-Realism in the Philosophy of Science, xvii–xxii.

many different places became the professor of philosophy at National South-west Associated University at Kunming from 1940 to 1945. In this period, he lectured on the philosophy of the Vienna Circle, and Schlick's in particular. He published many papers on this topic in Chinese philosophical journals and compiled these papers into a small book entitled *The Philosophy of Vienna Circle* which was published in 1945. In these writings, Hong systematically introduced the philosophy and scientific world-view of the Vienna Circle, criticized the traditional metaphysics, Kant's Apriorism, Husserl's and Scheler's phenomenalism, Windelband's and Rickert's view of 'spiritual' science, Mach's positivism and Feng Youlan's new Confucianism, and made a significant contribution to the dissemination of logical empiricism and analytical philosophy in China.

As to Mach's positivism, Hong Qian pointed out: Although logical empiricism inherited Mach's positivist spirit against metaphysics, there is a fundamental difference between them. According to Mach's positivism, the reality of material bodies is a complex of sensations, but logical empiricists think, "since there are possibilities that all scientific abstract formalism can be tested by the given, their reality is not less than that of tables and chairs which can be felt. Therefore, the members of Vienna Circle said: 'the atom is not a logical construct', not a way of 'thinking', but it is absolutely real."[1]

Hong also pointed out:

> The opposition to metaphysics by the logical empiricism is slightly different from that of traditional positivism. Logical empiricists negate the role of metaphysics in the theory of knowledge, but do not deny its significance in practical life.

Hong agreed with Schlick:

> that metaphysical *philosophemes* are conceptual poems: in the totality of culture they play, in fact, a role similar to that of poetry; they serve to enrich life, not knowledge.[2]

Metaphysics is meaningful to ethics and the philosophy of life, so it should not be eliminated.

Hong's book *The Philosophy of Vienna Circle* is similar to Ayer's *Language, Truth and Logic* in some aspects. Both were intended to introduce the philosophy of the Vienna Circle to readers of their native lands. But China lacked the tradition of empirical and analytical philosophy, while in contrast Britain was strong in these. Compared to Ayer who paid more attention to explicate his own viewpoint, Hong's book was more faithful and comprehensive. Ayer's book was republished and reprinted again and again, and translated into many different languages (including Chinese in 1980) and thereby became a very popular, basic philosophical reading in the world. Hong's book, after the 1st edition of 1945, owing to political and ideological reasons, could not be republished in mainland China until 1989, and then it was again welcomed as a valuable academic treatise. Due to the differences of cultural traditions and political circumstances, how different was what happened to these two books.

Hong went to Oxford in 1945 where he was elected a research fellow at New College. But events in China impelled him to return home in 1947, and while he was Professor and Director at the Department of Philosophy, Wuhan University, he published 'Moritz Schlick and Modern Empiricism' in *Philosophy and Phenomenological Research*, vol. 9, no. 4 (1949). In this paper, Hong introduced the debate between Schlick on one side and Neurath with Carnap on the other side on the problem of the foundation of knowledge, and criticized Carnap's viewpoint of reductionism and protocol sentences. It was two years before the publication of Quine's paper 'The Two Dogmas of Empiricism'.

After the People's Republic of China came into existence, Hong Qian moved to Beijing in 1950 and became the Professor and Head of the Philosophy Department, Yenching University in 1951 [Beijing University]. From 1952 to 1965, he was the Professor and head of the Seminar for History of Foreign Philosophy of the Philosophy Department, Beijing University and then was the director of the Institute of Foreign Philosophy of Beijing University until his retirement in 1987.

Since logical empiricism was condemned as a reactionary idealist philosophy serving imperialism, Hong could not continue his teaching and writing on logical empiricism. In the first half of 1957, encouraged by the Double Hundred Policy, Hong wrote two papers to introduce Mach's philosophical thought and Kant's *Allgemeine Naturgeschichte und Theorie des Himmels*. He suggested to leaders of the Chinese Communist Party to implement the Double Hundred Policy fully, "needn't overemphasize the leadership of Marxism, needn't be afraid of idealism", "pay more attention to the study of Western philosophy". In the summer of 1957, the Anti-Rightist Struggle broke out, Hong was criticized together with other famous philosophers such as Jin Yuelin, Feng Youlan and He Lin. The atmosphere of free contending soon completely vanished. The only thing Hong Qian could do was to keep his silence and do some translation and editing work.

Then Hong was made editor-in-chief of a series of western philosophy from the ancient Greeks to contemporary western philosophy (in 5 vols.) which became the basic materials for students of Philosophy Departments in mainland China.

But, Hong Qian insisted on reading the western philosophical journals and literatures, kept communications with members of Vienna Circle such as Carnap, Feigl etc. until the Cultural Revolution (1966), and followed the recent advances of logical empiricism and analytical philosophy in the Western world.

During the Cultural Revolution, like many other philosophers and scientists, Hong was criticized as a 'reactionary bourgeois academic authority'. His house was searched, rare copies of his early writings, letters from members of Vienna Circle, and some books were confiscated and destroyed.

Perhaps due to the fact that some famous foreign philosophers wrote to the leaders of the Chinese government and expressed their concern about the safety of Hong Qian, the revolutionary committee of Peking University did not force him to join the May 7th Cadre School to do physical labor, but asked him to attend political study routinely. In 1973, in order to coordinate the study of Lenin's *Materialism and Empirio-Criticism*, Hong Qian and two other professors were ordered by the Philosophy Department to translate the related paragraphs of Mach's *Die Analyse der Empfindungen* into Chinese. In 1975, the abridged translations of Mach's book accompanied by an official preface were published by the Commercial Press. (In 1986, Hong Qian and his co-translators translated Mach's whole book and published it without the official preface.)

When looking back upon the three decades since 1949, Hong Qian often said to us regretfully, "lacking the indispensable research materials, without the academic circumstances for free discussion, and with no international academic communication, I did little research work in these three decades. Much time vainly slipped by."

After the end of the 1970's, with the opening to the West, Hong Qian was again allowed to visit the outside world. Though by now in his seventies and relatively frail, Hong Qian became, once more, an indefatigable traveller, charming his hosts with his philosophical acumen and acerbic wit. Through the 1980's, he received visiting fellowships at Queens College, Oxford (1980, 1984) and Trinity College, Cambridge (1982), contributed to symposia on Wittgenstein (1980) and Schlick-

Neurath (1982) in Australia, and lectured in Tokyo (1986) and Hong Kong (1988). In 1984, Vienna University awarded him a second, honorary doctorate on the 50th anniversary of his first one. He also became research professor and member of the academic committee of the Philosophy Institute, Chinese Academy of Social Sciences since 1978, honorary President of the Chinese Society of Contemporary Foreign Philosophy since 1980. His great prestige was a crucial factor in establishing the Sino-British Summer School of Philosophy in Beijing in 1988 with Hong and Sir Alfred Ayer as honorary presidents. Sir Peter Strawson became an Honorary President after Ayer's death in 1989.

In his last decade, Hong's academic contributions reached a second peak. He edited and published collected papers on logical empiricism (vol. 1 in 1982, vol. 2 in 1984), enlarged and reedited the *Anthology of Contemporary Western Philosophy*. He wrote and published 17 papers such as: 'Wittgenstein and Schlick' (*Proceedings of 5th International Wittgenstein Symposium*, 1981), 'Moritz Schlick und der Logische Empirismus' (*Grazer Philosophische Studien* **16/17**, 1982), 'Remarks on Affirmations' (*Synthese* **64**, 1985), "On Logical Empiricism" (*Zhexue Yicong*, 1987), 'Some Problems about Logical Empiricism' (*Journal of Dialectics of Nature*, 1989), 'Ayer and the Vienna Circle' (*The Philosophy of A. J. Ayer*, 1992), 'Rudolf Carnap' (*21th Century*, 4, 1992), etc. In these papers, Hong Qian discussed the historical sources of the Vienna Circle, the conditions of its rising in Austria during the 1930's and its dissemination in English speaking countries; commented on the philosophical viewpoints of Schlick, Ayer, Carnap, Feigl, Kraft and others; reviewed the changes and developments of logical empiricism in recent decades, and its present situation and new trends; expressed his skepticism about foundationalism. His works are valuable resources for the study of the philosophy and history of the Vienna Circle and highly valued by the academic world.

In respect to the criticism of logical empiricism in the recent half-century, Hong Qian paid much more attention to Quine's criticism than to that of Popper, Kuhn and Feyerabend. Especially, Hong disagreed with what Popper had said, that he had killed logical empiricism. Instead, Hong believed that logical empiricism is still alive and that there was a trend to its rejuvenation in recent decades. Nevertheless, Hong acknowledged the weakness of logical empiricism. In his article entitled 'Philosophical information from travel in Europe' (in 1980), he wrote:

> No need for reticence, since the beginning of the Vienna Circle, most schools of contemporary analytical philosophy have not paid sufficient attention to ethics, have not placed ethics in the proper position in philosophy. Just as someone pointed out: a complete philosophical system, needs not only an integral part of theoretical philosophy, but also an integral part of practical philosophy. For instance, in Kant's philosophy, there are three great treatises of *Critique*; in Marxist philosophy, there are dialectical materialism and historical materialism. In view of this, it is no wonder that Russell said emotionally: Strictly speaking, in this kind of philosophy like logical positivism, actually there is no philosophy, but only methodology.[3]

Hong hoped that analytical philosophers including logical empiricists will pay more attention to practical philosophy and ethics.

Hong Qian was really very sad after the Beijing event in June 1989. He deeply worried that it would block China's international communications in philosophy which had just opened. It was Hong's principal advice that helped the Sino-British Summer School of Philosophy to be revived last summer. This Beijing International Conference on Philosophy of Science was also prepared under his energetic support. So Hong Qian's death is really an irremediable loss to Chinese philosophical circles and the philosophy of science in particular. We will cherish the memory of him forever.

Zhongguancun,
Beijing 100080,
China.

NOTES

[1] Hong Qian (1989). *The Philosophy of Vienna Circle*, Commercial Press, Beijing, p. 62.

[2] Moritz Schlick (1979). *Philosophical Papers*, vol. 2, D. Reidel, Dordrecht, p. 111.

[3] Hong Qian (1990). *Collected Papers on Logical Empiricism*, Shanlian Shudian, Hong Kong, pp. 247–248.

RISTO HILPINEN

ON SOME FORMULATIONS OF REALISM, OR HOW MANY OBJECTS ARE THERE IN THE WORLD?

I

This paper has a very modest objective. The recent discussion of realism contains expressions and metaphors which I find difficult to understand, and I shall try to translate some of these expressions into a more comprehensible idiom. The arguments about realism – or about different 'realisms'[1] – often involve claims about the *dependence* (or *independence*) of the world on our theories and concepts. What is the nature of this dependence or independence? Some philosophers make a distinction between our conceptualizations of the world and "the way of the world *really* is" (italics mine), or at least attribute such a distinction to other philosophers.[2] I assume that this distinction is not meant to be the same as the familiar distinction between a representation and its object.[3] Do scientists not try to represent the world the way it really is? (The representation of the world as something other than what it really is would seem to be misrepresentation.) The metaphors of 'carving' and 'cutting' are common in this context: Hartry Field speaks about the "carving up of the noumenal dough" by means of various "cookie cutters",[4] whereas Hilary Putnam has argued that this metaphor is a misleading characterization of scientific representation, that is, a misleading summary of what he regards as the correct account, his "internal" or "pragmatic" realism.[5] The target of Putnam's criticism, "external" or "metaphysical" realism, includes, according to him, the thesis that "the world consists of some fixed totality of mind-independent objects".[6] In this paper I am mainly interested in this thesis and its negation. I shall not discuss other aspects of what Putnam calls "external realism", viz., the assumptions of correspondence (i.e., the correspondence theory of truth) and uniqueness (the uniqueness of a complete and correct description of the world).[7]

R. S. Cohen, R. Hilpinen and Qiu Renzong (eds.), Realism and Anti-Realism in the Philosophy of Science, 1–10.

II

In his Paul Carus Lectures Putnam says:[8]

(P1) How many objects there are in the world is relative to the choice of a conceptual scheme.

According to Putnam, this statement may be regarded as an expression of a "relativistic" view:[9]

> How can one propound this sort of relativistic doctrine and still claim to believe that there is anything to the idea of 'externality', anything to the idea that there is something 'out there' independent of language and mind?

Sentence (P1) can be paraphrased as follows:

(P2) The number of objects in the world is relative to the choice of a conceptual scheme.

Putnam also speaks about the dependence and independence of objects and facts on our concepts. This suggests that the expression "relative to" in (P2) can be replaced by "dependent on", that is, that (P2) (and hence (P1)) can be paraphrased as

(P3) The number of objects in the world depends on the choice of a conceptual scheme.

If (P3) is part of Putnam's doctrine of 'internal' or 'pragmatic' realism, the doctrine he calls "metaphysical realism" presumably involves the claim

(M1) The number of objects in the world is independent of the choice of a conceptual scheme.

(M1) seems to follow from the thesis of metaphysical realism mentioned earlier,

(M2) The world consists of some fixed totality of mind-independent objects.

The expressions 'mind-dependence' and 'dependence on a conceptual scheme' are often used interchangeably in the discussion of realism, and they will be used in this way also in the present discussion. In this context, a 'mind' consists of a system of concepts by means of which the reality is represented and understood. Sometimes philosophers have confused this sense of 'mind' with its other meanings; for example, in his discussion of realism Robert Nola characterizes *after-images* as mind-dependent entities.[10] After-images may be called 'psychological' objects, but this does not make them necessarily "mind-dependent" in the present sense of the word.

(P3) and (M1) appear to be propositions about a number, viz., the number of objects in the world: (P3) seems to entail, e.g., the following:

(P4) There is a number n such that n = the number of objects in the world, but the number of objects in the world would not have been n if our choice of a conceptual scheme had been different from what it actually was.

Here 'n' is assumed to be a rigid number designator; (P4) says that the expression 'the number of objects in the world' refers to different numbers in different possible worlds, and this variation depends on variation in our conceptual schemes.

If (P4) were a well-defined proposition, it would express a truly relativistic doctrine. For example, the much discussed doctrine of linguistic relativity can be expressed in a form analogous to (P4):

> If a given speaker's language were (significantly) different from what it actually is, his "picture of the world" (world view) would differ from his actual view.[11]

But (P4) is of course not what Putnam means by 'internal' or 'pragmatic' realism. In *Representation and Reality* Putnam asks us to consider an example in which a person is taken to a room and asked, "How many objects are there in this room?".[12] He argues that the question has no clear and unambiguous answer: *there is no such thing* as the number of objects in a given room. This can be expressed by saying that the expression 'object' is not an *individuating expression* or a *sortal* expression.[13] To use Frege's words, the concept of an object "does not isolate in a definite manner what falls under it".[14] Putnam expresses this indirectly by saying that the answer to the question about the number of objects in the room "depends on what one means by 'object'", and that "there are many ways of using the notion of an object".[15] These characterizations are misleading, because they suggest that the expression 'object' is somehow ambiguous.[16] To say that there are different "ways of using the notion of an object" means in this context only that the expression 'object' can be joined together with different individuating descriptions.

The concept of an object, unlike the concept of a chair or the concept of a horse, is not an individuating concept, and therefore the description 'the number of objects in the world' (or 'the number of objects in this room') is not a proper definite description. Consequently the expression 'the number of objects in the world' does not denote anything (in any possible world), and sentences (P1)–(P4) and (M1) cannot be understood literally as statements about the number of objects in the world. Literally interpreted, they should be regarded as false (if we accept Russell's account of definite descriptions) or meaningless.

Charitably interpreted, (P1) may be regarded as an indirect or roundabout way of expressing the following 'grammatical' (semantical) fact about the use of the expression 'object':

(P5) The expression 'object' is not an individuating expression.

According to this interpretation of (P1), the doctrine which Putnam calls "metaphysical realism" involves a *semantical* mistake about the world 'object', and Putnam's "pragmatic" or "internal" realism, as far as (P1) is concerned, merely reminds us that 'object' is not an individuating expression. Putnam's example about the number of objects in a room supports only (P5), but no further "metaphysical" conclusion. In *Reason, Truth and History* Putnam expresses this view as follows:[17]

(P6) *What objects does the world consist of?* is a question that it only makes sense to ask within a theory or description.

According to the observations made earlier, this is true, but it does not express an interesting metaphysical doctrine: (P6) should be regarded as an uncontroversial observation about the semantics of the word 'object'. If there are "metaphysical realists" who seem to deny (P6), they are either semantically confused or are using some technical or special notion of Object; perhaps they have in mind some notion of metaphysically basic or "ultimate" Object with its own individuating characterization.

III

The word 'object' is not an individuating expression, but if *F* is any individuating expression, 'object which is *F*' or '*F*-object' is also an individuating expression: such an expression is synonymous with *F*. If the expression 'object' in (P3) is supplemented by some individuating expression *F*, we get

(P7) The number of *F*-objects in the world depends on the choice of a conceptual scheme.

For example, for alligators we obtain the following thesis of "internal realism"

(P8) The number of alligators in the world depends on the choice of a conceptual scheme.

This claim seems false. It makes good sense to speak about the number of alligators in the world, and for alligators the relativized thesis (M1), the thesis of relativized metaphysical realism, seems to hold true.[18] The falsity of (P8) can be seen more clearly if it is rewritten as follows:

(P9) If our conceptual scheme were different from what it actually is, the number of alligators would be different from the actual number of alligators.

(P9) is clearly false. (P8) may be regarded as true only if it is not understood as a statement about the number of alligators, but rather as a statement about (the extension of) the word (or concept) 'alligator'. Let Ext(G) be the extension of an expression (or concept) G, let $\text{Ext}_{\text{actual}}(G)$ be the actual extension of G (the extension of G in the actual world), and let $n(A)$ be the cardinality of a set A. Thesis (P9) must be distinguished from the following proposition:

(P10) If our conceptual scheme were different from what it actually is, n(Ext('Alligator')) would be different from $n(\text{Ext}_{\text{actual}}(\text{'Alligator'}))$.[19]

(P10) is not equivalent to (P9): (P10) may be true, even though (P9) is false. If "internal realism" involves the acceptance of (P9) (and not merely that of (P10)), it amounts to the view that if a dog's tail were called a leg, dogs would have five legs.[20]

According to the argument given above, theses (P3) and (M1) are meaningful only when relativized to objects "under" some individuating description, as shown in (P7). If the doctrine of realism depends on the truth of such theses, it is necessarily realism with respect to *F*-objects, where *F* is some individuating description. Thus the following thesis is not meaningless (or semantically confused) in the same way as (M2):

(M3) The world contains a fixed totality of mind-independent *F*-objects,

where *F* is some individuating description. By generalizing with respect to individuating concepts, we obtain from (M3) the following doctrine of "absolute metaphysical realism":

(M4) For any individuating concept *F*, the world contains a fixed totality of mind-independent *F*-objects.

By a similar generalization we get from (P7) the following thesis of "internal" realism:

(P11) For any individuating concept *F*, the number of *F*-objects in the world depends on the choice of a conceptual scheme.

It is not quite clear what 'mind-independent' means here. If the unavoidable relativity of any "fixed totality" to an individuating concept is regarded a form of "mind-dependence", (M4) is a self-contradictory thesis and (P11) becomes a conceptual truism. But this is not a plausible interpretation of 'mind-dependence' (and 'mind-independence'), since it is possible to distinguish kinds of objects whose character is "constituted" by people's beliefs, e.g., chess pieces and traffic signs, from other kinds whose character is largely independent of what people believe about them, for example, alligators.[21] Thus both (M4) and (P11) are clearly false. To ask questions about existence and identity, we need individuating concepts or descriptions, and for some such descriptions the correct (true) answers are objective in the sense that they are independent of people's beliefs, but in other cases the answers are genuinely "mind-dependent".

IV

When philosophers have discussed realism relativized to individuating descriptions (sorts or kinds), condition (M3) has usually not been regarded as a criterion of realism about *F*-objects. Sometimes realism about *F*'s has simply been regarded as the view that there exist *F*'s. For example, according to Mark Heller,[22] to be a realist about a some particular thing or kind of thing is to believe that thing or kind exists (or that kind has members).

Heller calls this form of realism "existential realism". This is not an interesting sense of 'realism': according to Heller's definition, a person who believes that there are swans and that all swans are white, is a realist about swans, but anti-realist (or non-realist) about black swans. Moreover, existential realism in Heller's sense is compatible with mind-dependence in the sense suggested in the previous section: There are (i.e., there exist) traffic signs and chess pieces, but they are not mind-independent; they are constituted by people's beliefs.[23] If the requirement of "mind-independence" is added to Heller's definition, we get the following thesis of (existential) realism with respect to *F*-objects:

(R1) *F*-objects exist in the world in a mind-independent manner.

Michael Devitt has formulated an "absolutized" version of (R1) as follows:[24]

(R2) Tokens of most current common-sense, and scientific, physical types objectively exist independently of the mental.

To make thesis (R1) immune to the objection against Heller's definition of existential realism mentioned above, it should be reformulated as

(R3) The existence of *F*-objects in the world does not depend on the mental (or on the choice of a conceptual scheme).

In the same way, Devitt's "absolutized" version of (R1) should be understood as follows:

(R4) For most individuating concepts *F*, the existence of *F*-objects does not depend on the mental (or on the choice of a conceptual scheme).

Devitt's version of realism (reformulated as (R4)) is more cautious than (M4), and therefore more likely to be true, since it allows the possibility that some objects exist in a mind-dependent way, or, to use the terminology of conceptual schemes, it allows the possibility that there are objects whose existence "depends on our conceptual scheme". Such objects (for example, chess pieces) satisfy the following modified form of (P7):

(P12) The existence of *F*-objects in the world depends on the choice of a conceptual scheme.

But whether these objects or object kinds form a majority or a minority among all object kinds is hardly of great conceptual and philosophical interest. The interesting questions about realism concern the interpretation of the expression 'mind-dependence' used above and the truth of the relativized thesis (R3) for various concepts (or object kinds) *F*. If Putnam's internal realism makes everything mind-dependent on the ground that the identification of anything as an object requires an identifying concept, he bypasses a significant distinction between objects constituted by people's beliefs and those whose character and existence are independent of what is believed about them.

Departments of Philosophy,
University of Turku, Finland;
University of Miami, USA.

NOTES

[1] Cf. Susan Haack (1987). '"Realisms"', *Synthese* **73**, pp. 275–299.

[2] Mark Heller (1988). 'Putnam, Reference, and Realism', in *Midwest Studies in Philosophy XII: Realism and Antirealism*, ed. by A. P. French *et al.*, University of Minnesota Press, Minneapolis, pp. 113–127; see p. 115.

[3] The distinction between the world as it really is and as it is represented by our theories resembles Charles Peirce's distinction between the *immediate* and the *dynamical* object of a sign: The immediate object of a sign is the object "as the sign itself represents it", whereas the dynamical object is the "reality which by some means contrives to determine the sign to its representation"; see *Collected Papers of Charles Sanders Peirce*, vol. 4, ed. by C. Hartshorne and P. Weiss (1933). Harvard University Press, Cambridge, Mass., paragraph 4.436. According to Peirce, the immediate object of a sign is a representation rather than the reality represented by a sign.

[4] Hartry Field (1982). 'Realism and Relativism', *The Journal of Philosophy* **79**, pp. 553–567; see p. 561.

[5] Hilary Putnam (1987). *The Many Faces of Realism*, Open Court, La Salle, III. p. 19, and *Representation and Reality* (1988). MIT Press, Cambridge, Mass. pp. 113–114. According to Putnam, the cookie-cutter metaphor founders on the question, "What are the parts of the dough?" I don't think this is a convincing objection.

[6] Hilary Putnam (1981). *Reason, Truth and History*, Cambridge University Press, Cambridge, p. 49.

[7] *Ibid.*, p. 49.

[8] Hilary Putnam, *The Many Faces of Realism*, p. 32.

[9] *Ibid.*, p. 32.
[10] See Robert Nola, 'Introduction: Some Issues Concerning Relativism and Realism in Science', in *Relativism and Realism in Science*, ed. by R. Nola (1988). Kluwer Academic Publishers, Dordrecht, pp. 1–35; see p. 5.
[11] See John Carroll (ed.) (1956). *Language, Thought, and Reality: Selected Writings of Benjamin Lee Whorf*, The MIT Press, Cambridge, Mass., p. 221.
[12] Hilary Putnam. *Representation and Reality*, pp. 110–112.
[13] See E. J. Lowe (1989). *Kinds of Being*, Basil Blackwell, Oxford, pp. 9–10.
[14] Gottlob Frege (1950). *The Foundations of Arithmetic*, transl. by J.L. Austin, Basil Blackwell, Oxford, p. 66. See also Nicholas Griffin (1977). *Relative Identity*, Clarendon Press, Oxford, p. 40, and Peter Geach (1968). *Reference and Generality*, Emended Edition, Cornell University Press, Ithaca, pp. 38–39.
[15] *Representation and Reality*, pp. 113–114.
[16] In his discussion of sortal predicates Geach discusses the problem of counting the red things in a room, but he does not conclude from the impossibility of this task that the expression 'red thing' is ambiguous or can be used in different ways; according to Geach, the example shows only that 'red thing' is not what he calls a "substantival" (that is, a sortal) expression. See *Reference and Generality*, p. 39.
[17] *Reason, Truth and History*, p. 49.
[18] This is even more obvious in the case of dinosaurs and (say) galaxies, since in such cases we need not worry about the causal and indirect effects of our conceptual activities. (In this context "the *F*-objects in the world" should be taken to include all *F*-objects which have existed or will exist in the world.)
[19] According to the possible worlds semantics of conditionals, (P9) and (P10) can be formulated as follows:

(P9*) In the nearby alternative worlds where our conceptual scheme differs from that adopted in the actual world, the number of alligators differs from the actual number of alligators.

(P10*) In the nearby alternative worlds where our conceptual scheme differs from that adopted in the actual world, n(Ext('Alligator')) differs from n($\text{Ext}_{\text{actual}}$('Alligator')).

[20] I am indebted to Professor Alan Musgrave for this observation.
[21] C.S. Peirce's well-known definition of 'reality' is based on this distinction. According to Peirce,

that is *real* which has such and such characters, whether anybody thinks it to have those characters or not.

(*Collected Papers of Charles Sanders Peirce* (1934). vol. 5, ed. by C. Hartshorne and P. Weiss, Harvard University Press, Cambridge, Mass., paragraph 5.430.)
[22] Mark Heller, 'Putnam, Reference, and Realism', pp. 113–114.
[23] See also Robert Nola, *op. cit.*, pp. 1–35; see pp. 4–5; the definitions given by Nola also entail that one can be a realist about swans but an anti-realist about black swans.

[24] See Michael Devitt (1984). *Realism and Truth*, Princeton University Press, Princeton, p. 22. Devitt's reference to *physical* (as opposed to, for example, psychological) types here may be based on the confusion between the two senses of 'mind' and 'mental' mentioned in section II, viz., mind as a system of concepts and mind as the psychological.

RUDOLPH KREJCI

DISSOLUTION OF THE REALISM/ANTIREALISM PROBLEM

INTRODUCTION

This paper investigates the way to dissolve the adversary either/or position represented by realism and antirealism by suggesting the complementarity approach which was so successfully introduced by Niels Bohr in his revolutionary way of thinking applied to modern physics. Bohr himself hoped that one day the complementarity principle would become the backbone of everyone's education and would be applied in the entire spectrum of our society. However, Bohr's approach is regarded as revolutionary only in the western philosophico-scientific tradition. Outside of it there are similar philosophical, moral, and spiritual perspectives to be found in Taoism, Buddhism, Upanishads, and Jainism whose followers regard complementarity of Bohr as more-or-less commonplace.

The argument of this paper focuses on the priority of the question and the underlying assumptions which are responsible for polarity systems of answers in the form of the epistemological realism coupled with ontological materialism on one hand and epistemological antirealism and ontological idealism on the other hand. In this very adversary form it has become a major philosophical, scientific, and ideological argument of the western tradition. We seem to be at a crossroad; we can blindly continue to amass other either/or positions in our realism/antirealism conflict. If we decided to follow Niels Bohr's suggestion of complementarity we will discover that any two opposing contradictory aspects are always integral, inseparable parts of a whole. "One cannot bow in front of somebody without showing one's back to somebody else" [1].

Any consequent dissolution of the realism/antirealism problem can be successfully achieved by adopting a new conceptual context going radically beyond the Aristotelean tradition by exclusion of his *Tertium non datur* principle and replacing it by the principle of complementarity which can be symbolized as excluding $[p \vee q \equiv \sim(p \wedge q)]$ and replacing it by $[p \vee q \equiv (p \wedge q) \wedge r \wedge s \wedge \infty]$.

R. S. Cohen, R. Hilpinen and Qiu Renzong (eds.), Realism and Anti-Realism in the Philosophy of Science, 11–18.

I

> A great insight is a statement whose denial is another great insight.
>
> Niels Bohr

Philosophical and scientific problems manifest their intrinsic interconnectedness all through the history by articulations of peculiar mind-sets driven by mankind's curiosity. Some of them became central to views about the mystery of the universe and man's role in it. Their implications found, in turn, their manifestations in religion, arts, science, politics and in due time penetrated the entire theory and praxis of mankind. All through history such a philosophical problem was and still is the problem of realism based on assumptions that there is a world outside of the perceiver, governed by laws, we can know and communicate to others. Two forms of the western realism are notably represented in teachings of Plato and his pupil Aristotle, and both of them are responsible for a great part of the philosophical, scientific and religious cultural context of the West. In Whitehead's comment that the entire history of the West consists of a few footnotes to Plato we have probably the best defined historical role of realism. As this type of realism in all its forms is essentially dualistic, it will find its contradictory nemesis in antirealism challenging the realistic position by advocating a view that the world and everything in it is nothing but a temporary useful fiction helping us to organize our observations for our benefit, however, predestined to be replaced by another useful fiction in the future. Both of these positions represent a conflict of mind-sets within which corresponding self-referential realities are being constructed.

Psychologically the realistic thesis provides an ultimate foundation for attainment of certainty by getting away from oneself and finding the anchor in the world of things in their totality. As one can see, the entire cultural, religious, political and social striving was, and, to a great degree, still is, carried on within the realistic tenets of mankind. On the other hand, the historical position of the antirealism starting with Xenophanes' anthropomorphism and method of projective hypostatization bases all human striving in the man himself as an observer who, driven by his curiosity, continues his search and re-search and never reaches the ultimate finding! The search with its temporary finding becomes the only uppermost reward. Following Xenophanes, the best of the philosophical and scientific tradition became necessarily anthropomorphic, hypothetical, open-ended. It found its corresponding world-view in the concept of open universe, open-society, open-mind without ever being able to reach certainty by either logical, mathematical, scientific, philosophical, or mystical ways.

Not until the 20th Century was it possible to depart from the traditional dualistic categories dictating our thought pattern with implied adversary relationships. The entire fabric of our western culture became saturated with this dualistic categorization manifested in most of its science, philosophy, religion, arts, politics, and social striving. When it entered the global scene it assumed the role of messianism attempting to disseminate its western mission all over the world. Given the view that there exists an absolute truth and by conviction that it was in possession of certain selected groups we have had the greatest historical examples of intolerance represented by the institution of the Inquisition and the 20th Century National Socialist, Fascist, and Communist ideologies. Realism always appealed to large collectives, while antirealism, on the other hand, was the conviction of individuals working in isolation from the masses.

II

> When the province of physical theory was extended to encompass microscopic phenomena, through the creation of quantum mechanics, the concept of consciousness came to the fore again: it was not possible to formulate the laws of quantum mechanics in a fully consistent way without reference to consciousness.
>
> Eugene Wigner [2]

It was a universal belief of the scientists of the 19th Century that the 20th Century would bring the completion of the Newtonian world-view by providing all the missing details. Materialism became not only the integral part of scientific method but the very condition under which science could be conducted. There were only rare exceptions among scientists and mathematicians who did not subscribe to this dogma [3].

The predictive power of the mechanistic science celebrated successes which silenced even the most vocal critics. Immanuel Kant himself declared the findings in mathematics, logic, and physics, as completed in principle, to which nothing substantial could be added. This very successful way of thinking in physical sciences inspired a speedy application to all other sciences such as biology, psychology, sociology, economics, anthropology, politics, and philosophy. The shift continued and persists practically up to now. Its political, social, and economical implications became responsible for much of the havoc of the 20th Century.

One of the most significant realistic and materialistic stands against the implications of the new advances in science at the end of the 19th Century and the beginning of the 20th Century is Vladimir I. Lenin's work *Materialism and Empiriocriticism*, regarded by his followers as an epoch making work of genius [4]. There were two major philosophical positions Lenin maintained in his desire to protect the dialectical materialism of his days. The first was the theory of knowledge of Marxism under attack from various streams of modern philosophy including the Social Democrats and the Russian Marxists – the second was the problem of the concept of matter which was under attack due to the implications of the discovery of radioactivity, which in turn released an avalanche of speculations suggesting "dematerialization of the atom" and "disappearance of matter".

Lenin was the first one to recognize these new emerging developments as mortally endangering the very realistic and materialistic foundation of Marxism in both theory and praxis. Truly in Marx's way Lenin attacks, under the name of 'physical idealism', scientists and philosophers, such as Mach, Poincaré, Avenarius, Duhem, Pearson, his Russian adversaries Lopatin and Bogdanov and a host of others. Lenin's defense of materialism and realism with his epistemology of copy-theory laid down the 20th Century position of the Soviet School of Marxism (Leninism-Stalinism) against the idealistic interpretations of Einstein's theory of relativity with its equivalence of mass and energy and of quantum mechanics of the Copenhagen School.

On the basis of Lenin's restatement of the 19th Century realistic and materialistic dogma it was rather consequential to reject the philosophical concepts of complementarity and uncertainty-relations due to their denial of causality and objectivity of micro-phenomena. The Copenhagen 'idealistic interpretations' was thus rejected because it regarded the wave-function describing the state of the micro-particles not as an objective state of affairs, but merely as an expression of the knowledge of the observer where the wave is not the property of the objects of the microworld but merely a wave of probability.

As we can see it in retrospect, Lenin's crusade against the spirit of the modern way of thinking which was manifested in new scientific and philosophical approaches, helped to create a closed system of thinking, which in turn produced a closed society unable to compete with the rest of the world. Its recent abdication from the stage of modern history of mankind must be regarded as an overdue disappearance of dysfunctional philosophy in both theory and praxis.

III

> Physics does not deal with physics. Physics deals with what we can say about physics. What we can say depends in turn on what we can agree about; and that depends on clicks of a counter, irreversible acts of amplification, indelible records.
>
> John Archibald Wheeler [5]

The above quotation is from Professor Wheeler's 'working paper' series of three lectures delivered by the author in Beijing, Hefei, and Shanghai in October 1981. It was almost 50 years ago, when in 1932 Niels Bohr, Professor Wheeler's great teacher and mentor, returned from his visit to China, inspired by the great cultural tradition of its past. The principle of complementarity formulated by Bohr five years before his pilgrimage to China acquired thus a new symbol represented by the Chinese YIN-YANG accompanied by the Latin motto *non contraria sed complementa*, and the hope that one day this very principle would become the central core of our entire educational system in the West. Professor Wheeler, one of the last surviving giants of Bohr's school could not help but come and report to you the accomplishments which followed Bohr's legacy during the remainder of the 20th Century. In light of this new tradition we can observe a slow process of introducing complementarity as a way of thinking into all other aspects of our contemporary modes of thought. Similarly, we see the implications of Heisenberg's principle of uncertainty for the historical striving of mankind to escape from itself into some postulated certainty out there, and with it the nonetheless stupendous implications of Gödel's incompleteness theorem which proved the openendedness of human thought as against the axiomatic closed mind. In light of these innovative ways of thinking I dare to argue for the dissolution of the realism-antirealism problem as one of the important metaphysical underpinnings of the historical and cultural tradition of the West which is entirely foreign to the traditional heritage of China and the great part of the Orient.

Professor Wheeler's 'Chinese' quotation provides a telling description of the new view of the nature of the reality of the universe as no more "sitting out there" independent of any act of observation. Rather, 'reality' is directly the co-production and by-product of the observer's intentional total tools, thus making observer's actions necessarily participatory actions. (*Esse est participare.*) There are many implications of this turn around, one of them is that we cannot have any information unless we have acquired it by some specific method accompanied by detailed disclosure of both – what we report and how we arrived at it. The observer is ourselves – thus we have to start and end with him (or her) – there seems to be no escape any more. Again there are necessary implications. The most important ones are that the traditional dualistic division of the subject and object, inner and outer world, body and soul, good and evil, true and false, right and wrong, paradise and hell are no longer adequate concepts for a more sophisticated description of our new 'realities'.

A long time before the impact of the above discussed ideas was felt in our century, there was an important discovery of the German physiologist Johannes Müller (1801–1858) who formulated the idea that the world as we sense it is our own construction – known as 'Müller's law of specific nerve energies'. An important conclusion of this law is that our sentences do not mirror the external stimuli but only the state of the nerve being stimulated which provides us with the justification that our system of cognition is self-referential and self-explicative [6]. Müller's law thus anticipates very important epistemological consequences of the new way of thinking in the 20th Century, expressed through principles of complementarity and uncertainty relations and the incompleteness theorem.

CONCLUSION

This paper attempted to expose the cultural, philosophical, and scientific assumptions behind the historical conflict of realism and antirealism. They were responsible for a state of petrification into dualistic categories of adversary relationships of our traditional way of thinking applied to the entire interpretation of reality in both theory and praxis. It was the radical departure from this traditional way of thinking due to our new scientific advances at the beginning of the 20th Century which forced us to rebuild our world-view during the 20th Century. It has been proved over and over again that deep-rooted prejudices of human thinking die hard as it was the case with false hypotheses about ether, phlogiston, or pure race, survival struggle, class struggle, Oedipus complex, geocentricity of the solar system, phrenology and so forth. The old world-view was a result of questions we asked and extralogical strivings for certainty – it helped us to collect facts we desired, and preferred interpretations we were pleased with. The old world-view was self-referential and self-explicative without being aware of it.

In view of the above insight many philosophical problems which resulted in adversary positions will disappear because a new world-view does not provide the soil in which they can flourish. Among these will be notably our realism/antirealism problem and hosts of others such as materialism and idealism, subjectivism and objectivism, mind-body problem, individual-collective and any other adversary application of categories of exclusion.

The transition from one to another world-view is the best testimony of humanity triumphing over its own prejudices and limits and reaching its higher level of self improvement on its way *ad summum*.

$$(p \vee g) \equiv \sim(p \wedge g)$$

is rather

$$p \vee q \equiv p \wedge q \wedge r \wedge s \wedge \infty$$

Department of Philosophy,
University of Alaska,
USA.

BIBLIOGRAPHY

1. Abbé of Galiana (1728–1787) quoted by John Wheeler, in his address at the ceremony conferring the Atoms for Peace Award on Niels Bohr, 1957.
2. Wigner, Eugene (1967). *Symmetries and Reflections*, Indiana University Press, Bloomington, p. 189.
3. Isaac Newton himself hoped to explain with his mechanistic laws only "all corporeal things", still maintaining an existence of non-corporeal ones. Mathematicians and logicians due to their *a prioristic* approaches were outside of the realist-materialistic orbits.

4. Lenin, Vladimir I. (1952). *Materialism and Empiriocriticism, Critical comments on a Reactionary Philosophy*, 1909 (originally published), Moscow.
5. Wheeler, John Archibald (1982). 'Physics and Austerity – Laws without Laws', working paper, Center for Theoretical Physics, University of Texas, Austin, Texas, p. 32f.
6. Müller, Johannes (1834–1840). *Handbuch der Physiologie des Menschen*. "Sensation consists in the sensorium's receiving through the medium of the nerves, and as a result of the action of an external cause, a knowledge of certain qualities or conditions, not of external bodies, but of nerves themselves." Müller's quotation is in E. Boring (1950). *A History of Experimental Psychology*, 2nd ed., Appleton, New York, p. 82.

ALAN MUSGRAVE

REALISM, TRUTH AND OBJECTIVITY*

Bill Lycan once called me a mad-dog realist. Whereupon Deborah Mayo and Dick Burian presented me with a large wooden sign bearing the legend "BEWARE OF THE mad-DOG realist". I am very proud of this sign and I keep it in my office back in New Zealand. I thought of bringing it to this conference. But I decided not to. You see, the sign is misleading, I am not a mad-dog realist at all, but rather a pretty tame one. More rabid realists than me are common, as we will see.

1. REALISM AND TRUTH

I subscribe to the old-fashioned idea that scientific realism is first and foremost a thesis about the aim of science. It says that the aim of a scientific inquiry is to discover the truth about the matter inquired into. This incorporates a semantic thesis (inquiry results in true or false statements about the world) and an axiological thesis (science aims for true statements). Both are controversial: some antirealists deny the semantic thesis, others the axiological thesis.

This cannot be the whole realist story, of course. To what extent has the realist aim actually been achieved? It is now fashionable for realists to give strong positive answers to this question. What Putnam (before he abandoned realism) or Boyd or Leplin would like to say is that current science is all true. But history intrudes and they hedge this in various ways: they say that most of current science is true (supposing that the notion of 'current science' is unproblematic, which it is not); or they say that most of current science is true or nearly true (supposing that the notion of 'near truth' is unproblematic, which it is not); or they say that most of current mature science is true or nearly true (supposing that the notion of 'maturity' is unproblematic, which it is not). Furthermore, they make these strong claims definitional of scientific realism: you cannot be a scientific realist unless you think that all (or most) of current science (or current mature science) is true (or nearly true).

Similarly strong claims are made about the existence of the 'theo-

R. S. Cohen, R. Hilpinen and Qiu Renzong (eds.), Realism and Anti-Realism in the Philosophy of Science, 19–44.

retical entities' postulated by current science – or equivalently, about how the 'theoretical terms' of current science succeed in referring to real things in the world. Such claims also become definitional of scientific realism: you cannot be a scientific realist unless you think that all or most current theoretical entities really exist, that all or most current theoretical terms really do refer.

Thus construed, scientific realism is not a thesis about the aim of science at all. It is a substantive ontological or metaphysical thesis. It is the thesis (I omit the hedgings for simplicity) that there really are such entities as current science claims there to be and that what current science tells us about such entities is true (see Leplin, 1982, p. 88 & p. 90). This is mad-dog realism proper.

Some try to soften it by harping on reference but giving up on truth. We are to believe in scientific entities (believe that theoretical terms refer), without thinking true any theory about those entities. We are to be 'entity-realists' without being 'theory-realists'. Or so say Nancy Cartwright (1983), Ian Hacking (1983), and Michael Devitt (1991) – if I understand them.

This is incoherent. To believe in an entity, while believing nothing further about that entity, is to believe nothing, I tell you that I believe in hobgoblins (believe that the term 'hobgoblin' is a referring term). So, you reply, you think there are little people who creep into houses at night and do the housework. Oh no, say I, I do not believe that hobgoblins do that. Actually, I have no beliefs at all about what hobgoblins do or what they are like. I just believe in them. (Compare contemporary believers in God.)

Cartwright and Hacking, despite their entity-realist rhetoric, do not actually adopt such a position. Their belief in some entity X always involves as well belief in some X-theory, in Cartwright's case a theory about what X's do, in Hacking's case a theory about what scientists can do with X's. You cannot intervene in nature and do things with electrons unless you believe (think true) all sorts of things about the properties and behaviour of electrons. Nor, despite what Devitt says, does supposing merely that electrons exist enable us to "give good explanations of the behaviour and characteristics of observed entities" (Devitt, 1991, p. 108). "Electrons exist", by itself, can explain nothing whatever about the behaviour and characteristics of observed entities. To do any explaining we need an electron-theory which says that electrons exist, to be sure, but which says much more besides. And a successful electron-theory will do more than get the existence of electrons right, will do more than succeed in referring to electrons. For as Larry Laudan has correctly urged (1981, p. 25), a theory may be referential yet false, and because of that thoroughly unsuccessful. Reference is important for realists because they think it is a necessary condition for success. But it cannot be a sufficient condition – realists need truth as well.

It may be objected here that one can believe in X's and believe all sorts of things about X's, without having a 'fully-fledged' or 'global' X-theory – and that it is only 'fully-fledged' or 'global' X-theories that we have no business believing. To which I reply that for me any collection of beliefs about X's counts as an X-theory – though perhaps not as a 'fully-fledged' or 'global' X-theory, whatever that might be.

So you cannot soften mad-dog realism by harping on reference and giving up on truth. What of mad-dog realism itself? What of the idea that to be a scientific realist one must think current science true and current theoretical entities real? (I omit the hedgings for simplicity.) This is the latest manifestation of an old error: it erects current science into a metaphysic and ties scientific realism too closely to that metaphysic. Cartesians did it in the 17th century, Newtonians in the 18th century, determinists in the 19th century. Theories which did not square with the Cartesian, Newtonian, or deterministic metaphysic were deemed false, incomplete, or mere 'convenient fictions'. (Duhem rightly objected to the practice, but wrongly thought that the only way to save science from metaphysical interference was to adopt the view that convenient fictions are all that science seeks.) And now our own mad-dog realists tell us that you cannot be a realist unless you adopt the metaphysic of curved space-time and quarks and gluons (or whatever). Realism thus defined might be empirically refuted, as antirealists like Laudan are quick to see.

But scientific realism has survived the demise of Cartesian, Newtonian, perhaps even deterministic metaphysics. And it could survive the demise of some, most, even all, of current science. Realism would not be refuted if curved space-time or quarks or gluons went the way of phlogiston or the heavenly spheres. Or perhaps we should put it thus: epistemic realism would survive, while substantive realism anno 1990 would not.

Do not misunderstand me. Realists think that science should inform our metaphysical picture of the world, and I am no exception. Sensible realists are so informed, and suppose that some scientific entities really do exist and that some of what science tells us about them is true. Again, I am no exception. You would have to be a pretty pessimistic realist to think that science has discovered no entities whatever and come up with no truths whatever about them. Such pessimists may exist. Popperians, gripped of all things by the pessimistic induction, sometimes tell us that all scientific theories are false. This is a ludicrous proposition and it is backed by an argument which is as bad as an argument can be: its premise ("All past theories are false") is preposterous and its conclusion ("All theories are false") does not follow.

Mad dog realists go to the other extreme: they are besotted with current science and think that it has at last got to the bottom of things. But current science includes the current frontiers of science, the places where the current scientific action is, the places where scientists do not know and are trying to find out. Scientists tend to be pretty cagey about entities and theories at the frontier – realist philosophers of science would do well to follow them in this. It is well-entrenched and well-tested past science that should inform our metaphysic. We should be more confident about atoms and molecules than we are about electrons, and more confident about electrons than we are about quarks and gluons. Realism about the entities and theories of current science should be rather guarded. And whether guarded or not, it should not be seen as definitional of scientific realism.

The maddest of the mad-dog realists go even further. They say that only the entities of current science exist, or 'really exist'. When science found out that tables and chairs were composed of atoms and molecules, it found out that tables and chairs are unreal or do not really exist. Remember Eddington's tale (1982, pp. xi–xiv) of the 'two tables', the 'table of common sense' and the 'table of physics', and his question "Which table is the real table?". Remember Wilfrid Sellars' tale (1963, p. 126) of the 'manifest' and the 'scientific' images of the world and his insistence that if the latter is true the objects of the former "*do not really exist – there are no such things*". Eddington's mistake might be called the 'explaining-is-explaining-away fallacy'. If science finds out that atoms are composed of quarks and gluons, it will not have found out that atoms do not really exist. If social science finds out that social wholes can be explained individualistically, it will not have found out that they do not exist.

Eddington's mistake is, curiously, the mirror-image of the positivist mistake. Positivists argued that since tables and chairs are real, atoms and molecules (not to mention quarks and gluons) cannot be. Sensible realists (lap-dog realists?) resist both mistakes. Tables and chairs are real, and so (it is reasonable to suppose) are the atoms and molecules of which they are made. Atoms and molecules are real, and so (it might be reasonable to suppose) are the quarks and gluons of which they might be made. Commonsense realism about tables and chairs neither precludes (pace positivists) nor is precluded by (pace Eddington) scientific realism about atoms and molecules.

This is not, of course, to say that science never conflicts with common sense nor later science with earlier science. We think of a table as a solid object. Perhaps (though I doubt it) this is to think that the space occupied by a table is continuously filled with matter. If so, then the atomic theory of matter conflicts with this commonsense belief, and if we accept the former we must reject the latter. But this is to change a belief about tables, not to start thinking that the 'table of common sense' does not exist. (The 'table-of-common-sense' and the 'table-of-physics' are ersatz entities dreamt-up by philosophers. Once you start taking them seriously, nothing can stop you from sliding into conceptual idealism. But I anticipate.)

Lap-dog realism differs from mad-dog realism in another respect that I digress briefly to mention. Some lap-dog realists are suspicious of platonic realism, realism about abstract entities. They would like to find a way to drive a wedge between realism about some of the entities posited by science and realism about the abstract entities posited by mathematics and other disciplines. Again, some lap-dog realists have sympathy with nonrealist accounts of morals. Can lap-dog realists be discriminating in this way? In particular, does not a realist theory of truth commit one willy-nilly to all kinds of realisms, some of them suspect in the extreme? Which brings me to my next section.

2. TRUTH

I return to semantic realism (scientific theories are true or false statements about the world) and axiological realism (science aims for true theories). As I said, both are controversial. What theory of truth do realists have in mind when they say that the aim of inquiry is truth?

It is obvious to me, and to others, that realism assumes some version of the classical or objective or correspondence (or we might add, realist) theory of truth. To say that the aim of inquiry is truth in any other sense of the term 'truth' is not to advocate realism at all. For example, suppose you go in for an 'empirical adequacy theory of truth'. Then to say that science aims for truth is to advocate, not realism, but Bas van Fraassen's constructive empiricism. Or suppose you go in for some sort of pragmatist or instrumentalist theory of truth (whatever that is). Then to say that science aims for truth is to advocate, not realism, but some sort of pragmatism or instrumentalism.

Arthur Fine sees the realism-antirealism dispute in a perverse way. Both realists and antirealists think that truth is the aim. They disagree only about the meaning to be attached to the term 'true': realists go in for a realist conception of truth, antirealists for some antirealist conception of truth. One can see the dispute this way – but it is not the clearest way to see it. Nor do antirealists see themselves as giving esoteric antirealist readings to the term 'true'. Van Fraassen would not bar an 'empirical adequacy theory of truth'. He is quite clear about the difference between truth and empirical adequacy and his best antirealist arguments rely upon it. So let us set aside Fine's perversity and say that realists assert and antirealists deny that truth is the aim.

The most fundamental objection to scientific realism concerns the realist conception of truth. What is this conception and what are the problems, or alleged problems, which beset it?

It is well to begin with Tarski, who thought that his semantic conception of truth was a version of the classical correspondence theory. Most philosophers think otherwise, as we will see. It is well to begin, in particular, with Tarski's famous requirement that a materially adequate definition of truth for a language L must entail, for each statement S of L, a statement of the form

(T) The statement S is true in L if and only if p

where S names the statement of L and p translates that statement into the language in which the truth-definition is formulated. This is the famous T-scheme or equivalence thesis. It is well to begin with it, not of course because it exhausts Tarski's semantic conception of truth, but because it exhausts most of the philosophical worries about that conception. Some philosophers think it trivial, others think it far from trivial. I begin with the latter.

There are those who invest the T-scheme with magical powers. They observe that it can be applied across the board, so to speak. Commonsense realists apply it to commonsense statements. Scientific realists apply it to scientific statements. But it can also be applied to statements about numbers, moral statements, modal statements, or whatever. If the T-scheme gives commonsense and scientific realists all they need from a theory of truth, won't it give all sorts of suspect realists all they need too? If commonsense instances of it yield realism about tables and chairs, and scientific instances of it yield realism about genes and electrons, will not arithmetic instances of it yield platonic realism about natural numbers, and moral instances of it yield moral realism, and modal instances of it yield modal realism?

This worry is quite groundless. You might as well worry that since the T-scheme can be applied to "Margaret Thatcher gives me the creeps", we must all be creeps-realists and believe in a mysterious entity (the creeps) which Margaret Thatcher gives to me and simultaneously to others too. The T-scheme by itself yields no realist claim. We all avoid creeps-realism by saying that although "Margaret Thatcher gives me the creeps" is true (which it is), it is an idiom which is not to be taken at face-value for logico-philosophical purposes. We replace it with an equivalent non-idiom (say, "Margaret Thatcher makes me nervous") and avoid ontological commitment to the creeps. Similarly, one sceptical of moral realism might refuse to take "Eating people is wrong" at face-value for logico-philosophical purposes – which is just what emotivists, prescriptivists, and the like do. Alternatively, one can go for an error-theory: take moral statements at face-value and say that they are all false. This is what John Mackie and my colleague Charles Pigden do to avoid moral realism. Hartry Field does a similar thing to avoid arithmetical platonism: he takes "Two plus two equals four" at face-value and says that it is false, not because two and two add up to some other number than four, but because there are no numbers for the numerals 'two' and 'four' to be names of. Similarly with scientific realism: some antirealists (the instrumentalists) refuse to take scientific statements at face-value, other antirealists (the positivists) take them at face-value and say that they are all false because there are no theoretical entities. Similarly, finally, with commonsense realism: idealists either refuse to take commonsense statements at face-value or take them at face-value and say they are false because there are no tables and chairs.

In fact, antirealists usually go in for a mixture of these strategies. They first tell you that a certain kind of statement is false if taken at face-value or realistically. They then soften that conclusion by telling you what such statements "really mean", that is, how they are to be taken for logico-philosophical purposes. Think of phenomenalist 'translations', so-called, of statements about physical objects, or of emotivist 'translations', so-called, of moral statements.

So the idea that the T-scheme by itself has magical powers and begs all kinds of metaphysical questions is quite mistaken. But if it does not by itself yield realism about tables and chairs, electrons, numbers, moral qualities, the creeps, or anything else, what is its importance for realism? The answer is that it makes realism about all these things possible. This is its importance for realism, this is why it is the theory of truth that realists need. Antirealist conceptions of truth make realism impossible, they leave no room for it. Realists need the T-scheme – but they also need much more. To be a realist about X's (whatever X's might be) you must take statements about X's at face-value or realistically, apply the T-scheme to them, and accept some of them (appropriate ones: "There are no X's" will not do) as true.

So much for those who invest the T-scheme with magical powers. Most philosophers have quite the opposite view – that it is trivial. Some think it trivial because circular: the same statement appears on each side. This is a misunderstanding: on the left is a statement about a statement, on the right a statement about something else. The T-scheme is called a 'disquotational' scheme and is said to encapsulate a 'disquotational' or 'deflationary' or 'redundancy' view of truth. But an instance of T will not be 'disquotational' if something other than a quotation-mark name of a statement appears on the left-hand side. And 'disquotation' will result in nonsense if L is a different language from the one in which the instance of T is formulated. (Consider: "The statement 'La neige est blanche' is true in French if and only if la neige est blanche".) Again it is said that for Tarski sentences construed as meaningless strings of symbols are the truth-bearers. But as the adjective 'semantic' and the reference to translation make plain, the truth-bearers are meaningful sentences or statements. To set up an instance of T you must know what the referring terms in S refer to, and you must know what the predicates applied to those terms mean in order to translate them into the language you are speaking.

It is true that some instances of the T-scheme are 'disquotational', like:

(1) The statement "Snow is white" is true in English if and only if snow is white.

Confining himself to such instances, Alan Chalmers once worried that the T-scheme is trivial in that it merely describes how one bit of language relates to another bit of language, not how language relates to the world. And he suggested that some notion of truth-as-correspondence must be added to Tarski's theory to fill the gap. (Chalmers, 1976, p. 121. In the second edition of his book Chalmers corrected this mistake.)

This is a queer worry. Instances of the T-scheme do relate language to the world. True, if one is to speak about the way language relates to the world, one must use language. But this is no deep truth – rather, it is a pallid truism. Sweeney had it right:

> I gotta use words when I talk to you
> But if you understand or if you don't
> That's nothing to me and nothing to you
> We gotta do what we gotta do.
> (T. S. Eliot, 1963, p. 135).

The fact that "I gotta use words when I talk to you" does not mean that I am trapped inside language and can only ever talk about words. To think otherwise is to ignore the hard-won philosophical distinction between using a word and mentioning it. (Wittgenstein ignored this distinction in his *Tractatus* and concluded that the way language relates to the world cannot be 'said' but can only be 'shown'. Wittgenstein's 'logocentric predicament' is actually old psychologistic wine poured into new linguistic bottles. The British empiricists thought that thinking consists in having a stream of 'ideas', and concluded that all we ever think about are our own ideas.)

What of Chalmer's idea that 'truth as correspondence' is needed to fill some gap which Tarski leaves unfilled? Arthur Fine has a similar idea. Tarski's semantic theory of truth is one thing and the correspondence theory another. The correspondence theorist provides (or seeks to provide) something which Tarski does not, a general account of the 'correspondence relation' between language and reality, a theory of the way the two can 'match up' so that truth results. Fine says that such a theory "would explain what *makes* the truth true" (Fine 1984a, p. 97). Armed with an account of the relation which all truths bear to reality, the correspondence theorist will know what all truths have in common, their essence, what makes them a natural kind. And here Fine gets sceptical: he doubts that truths form a natural kind (Fine, 1984b, p. 56).

I share the scepticism. So, I believe, would Tarski. For Tarski it is meaningful linguistic items that are true or false. Languages are conventional human devices, suited to different purposes. The truths they contain are many and various, and so are the ways they correspond to reality. Why suppose that they share an essence and form a natural kind? It is surely better to suppose that Tarski captures 'the correspondence relation' as well as it can be captured, that there is no more to that relation than Tarksi gives us. As Michael Levin says:

> Tarski tells us that all true conjunctions have in common the truth of each conjunct, that each true existential generalisation is such that its matrix is satisfied by at least one sequence, and so on. To be sure, at the level of the basis clauses, the definition goes strongly extensional, but that is the way it ought to go. The truths "Ron Reagan is a man" and "This tulip is red" are not shown to have much in common: each is a matter of a different object satisfying a different open sentence. But *do* they have more in common than being a man has in common with being red, which is to say, very little? I cannot see that they do, or at least more than Tarski gives them (Levin, 1984, p. 126).

On this view, one thing Tarski taught us is that the essence of truth is a chimera, that an essentialist correspondence theory is out. Or if we must talk of essences, we might talk thus: Tarski gives us the essence of the correspondence theory without giving truth an essence.

But it does not matter if we talk otherwise. Suppose that a correspondence theory worth its salt must give truth an essence, must provide an account of the correspondence relation which all truths bear to reality. Then Tarski's theory of truth is not a correspondence theory. Indeed, in view of all the dust philosophers have raised down the ages with the word 'correspondence', only to complain afterwards that they cannot see, perhaps we had better drop that word altogether in this context.

Tarski's theory of truth, or rather Tarski's T-scheme, has been accused of being trivial in yet another sense. It does not tell us which statements are true and which are false, nor does it say anything about how we might determine this. It is a non-epistemic conception of truth. So it is, and so it should be. The questions "What is truth?" and "What is true?" are not the same question at all. The trouble is, many suppose, that a non-epistemic conception of truth lets in the sceptic. Work out your epistemology, your way of sorting the true from the false. If truth is non-epistemic, the sceptic can still ask whether what your epistemology throws up is guaranteed to be true. To beat the sceptic we need an epistemic concept of truth, we need to define truth as whatever our epistemology throws up. And so we have a rash of epistemic truth-theories.

My stalking-horse will be the 'ideal limit theory of truth' recently proposed by Carl Matheson. According to this theory, "the truth is to be equated with the limit of ideal scientific practice: what is true is just perfect science pursued to its conclusion" (Matheson, 1989, p. 253). The claim is not that "All truths would as a matter of fact be part of ideal science", a claim which a mad-dog realist convinced of the omnipotence of scientific methods might endorse. Rather the claim is that this is a matter of (conceptual) necessity: "It is a *conceptual* truth, that all truths would be part of our ideal science" (Matheson, 1989, p. 253, note).

Matheson does not tell us what 'ideal science' or 'ideal scientific practice' is, nor does he justify his presumption that it will reach some sort of 'limit'. But never mind. One thing is clear: whatever account is given, it cannot be couched in terms of truth and falsehood. For to be told that truth is what ideal scientific practice reaches in the limit, and then to be told that ideal scientific practice is that which in the limit reaches truth, is to be told nothing. If we are to have a non-trivial epistemic conception of truth, then the epistemology must be defined in austere truth-free terms. For then we will be told something non-trivial when we are told "This is what ideal scientific practice is – and truth is by definition what it comes up with in the limit".

This opens the theory up to Moore's 'Open Question Test'. However ideal scientific practice is to be defined, it remains an open question whether what it yields in the limit is true. The sceptic does not contradict herself when she says that it is possible that ideal scientific practice pursued to its limit might issue in falsehoods.

Besides, the anti-sceptical virtues of epistemic conceptions of truth are entirely spurious. To recognize the truth I am supposed to recognise what ideal scientific practice is and recognise that on some issue it has reached it 'limit'. What some 'ideally rational inquirer' (is this God, I wonder?) will be coherently believing at the end of time is just as inaccessible to you or me as is truth-as-correspondence. In fact, of course, it is much more inaccessible.

Consider, finally, Matheson's assertion that "It is a *conceptual* truth, that all truths would be part of our ideal science". What is the meaning of the word 'truth' (as opposed to the word 'truths') in this assertion? Applying the ideal limit conception of truth to it yields "It is a conceptual part of ideal science, that all truths would be part of our ideal science". The contention is, then, that this particular statement (the statement of the ideal limit theory of truth) has been or will be thrown up by ideal science pursued to its limit. I defy Matheson or anyone else to locate this statement in the writings of any reputable scientist, or to give some reason for thinking that it will occur in the writings of any reputable scientist of the future. No, the ideal limit theory of truth is not science but philosophy – and bad philosophy to boot.

So much for epistemic conceptions of truth, of which the ideal limit theory is my exemplar. Returning to Tarski's T-scheme, it has been accused of triviality in yet another sense. It is said (by Davidson, Devitt, and others) that the T-scheme is trivial in the sense that it is common to all conceptions of truth. This is not true. According to the T-scheme, what makes the statement "Snow is white" true in English is the fact that snow is white. According to the ideal limit theory of truth, what makes it true is the fact that ideal science will subscribe to it in the limit. According to the empirical adequacy theory of truth, what makes it true is the fact that it is empirically adequate. According to the pragmatist theory of truth, what makes it true is the fact that it is "expedient in the way of belief" (as William James put it). According to the coherence theory of truth, what makes it true is the fact that it figures in some coherent system of beliefs.

What happens if we combine the T-scheme with some epistemic conception of truth? To fix our ideas, let us apply both to the statement that electrons exist. The former gives us:

(1) The statement "Electrons exist"is true in English if and only if electrons exist.

The latter gives us:

(2) The statement "Electrons exist" is true in English if and only if the statement "Electrons exist" satisfies epistemic condition E.

It does not matter what epistemic condition E is, except that to avoid circularity it must not involve "Electrons exist" being true in English. Notice that (2) asserts the (necessary?) equivalence of two statements about the statement "Electrons exist". It says nothing about the world, nothing 'metaphysical'. Hence Putnam talks of 'internal realism' as opposed to 'metaphysical realism'.

Now (1) and (2) yield, by transitivity of the bi-conditional:

(3) Electrons exist if and only if the statement "Electrons exist" satisfies epistemic condition E.

Surely (3) is, or might be, false. Electrons existing (or equivalently "Electrons exist" being true in English) is no part of the specification of E. However our best methods for seeking truth are to be specified, they are not guaranteed to reach the truth. There is nothing incoherent or unintelligible about the idea that our best methods might lead us to think that electrons exist when they do not. At least, this is so unless we suppose that a statement's satisfying condition E (which specifies those methods) somehow brings it about that electrons exist. This 'internal realism' is, of course, a form of idealism. It conflicts with what science teaches about electrons, for example, that they existed before any sentient creatures able to theorise about them had evolved. You cannot combine an epistemic truth-theory with the T-scheme and retain realism about the world and its contents being (largely) independent of our epistemic activities.

Michael Devitt, following Stephen Leeds, thinks the T-scheme trivial in that T-schemes for a language are two-a-penny. Given a bunch of words and a bunch of things, we can set up a great many 'reference-schemes' between the words and the things. One reference-scheme says "The English word 'snow' refers to snow", another says "The English word 'snow' refers to grass", and so forth. Each of these reference-schemes yields a Tarskian T-scheme for the language in question. And each T-scheme satisfies the equivalence thesis (Devitt, 1984, pp. 30–31). But a T-scheme which assigned grass is green to the English statement "Snow is white" (the example is Devitt's) would not satisfy the equivalence thesis (and would not, by the way, be 'disquotational' either). "Grass is green' does not translate "Snow is white", because the English word 'snow' does not refer to grass. It might have so referred, of course, in which case the English language would have been different, the reference-scheme for English would have been different and the resulting T-scheme for English would have been different too. But as things are with the English language, 'snow' refers to snow, not to grass.

Is this naive? Can we say confidently that 'snow' refers to snow and not to grass? Can we speak confidently (as I just did) of *the* reference-scheme for English or any other language? Hilary Putnam thinks not. He has a 'model-theoretic argument against realism' (Putnam, 1978, 1980). It is this argument which probably lies behind the Leeds-Devitt point of view just considered. Let us consider the argument.

The confident assumptions just questioned make the realist theory of reference and truth radically non-epistemic. Those assumptions make room for the following possibility: there might be a theory that is empirically adequate, that satisfies any other epistemic desiderata we care to impose (is consistent, simple, elegant, or whatever), and yet is false. Such an 'ideal theory' might be false either because its theoretical terms fail to refer, or because it makes false claims about its theoretical entities. Either way, its falsity is empirically undetectable, since by assumption it is an empirically adequate theory. Putnam calls the view that an 'ideal theory' might be false 'metaphysical realism'. And he argues that metaphysical realism is mistaken.

The argument is this. Since an ideal theory will be consistent, it will (by the Lowenheim-Skolem Theorem) have a model, an interpretation in which it is true, whose domain is the natural numbers. We can map the natural numbers onto objects in the world, and use this mapping to define an interpretation in the world in which the theory is true. So the theory is true of the world. So metaphysical realism is mistaken.

The obvious objection to this is that the interpretation which Putnam has defined is not the intended interpretation of the ideal theory, that the terms of that theory do not refer to whatever they are made to refer to in Putnam's real world model. The arithmetical model, and the real world model induced by the mapping, serve to establish the consistency of the ideal theory but not its truth.

Putnam is unimpressed with this obvious objection. He finds the notion of an 'intended interpretation' fishy, because he finds any appeal to the 'intended' or real referent of a term fishy. Any attempt to spell out what the referents of our terms are is "just more theory". It can be added to the original theory, in the form of constraints on the interpretation of that theory, and the entire argument can be applied to the augmented theory. This will yield an interpretation in the world in which the original theory T and any constraints C on the interpretation of T are both true. This will be the 'intended interpretation' of the theory, or as near to its 'intended interpretation' as we can ever get.

The crucial flaw in the argument is that this is not the case. As David Lewis points out "The constraint is *not* that an intended interpretation must somehow make our account of C come true. The constraint is that an intended interpretation must conform to C itself" (Lewis, 1984, p. 225). In other words, an interpretation of T which also makes C come out true, which is all that Putnam's argument provides, is not an interpretation of T which conforms to C. This disposes of Putnam's paradox.

Lewis's point seems subtle but it is really quite obvious, as a simple little example will show. Let T be "Ravens are white" and let the actual world W* contain just five objects, a white swam (i_1), a black raven (i_2), a chair (i_3), a table (i_4), and a cat (i_5). You might think that T is false of W*, since W* contains a non-white raven i_2. Enter Putnam. Being consistent, T has a model in the natural numbers – actually, it has a model in the first five natural numbers. Let interpretation I assign prime numbers greater than two to the term 'raven' and odd numbers to the term 'white'. Clearly, "Ravens are white" is true in this interpretation. Now define a mapping between the first five natural numbers and the objects in W*, which takes n to i_n, and through this mapping an interpretation I* of T in W*. Clearly, T is true of W* under interpretation I*.

We protest that I* is not the intended interpretation of T. We protest, in particular, that the term 'raven' refers to i_2. Putnam replies that this constraint C is just more theory and invites us to add it to T, yielding "Ravens are white and the term 'raven' refers to i_2". This augmented theory is consistent, so there is an interpretation in the natural numbers (actually in the first five of them) in which it is true. Let this interpretation be as before for the terms 'raven' and 'white'; let it assign 1 to the term "the term 'raven' ", 5 to the term 'i_2', and let the pairs of numbers 1 and 3 and 1 and 5 stand in the relation 'refers to'. As before, use the mapping between the first five numbers and the objects in W* to define an interpretation in W* in which the augmented theory T&C is true.

Now the crucial point is that this interpretation does not conform to constraint C. Constraint C requires that the term 'raven' refer to i_2. In the interpretation Putnam defines, the term 'raven' does not refer to i_2. The constraint is true in that interpretation, but the interpretation does not conform to the constraint.

So far we have had Putnam allowing us to formulate constraints on the interpretation of theories and arguing (invalidly) that they do not help the 'metaphysical realist'. But the position to which Putnam is led by this argument does not allow us such constraints in the first place. He says that statements like "The term t refers to the object o" presuppose a 'magical' theory of reference which sticks words onto things with some sort of 'cosmic glue'. He says that we cannot 'single out' any sort of correspondence between words and things: to do so we would need independent access to the things, which we do not have. For ""objects do not exist independently of our conceptual schemes. We cut the world up into objects when we introduce one or another possible scheme of description" (Putnam, 1981, p. 51). Putnam calls this 'internal realism'.

It is an odd doctrine. Consider the Putnamesque possibility of two 'ideal theories' which give utterly different and mutually incompatible accounts of the way the world is. (One might be a particle theory, the other a field theory – the details do not matter.) Having abandoned metaphysical realism, Putnam thinks both theories are true of the world. But utterly different and mutually incompatible theories can both be true of the world only if the notion of truth is relativised to each of them – which relativises the notion of 'the world' to theory also. There is no real world independent of our theorising – or if there is we have nothing to do with it. The world we theorise about is a world-as-theorised-about-by-us – and there may be more than one of them. This is not realism of any kind – it is idealism. It spells trouble not just for esoteric doctrines like Putnam's metaphysical realism – it makes nonsense of scientific realism and commonsense realism too.

Before I argue this, let me say what fun it is to see the antirealists squabbling amongst themselves. Van Fraassen's chief argument against realism is that realists, being truth-freaks, will want to know which of two incompatible 'ideal theories' is true. Evidence cannot help them (being 'ideal' the theories will be empirically equivalent and empirically adequate). So realists must choose on non-evidential or metaphysical grounds. Realism lets in metaphysics, metaphysics is bad news, so realism is bad news. I will not discuss this argument now (I have done so elsewhere). I mention it only to observe that if Putnam's model-theoretic argument is correct, van Fraassen's argument cannot get off the ground. Any empirically adequate theory might be deemed 'ideal'. Any 'ideal theory' is, by Putnam's argument, true. So the realist does not have to choose between two ideal theories at all, but can accept them both. But only at a very heavy antirealist price, as we will see in the final section of my paper.

3. OBJECTIVITY

A Dutch sociologist (Gerard de Vries) once told me in all seriousness that when the concept 'person with an IQ two standard deviations above the mean' was invented new entities were brought into being. So it turns out (I replied) that there are two ways of making babies, the way we all know and love, namely love, and this new way, psychological theorizing! I was told not to be silly, that the new entities were not babies. I tried to find out what they were but I did not succeed.

My ex-friend the Dutch sociologist was a conceptual idealist. And conceptual idealism, a view which I find quite absurd, is arguably the dominant metaphysical position of our age among prominent philosophers. Putnam's 'internal realism' is a species of it. And there are many other species of it. Remember all that guff about 'the social construction of reality'. Recall Kuhn's suggestion that scientists imbued with different paradigms 'inhabit different worlds'. Think on Nelson Goodman's 'ways of world-making' and his extraordinary contention that "We can have words without a world, but no world without words or other symbols". Think too on Richard Rorty's 'the world well lost'. Consider Simon Blackburn's view that we can 'project', "speak and think as though there were a property of things" when there is no such property, and yet make no mistake (Blackburn, 1984, p. 171). An extraordinary state of affairs!

Conceptual idealism was ushered into the world by Kant, who distinguished between the phenomenal and noumenal 'worlds', between things-in-themselves and things-as-experienced-by-us. There is the Moon-in-itself (call it Moon_i), about which we know nothing, and there is the Moon-as-experienced-by-us (call it Moon_e). Moon_i is not the same object as Moon_e – if it were the distinction would have no point and we could rest content just with the Moon, without subscript, unhyphenated. For example, Moon_i is not located in space and time, these being 'forms of sensibility' in which only Moon_e is located. Nor does Moon_i cause or help cause moon-experiences in humans down on earth, causality being a 'category of the understanding' which applies only in the phenomenal world. Some of Kant's immediate followers realised that Moon_i is nowhere, at no time, and does nothing, concluded that it was an idle metaphysical posit, and did away with it altogether to become fully-fledged idealists. I do not blame them. My Kantian friends tell me that I misunderstand, that Kant was an 'empirical realist', and not an idealist at all but only a 'transcendental idealist'. And I am reminded of how fond Bishop Berkeley was of presenting himself as a defender of commonsense realism.

Kant, like Berkeley before him, thought that there was only one 'phenomenal world': he assumed that humans all have the same immutable set of basic concepts with which they structure incoming stimuli. Contemporary philosophical wisdom has outgrown that assumption. Modern conceptual idealism is Kantian idealism relativised to conceptual or linguistic scheme. Concepts and languages vary and change. There is not a unique 'phenomenal world' or world-as-conceived-by-humans at all. The world-as-conceived-by-the Aristotelian differs radically from the world-as-conceived-by-the-Newtonian. The world-of-the-Eskimo is not at all the world-of-the-Kalahari-bushman. This gets really exciting if we drop human chauvinism and consider non-human animals too: the world-of-the-chimpanzee is not at all the world-of-Albert-Einstein, and both are worlds apart from the world-of-the-honeybee. (Does Kant think that non-human animals have experiences? If he does, does he think that they possess all those categories of the understanding deployment of which is a condition of the possibility of all experience? It is pretty implausible to suppose that chimps and honeybees and flatworms structure incoming stimuli the same way humans do. It is even more implausible to suppose that they have no experiences at all. The upshot seems to be that chimps can have experiences without deploying the Kantian categories. But if chimps can do without the Kantian categories, why can't we do without them too?)

Of course, all this idealist talk of different worlds-as-experienced (or conceived or talked about)-by-X's need not be taken seriously. We can see it just as a fancy way of drawing attention to the diversity of experience or concepts or talk of the world. On this view, hyphenated entities like the Moon-as-experienced-by-humans are all ersatz entities. The Moon-as-experienced-by-humans is just the Moon – and similarly for all other hyphenated entities, including the Kantian Moon-in-itself. And a profundity such as "The Moon-as-conceived-of-by-Aristotelians was perfectly spherical" is just philosopher's gobbledy-gook for "Aristotelians thought that the Moon was perfectly spherical".

Conceptual idealists do take seriously talk of different worlds-as-experienced (or conceived or talked about)-by-X's. Given their metaphysic, the realist idea that our experience or thought or talk might correspond with a reality independent of experience or thought or talk is seen to be thoroughly naive. The things we experience or think about or talk about are partly of our own experiential or conceptual or linguistic making. What arguments lead to this extraordinary metaphysic?

To begin with there is the idea that what I can see (or say) depends partly on the concepts or words that I possess. Now it is trivially true that a being lacking the typewriter-concept (the word 'typewriter') cannot say that there is a typewriter on the table. Assuming that a being which cannot say that P cannot see that P either (an assumption I will not here dispute), it follows that a being lacking the typewriter-concept (the word 'typewriter') cannot see that there is a typewriter on the table either. But it does not follow that such a being cannot see the typewriter on the table. The Kalahari bushman may see the typewriter perfectly well, as evidenced by his response to the request (couched in Kalahari-bushman-ese) to pass us that thing over there. The cat may see the typewriter perfectly well too, as evidenced by her not bumping into it when the mouse she is chasing hides under it. This argument wrongly conflates seeing-that with seeing. (In between these there is also seeing-as: the bushman or the cat may see the mouse as food, and the typewriter as non-food, even though they lack the mouse-concept and the typewriter-concept.)

[Digression: It is not, in fact, "trivially true that a being lacking the word 'typewriter' cannot say that there is a typewriter on the table". A French-speaker who knows no English and who therefore lacks the word 'typewriter' can say this – in French. Such a being lacks the word 'typewriter' but possesses the typewriter-concept, since French has a word for typewriters too. I ignore this complication, and the knotty problems involved in it, as being tangential to my main concerns. One can also dispute the assumption which for simplicity I granted just now, the assumption that a being which cannot say that P cannot see that P either.]

The conflation mentioned a paragraph back between seeing-that and seeing leads to the view that beings possessed of different 'conceptual schemes' or languages or theories literally see different worlds. This is nonsense. The Aristotelian and the Copernican, watching the sunrise, see the same thing. What might be true is that the Aristotelian says of the sunrise "I see that the sun is still orbiting the earth", while the Copernican says "I see that the earth is still rotating on its axis". The profundity "The limits of my language are the limits of my world" is false. What is true is the triviality that my language limits what I can say of the world. Or consider the anthropological profundity "The world of the Kalahari bushman contains no typewriters". Is this the hidden platitude that, Kalahari language or concepts being what they are, the bushman cannot see or say that a typewriter is on a table? Or is it the gratuitous paradox that typewriters fail to interact causally with the visual system of the Kalahari bushman? Since when did reflected light get to be so discriminating or the bushman's eyesight so weak? Or is it the even sillier view that the (assumed) absence of some concept or word makes certain physical objects disappear? What magical causal powers are being invoked here?

Arthur Fine thinks that the realist 'metaphysical picture' of an objective, independent, external world is misguided. It is misguided, he says, because of "the obscurity of the correspondence relation and the inscrutability of realist-style reference". He elaborates:

> The problem is one of access. The correspondence relation would map true statements (let us say) to states of affairs (let us say). But if we want to compare a statement with its corresponding state of affairs, how do we proceed? How do we get at a state of affairs when that is to be understood, realist-style, as a feature of the *World*? . . . A similar question comes up if we move to reference . . . , for there again what the realist needs by way of the referent for a term is some entity in the *World* The difficulty is that whatever we observe, or, more generously, whatever we causally interact with, is certainly not independent of us. This is the problem of *reciprocity*. Moreover, whatever information we retrieve from such interaction is . . . information about interacted-with things. This is the problem of *contamination* How then, faced with reciprocity and contamination, can one get entities both independence and objective? Clearly, the realist has no direct access to his *World* . . . (Fine, 1986, p. 151).

What exactly is the problem here? Suppose somebody says "The Moon is full tonight" and I look up into the night-sky and ascertain that the statement is true. (I use a humdrum commensical example rather than an esoteric scientific one because if there is a problem here it is a quite general one which will afflict the commonsense realist metaphysic just as much as the scientific realist metaphysic.) I have access to both terms of the so-called 'correspondence relation': my linguistic competence gives me access to what was said, my eyes give me access to the Moon out there in the world. Fine does not think that we lack either linguistic competence or sensory awareness. Nor is his worry the traditional sceptical one – that my perceptual belief that the Moon is full is fallible, because I might have been the victim of some illusion or hallucination. No, Fine's worry is that reciprocity and contamination mean that I have no access to the Moon at all.

Reciprocity was supposed to show that the Moon is not independent of us because we can see it or otherwise causally interact with it. But implicit in this is a silly account of independence: an object will be independent of us only if we cannot see it or otherwise causally interact with it. The only independent entities in this sense will be platonic entities, which do not exist in space and time and which have no relations causal or otherwise with beings like us which do exist in space and time. No realist, commonsense or scientific, thinks that the only independent entities are platonic entities and the only independent reality the platonic realm. When a commonsense or scientific realist says that the Moon is (largely) independent of us, she means that it is non-mental, continues to exist when we are not looking at it, existed long before we did, and so forth.

What of Fine's 'problem of contamination'? That was supposed to show that when we see that the Moon is full, we gain information about an interacted-with-Moon, not about an objective Moon out there in the World. Is this interacted-with-Moon then subjective and inside our heads, like the 'Moonish-sense-datum' that was once thought to be the immediate object of perception? Realists are perceptual realists and have rightly discarded that view: perceiving external objects may involve us in having perceptual experiences, but it is the external objects that we perceive nonetheless. (Hallucinators, who are commonly said to be 'seeing things', do not see things at all, they just may mistakenly think that they do!)

Fine may not be seeking to revive the theory of sense-data – but he arrives at something pretty close to it. His interacted-with-moon, his Moon-as-observed-by-humans, is not 'objective' in the sense that it is somehow partly constituted by the Moon-concept which is our invention. (The Moon-in-itself is, of course, no better off in this regard!) Like Putnam, Fine thinks that objects do not exist independently of conceptual schemes, that we carve up the world into objects when we introduce a conceptual scheme, and that having carved-up the world we cannot partake of it. (Evidently, the world is not like the Sunday joint!) The Moon which we observe, talk about, theorise about, is an object that we create by observing or talking or theorising. And this is said, believe it or not, to be part and parcel of a 'natural ontological attitude', which accepts the results of science as true on a par with 'homely' commonsense truths. Well, some pretty mundane and well-entrenched bits of science tell us that the Moon (not some hyphenated Moon, not even the Kantian Moon-in-itself, just the Moon) is objective and independent of us: it existed long before we did, was not created by us, and so forth. Someone with a natural ontological attitude, someone who accepts these bits of science as true, will have precisely the realist 'metaphysical picture' which Fine rejects.

Although he does not know it, Fine is heir to a long and distinguished philosophical – that is, idealist – tradition. The giveaway is the hyphenated entity, the 'interacted-with-thing'. Fine implicitly argues as follows: "We cannot find out about things without interacting with them. Therefore, we cannot find out about things-as-they-are-in-themselves". What can we find out about, then? Why, interacted-with-things, of course. The argument is invalid. It is one of the countless versions of what David Stove (1991) calls 'The Gem', an argument which has converted most philosophers to idealism from Berkeley to the present day. The general form of The Gem is: "We can X things only if condition C, which is necessary for us to X things, is met. Therefore, we cannot X things-as-they-are-in-themselves." For X you can substitute 'know', 'perceive', 'think of', 'talk of ', 'refer to', and so forth. For the necessary condition C you can substitute Berkeley's 'have in mind', Kant's 'bring under the categories of the understanding', Fine's 'interact with', and so forth. Gems, implicit and explicit, take many forms.

Conceptual idealism takes many forms, too. John Mackie had an 'error theory' of moral judgements, according to which we 'project' moral properties onto the world. Simon Blackburn liked the idea of 'projection', but did not like the idea that it involves us in error. So he decided to square the circle, and say that we can 'project' things onto a world which lacks those things and make no mistake. The project of projecting things has recently been extended to science, and conceptual idealism is again the result.

Richard Jennings does not like the scientific realist's view that "there are in the world theoretical (i.e. unobservable) properties and entities which are independent of our theories (and thus do not change when our theories change)" (1989, p. 243). Instead, we are to have 'scientific quasi-realism': "theoretical properties and entities are projected onto the world, they are the offspring of our theories . . . they are things in the projected world rather than in some inaccessible world with which we are utterly unacquainted" (1989, pp. 235–6).

Once upon a time many people believed in witches and subscribed to a body of theory (call it witch-theory) about them – they projected the predicate "is a witch" onto the world. For a while witch-theorists were quite successful: suspected witches were identified, tested for witch-hood in accepted ways, and when they tested positive disposed of in accepted ways too. Despite these 'successes', realists think that witch-theory was false and that there were not really any witches. But quasi-realists think this metaphysically naive: given a consistent witch-theory, there is a 'projected world' of which that theory is true; so there were witches after all, not inaccessible noumenal witches perhaps, but projected witches. Or as Alan Chalmers' once put it (1976: p. 131): "In medieval Europe, witches really inhabited the common-sense world while in modern times they do not . . .".

A sixteenth-century witch hunter prone to bang the table and insist that there really are witches, would not thank Jennings for telling him that what he believes is true, but only of 'projected witches' in some 'projected world'. For the witch-hunter's project was to find the real witches in the real world. Contrary to what Jennings says, no ordinary sense of the term 'true' is involved in quasi-realism. for quasi-realist truth does not seem to admit of an opposite. Suppose the witch-hunter armed with his consistent witch-theory confronts an opponent armed with a consistent no-witch-theory. Is there a truth of the matter, as plain folk think? Jennings thinks not: both theories are true, but of different 'projected worlds': ". . . theories with which we no longer agree . . . are seen as having projected different properties and entities onto the world, and based on their projections, they spoke truly. We now, of course, project differently . . ." (1989, p. 239). The realist question "But are there any real witches in the real world?" is naive and the real world it asks about is inaccessible to us. "There are witches" was true of the projected world of the witch-hunter, "There are no witches" is true of the projected world of today. And that is all we can say.

Jennings claims that his "quasi-realist account of realist talk offers a more reasonable account of scientific talk than the realist account" (1989: p. 239). Stuff and nonsense. Science has discovered that there are in the world no witches, heavenly spheres, phlogiston-rich substances, or caloric fluids. (These negative discoveries stand no matter how sceptical we may be about whether science has any positive discoveries to its credit.) An account of "scientific talk" which denies that science has these negative discoveries to its credit is quite unreasonable.

As for the quasi-realist metaphysic itself, think of the extraordinary claim that "the theoretical things we talk about in our science are the products of our talk" (1989, p. 238). So our talk produces electrons, black holes, genes, or whatever. From which it follows (assuming that the word 'produce' means what it ordinarily means) that there were no electrons, genes, black holes or whatever before people started talking of them. This flies in the face of everything science teaches us about such entities. Science teaches us, for example, that black holes (if there are any) are located far out in space and cannot be produced by anyone's thought or talk down here on earth. If the quasi-realist metaphysic is true, then science is all false. Scientific entities are not 'projected entities', because projected entities only "seem to belong to the outside world" (1989, p. 245) but do not really belong there. Projected entities are mental or linguistic entities, because mere thought or talk can produce them. Quasi-realism is just conceptual idealism, and conceptual idealism is just Bishop Berkeley's metaphysic dressed-up. We should be grateful to Jennings for making this abundantly clear.

The ideal limit theory of truth, discussed in my previous section as an exemplar of epistemic truth-theories, also has relativistic and idealistic implications. Why suppose that ideal scientific practice will converge upon a single limit? Might it not converge upon two different and seemingly incompatible limits, say an ideal particle theory and an ideal field theory? Might not Martian science pursued to its ideal limit be very different from our science pursued to its ideal limit? Matheson canvasses both possibilities and embraces them: "Thus the ideal limit theory might better be called the ideal limits theory. According to it, your community's limit might not be my community's limit; what is true for you might not be true for me. In allowing for the possibility of relativism, the ideal limit theory contradicts global realism" (1989, pp. 253–4). Relativism and idealism go hand-in-hand. Seemingly incompatible ideal limit theories are, according to the ideal limit theory of truth, both true. Incompatible theories cannot both be true of the world. So each must be true of its own world, each must be true of an ersatz world-as-it-is-according-to-my-community's-ideal-theory. Matheson sees and embraces the relativistic consequence of his theory of truth – he does not see the idealist consequence.

Brian Ellis sees neither consequence very clearly. He calls himself in 'internal realist' and embraces something like the ideal limit theory of truth – "truth is a kind of limit notion of reasonable belief" (1985: p. 68). He keeps saying that an internal realist can believe that reality is independent of what we think or say. But in the end the pressure tells and we have ersatz or hyphenated entities again: "The way the world is relative to the sorts of beings we are. That is one of the consequences of internal realism . . . according to the internal realist, there is no way that the world is absolutely, only ways in which it is relative to various kinds of beings" (1985, p. 71). Here the hyphenated entities are the-way-the-world-is (noumenal) and the-way-the-world-is-relative-to-us-humans (phenomenal). Ellis finally gives up the realist ghost: "There is not and cannot be any absolute truth, and therefore there cannot be any way that the world is independently of how we, or some other kind of creature, would evaluate its beliefs about it" (1985, p. 72). Goodness me, science teaches (speculation about God and aliens aside) that the world existed long before sentient creatures did. Is there no way that a world without sentient creatures is? Did the world have to wait for sentient creatures to evolve before it had any objects in it or any structure? How, in an objectless and structureless world, did sentient creatures evolve?

Department of Philosophy,
University of Otago,
New Zealand.

NOTE

* An earlier version of this paper was read at a conference on Objectivity held at La Trobe University, Melbourne, Australia, on 20–22 April 1990.

BIBLIOGRAPHY

Blackburn, S. (1984). *Spreading the Word*, Clarendon Press, Oxford.

Cartwright, N. (1983). *How the Laws of Physics Lie*, Oxford University Press.

Chalmers, A. F. (1976). *What is This Thing Called Science?*, University of Queensland Press.

Devitt, M. (1984). *Realism and Truth*. Basil Blackwell (Second edition 1992).

Eddington, Sir A. S. (1928). *The Nature of the Physical World*, Cambridge University Press.

Eliot, T. S. (1963). *Collected Poems 1909–1962*, Faber and Faber.

Ellis, B. (1985). 'What Science Aims To Do', in P. M. Churchland and C. Hooker (eds.), *Images of Science*, University of Chicago Press, pp. 48–74.

Fine, A. (1984a). 'The Natural Ontological Attitude', in J. Leplin (ed.), 1984, pp. 83–107.

Fine, A. (1984b). 'And Not Anti-Realism Either', *Nous* **18**, pp. 51–65.

Fine, A. (1986). 'Unnatural Attitudes: Realist and Instrumentalist Attachments to Science', *Mind* **95**, pp. 149–179.

Hacking, I. (1983). *Representing and Intervening*, Cambridge University Press.

Jennings, R. (1989). 'Scientific Quasi-Realism', *Mind* **98**, pp. 225–245.

Laudan, L. (1981). 'A Confutation of Convergent Realism', *Philosophy of Science* **48**, pp. 19–49.

Leplin, J. (1982). 'The Historical Objection to Scientific Realism', *PSA 1982* **1**, pp. 88–97.

Leplin, J. (ed.) (1984). *Scientific Realism*, University of California Press.

Levin, M. (1984). 'What kind of explanation is truth?', in J. Leplin (ed.), 1984, pp. 124–139.

Lewis, D. (1984). 'Putnam's Paradox', *Australasian Journal of Philosophy* **62**, pp. 221–236.

Matheson, K. (1989). 'Is the Naturalist Really Naturally a Realist?', *Mind* **98**, pp. 247–258.
Putnam, H. (1978). *Meaning and the Moral Sciences*, Routledge & Kegan Paul.
Putnam, H. (1980). 'Models and Reality', *Journal of Symbolic Logic* **45**, pp. 464–482.
Putnam, H. (1981). *Reason, Truth and History*, Cambridge University Press.
Sellars, W. (1986). *Science, Perception and Reality*, Routledge and Kegan Paul.
Stove, D. C. (1991). *The Plato Cult and Other Philosophical Follies*, Basil Blackwell, Oxford.

ILKKA NIINILUOTO

QUERIES ABOUT INTERNAL REALISM

1. IS KUHN AN INTERNAL REALIST?

In his Presidential Address at the 1990 biennial meeting of the Philosophy of Science Association (PSA) in Minneapolis, Thomas S. Kuhn announced his project of writing a philosophical book on realism and truth. The audience included also members of the Society for Social Studies of Science (4S). To the disappointment of many sociologists of science, who have taken Kuhn as an authority in their campaign for relativist and anti-realist views, Kuhn now argued that the concept of incommensurability is in fact needed "to defend notions like truth and knowledge from, for example, the excesses of post-modernist movements like the strong program".[1] A "strong concept of truth" is needed, he argued, to understand the "normal rules of evidence" for accepting or rejecting theories.[2]

Kuhn thus seems to be a kind of 'realist' after all – to the surprise of some or many of his admirers. But what kind of realist? Kuhn himself notes that his account of the relationship "between the lexicon – the shared taxonomy of a speech community – and the world the members of that community jointly inhabit" clearly cannot be the one Hilary Putnam calls "metaphysical realism".[3] As Kuhn also follows Putnam in giving up the correspondence theory of truth,[4] it seems tempting to conclude that Kuhn's "Post-Darwinian Kantianism" might be a version of Putnam's "internal realism".

The suggestion that Kuhn's and Putnam's positions have important similarities is not a new one. In his insightful book, *Representing and Intervening* (1983), Ian Hacking characterizes both of them as "transcendental nominalists".[5] This description fits very well Kuhn's new account: the world itself is not mind-dependent, or our construction, but its structure or "taxonomy" is relative to a shared conceptual system or "lexicon" which changes with the passage from one community to another.

However, Kuhn is wisely cautious, and does not call himself an "internal realist". We shall see that it is indeed difficult to classify Kuhn's

R. S. Cohen, R. Hilpinen and Qiu Renzong (eds.), Realism and Anti-Realism in the Philosophy of Science, 45–54.

position in this philosophical taxonomy. What is more, I shall argue that Putnam's 1976 distinction between the two rival kinds of realism introduces a misleading dichotomy: a reasonable realist should accept neither metaphysical nor internal realism in Putnam's sense.[6]

2. PUTNAM ON INTERNAL REALISM

Perhaps Putnam's best formulation of the metaphysical-internal distinction is given in *Reason, Truth and History*.[7] The "externalist perspective" (or "God's Eye point of view") of *metaphysical realism* (MR) includes three theses:

(MR1) The world consists of some fixed totality of mind-independent objects.

(MR2) There is exactly one true and complete description of "the way the world is".

(MR3) Truth involves some sort of correspondence relation between words and external things or sets of things.

On the other hand, *internal realism* (IR) accepts the negations of MR1, MR2 and MR3:

(IR1) What objects does the world consist of? is a question that it only makes sense to ask within a theory or description.

(IR2) There is more than one 'true' description of the world.

(IR3) Truth is some sort of idealized rational acceptability.

In 'Why there isn't a ready-made world' (Putnam, 1983), Putnam reformulates or generalizes IR1 to the form

(IR1′) The world is not ready-made,

which denies that the world has any built-in structure. In other words, the world is carved or sliced into pieces (objects, essential properties, facts, causal relations, etc.) only relative to descriptions, theories, or conceptual schemes. In Raimo Tuomela's (1985) terms, IR1′ denies the ontological Myth of the Given.

While MR2 assumes that the whole truth about everything is uniquely determined, IR2 asserts the plurality of truths.

According to Putnam (1978), the existence of a fixed correspondence between words and mind-independent things would imply that

(MR3′) Truth is a radically non-epistemic notion.

Instead, IR3 claims that an epistemically "ideal" theory could not be false.[8]

Putnam clearly thinks that the MR-theses and IR-theses go together in natural "packages". It is of course possible to find, with some variations, more or less articulated formulations of MR (e.g., Wittgenstein's *Tractatus*, where the world, as the totality of facts, is pictured by the language) and IR (Putnam, 1981; Rescher, 1982; Tuomela, 1985).

However, it seems equally clear that theses MR1, MR2 and MR3 (and thereby their negations IR1, IR2 and IR3) are logically independent of each other. Therefore, they allow us to define, besides the 'pure' doctrines MR and IR, six other 'mixed' positions.

For example, the version of "internal realism" defended by Brian Ellis (1989) accepts an epistemic notion of truth (IR3), and the existence of various "epistemic perspectives" (IR2), but at least starts from the assumption that the world has a categorial ontological structure (MR1).

Before studying some of these mixed views, let us note that Putnam's dichotomy allows a realist to advocate either the correspondence theory of truth or the pragmatist theory of epistemic truth. It may be debated whether the redundancy theory (or disquotational theory) is sufficient for someone who wishes to be called a 'realist'.[9] But at least the nominalist Hartry Field (1982), who takes space-time points as the ultimate individuals, suggests that a metaphysical realist could combine MR1 with the redundancy theory. Kuhn (1991) in turn combines the internalist theses IR1 and IR2 with the redundancy theory. When Kuhn states that a conceptual scheme is not a "set of beliefs" – and that sharing such a scheme involves a common structure, not its embodiment in the head of an individual – I think he can be interpreted as denying that truth is an epistemic concept in the sense of IR3. This means that Kuhn's position cannot be classified as MR or IR.

3. THE WHOLE TRUTH ABOUT EVERYTHING

Field (1982) has argued that MR1 does not imply MR2 which "should not be taken as a component of any sane version of realism". Putnam's reply appeals to the following point: if by MR1 the world consists of a fixed set I or individuals and a set P of their properties and relations (and nothing else), then there is an ideal language L which speaks of I and P. Such a language L needs not to be denumerable. But the set Tr_L of true sentences of L, which is well-defined by MR3, is then the set of all truths, as MR2 requires.[10] Another way of formulating this argument is to say that, by MR1, the world is (or can be represented) by a set-theoretical structure W = (I,P). If L is a language which gives a complete description of W, then Tr_L is the theory of W (in the model-theoretical sense), i.e., $Th(W) = \{h \text{ in } L \mid W \models h\}$.

This debate seems to conflate different senses of 'exist'. Field is right that *we* need not actually have a unique complete language L, even if MR1 holds. If the structure W = (I,P) is infinitely complex (e.g., I contains all past, present, and future individuals, P contains n-order predicates and n-place relations for all finite n), then L would be an infinitary language which need not to be finitely definable at all. Further, the set of truths Tr_L in L may fail to be finitely or even recursively axiomatizable. To say with Putnam that such an ideal language "exists" is indeed a very strong idealization.

Grim (1991) has recently argued, by using diagonalization methods of set theory, that Tr_L for a given language L does not constitute a *set*. This is not quite conclusive, since the argument appeals to set-theoretical or metalinguistic truths about Tr_L. Moreover, there are versions of set theory which admit a universal set (i.e., the set of all sets). But at least Grim's claim adds a further obstacle to the derivation of MR2 from MR1.

But who then would be willing to accept the strong thesis MR2? Somewhat surprisingly, one answer is Wilfrid Sellars (1968) whose scientific realism starts from a criticism of the Myth of the Given (i.e., rejection of MR1) and concludes with the equation of truth as assertability relative to the ideal "Peirceish" conceptual system. In the Sellarsian realism, the scientific community reaches in the asymptotic limit the ultimate "God's Eye perspective". This means that Sellars combines the theses IR1, MR2 and IR3.

More generally, the combination of MR2 and IR3 is typical to most versions of epistemic, pragmatist or coherence theories of truth: a sentence is true if and only if it follows from, or coheres with, the set of all true beliefs.[11] These theories presuppose the existence of an ideal belief state which contains all truths about the world. If MR2 fails, then these epistemic theories of truth collapse.

Putnam has recently clarified his earlier discussion of truth as ideal acceptability (IR3) by emphasizing that the ideal state does not simultaneously concern any truth whatsoever – there is no single state of justification that would entail all truths.[12] This is important for Putnam, for otherwise his IR3 would conflict with his IR2.

Peter Unger (1975) has claimed that our understanding of truth assumes that a sentence is true if and only if it expresses the whole truth about everything or else is a part of the whole truth. As the concept of the whole truth about the world is inconsistent (i.e., Unger rejects MR2), Unger concludes that truth is impossible.

Unger's argument fails, however, to shake such theories of truth which do not rely on the fiction of the whole truth. Such accounts are given by non-epistemic notions of truth, in particular by the correspondence theory, which defines truth as a relation between a statement and the world, not between a statement and the set of true beliefs.

In the next section, I defend the claim that the best version of realism is neither "metaphysical" nor "internal" in Putnam's sense, but avoids the problematic thesis MR2 by combining the correspondence theory of truth (MR3) with the pluralist principle IR2 and with a refined version of IR1 which does not imply "conceptual idealism".

4. CORRESPONDENCE WITHOUT "METAPHYSICS"

A scientific realist accepts the minimal ontological assumption that there is an actual world independent of human minds, concepts, beliefs and interests. Let us call it THE WORLD.[13] If you wish, this assumption may be called "metaphysical". But all the evidence of science assures us that THE WORLD has existed long before the emergence of man on earth. Man is an evolutionary product of THE WORLD, and he has learned to interact with this reality and to transform it according to his needs. Against Kant (and later "conceptual idealists"),[14] we have reason to believe that THE WORLD has a physical spatio-temporal structure and obeys natural causal laws.[15] Thus, the mind-independent WORLD is a lawlike flux of causal processes.

For a realist, there must also be a legitimate sense of saying that electrons, planets, stones, trees, and dinosaurs existed before human beings invented languages or conceptual schemes. The suspicion that this is philosophically problematic arises easily from the following fallacy: the equivalence

(1) x exists at t iff the sentence 'x exists' is true at t

implies that the left side is false at those moments of time t, when a concept or a sentence referring to x does not yet exist. But the correct form of (1) is

(2) x exists at t iff the sentence 'x exists at t' is true.

The meta-linguistic sentence on the right side is temporally indefinite, and if it is true now it is true at any time. Even if the sentence 'Dinosaurs exist' was not yet invented at t_0 = 100 million B.C., it is nevertheless counterfactually true that 'Dinosaurs exist' would have been a true sentence at t_0, if someone had formulated it at t_0.

It is another matter that dinosaurs were *identified* as dinosaurs, i.e., as special kinds of physical objects, through a human language. To identify something, it has to be distinguished from other things; to identify an object as a dinosaur, at least some of its characters or properties have to be described. These identifications involve human activities (thinking, perceiving, using language, etc.), as Rescher (1982) convincingly argues. Rescher further asserts that to be *identifiable* is mind-involving as well, since the realm of possibility is mind-dependent, but this is less convincing. If an object x has *some* mind-involving properties (e.g., can be identified, seen, thought, etc., by human beings), x itself may nevertheless exist in a mind-independent way. The possibility of the identification of a physical thing, like a dinosaur or a chair, is indeed based on its mind-independent properties (location in space and time, causal continuity, qualities).[16]

We may thus make a distinction between *UFO's* ('unidentified flying objects') and *IFO's* ('identified flying objects'). Thesis MR1 is then true, and IR1′ false, in the sense that THE WORLD contains UFO's, which are not our constructions, or produced by us in any causal sense. But these UFO's are not "self-identifying objects" in a bad metaphysical sense:[17] they are potentially identifiable by us, as extended elements or "chunks" of the world flux, by means of continuity and similarity.[18] IFO's, on the other hand, are in a sense man-made constructions, objects under a description, and hence exist only relative to conceptual schemes.

Even if we reject the conceptual idealism ("all objects are IFO's"), our minimal realism is compatible with a version of IR1: THE WORLD does not contain self-identifying individuals, but can be categorized into objects in several alternative, overlapping ways relative to conceptual schemes. For example, depending on the choice of a suitable conceptual framework, THE WORLD can be 'sliced' or 'structured' to a system of momentary events, mass points, physical systems, etc.

The distinction between UFO's and IFO's may seem to resemble Kant's *Dinge an sich* and *Dinge für uns*, but there is a crucial difference in interpretation: an IFO is not a veil which hides a UFO from us, but rather a *partial truth* about a UFO. A UFO corresponding to an IFO is not defined negatively, as a propertyless bare particular, i.e., as nothing, but rather it is a complex composition of properties, and the IFO as it were gives a partial description of it (i.e., of those of its properties expressible in the chosen language L). In this sense, the existence and the properties of IFO's depend on the "reality", and knowledge about IFO's gives us truthlike information about the UFO's.[19]

Thus every interpreted language or conceptual system L, whose terms have a meaning through social conventions, "picks out" or determines a structure W_L, consisting of objects, properties, and relations, and exhibiting THE WORLD as it appears relative to the expressive power of L.[20] Such structures W_L are fragments or "versions" of THE WORLD. This view can be regarded as a formulation of the internalist principle IR1, since it allows the reality to be structured in many ways. It also denies the sort of metaphysical realism which assumes the existence of an ideal "Peirceish" language L such that THE WORLD = W_L.

It is important to point out that the concept of W_L is not epistemic: if a community shares the language L (to use Kuhn's phrase), W_L is not what they *believe* about the world, but rather "the way the world *is*" relative to L (to use Goodman's phrase). In other words, if a member of this community has false beliefs (expressible in L), then his "life-world" differs from W_L.

Furthermore, this account does not lead to relativism, since all structures W_L are fragments of the same WORLD – and therefore cannot be incompatible with each other.[21]

It might be objected that my suggestion leads back to unacceptable metaphysics. Martin Kusch (1991) has argued that my WORLD is something like Schopenhauer's *das Ding an sich*, a blind will: how can THE WORLD, which does not have any inherent categorical structure, "choose" the structure W_L?

The metaphorical talk about "choices" can be understood here as easily as in decision theory and game-theoretical semantics: in the game of exploring reality, the choice of the language L is my first move, and it is followed by "Nature's choice" of the structure W_L. The game continues by my attempt to study the secrets of W_L. And all true information about W_L, viz. about a fragment of THE WORLD, also tells something about THE WORLD.

As each W_L is a structure for language L, we can directly apply Tarski's model-theoretical definition of truth for sentences of L. For each L, we can define the class of truths in L:

$$Tr_L = Th(W_L).$$

This agrees with IR2, and does not imply the problematic MR2, since the union of all classes Tr_L over "all possible languages" L (whatever that might mean) is not well-defined. This approach also allows us to define an objective notion of truth about mind-independent reality: a sentence h in L is *true in THE WORLD* if it is true in W_L.

In this way we have combined IR1 and IR2 with MR3: our conception of truth is clearly a version of the correspondence theory, but it is not metaphysically suspectible in the sense feared by Kuhn and Putnam.[22] Hence, if Kuhn were willing to accept W_L as an explication of his notion of a world with a taxonomy, he could annul his rejection of the correspondence theory of truth.

Department of Philosophy,
University of Helsinki,
Finland.

NOTES

[1] See Kuhn (1991). p. 4. For my evaluation of the strong program, see Niiniluoto (1991).

[2] *Ibid.*, p. 9.

[3] *Ibid.*, p. 10.

[4] *Ibid.*, p. 6.

[5] See Hacking (1983), p. 109. According to metaphysical nominalism the world consists of a fixed class of individuals, while "transcendental nominalism" claims that individuals (and their classes) are "constituted" by human minds or languages.

[6] My first statement to this effect was given in Niiniluoto (1980). See also Niiniluoto (1984), pp. 177–178; (1987a), pp. 141–143.

[7] See Putnam (1981), p. 49. In Putnam (1978), p. 123 internal realism was defined as an empirical theory which explains the convergence and success of science.

[8] Cf. Putnam (1978), p. 125; (1983) p.xvii. To say that truth is non-epistemic does not imply that truth is entirely inaccessible to us (cf. Niiniluoto, 1984, p. 178). Rather it means

that the concept of truth cannot be defined, or co-extensively characterized, by epistemic concepts. Tuomela's (1990) internal realism claims truth to be epistemic in a much weaker sense: talk of truth presupposes knowledge about meanings. But Tuomela (1985) has also argued that, by the *scientia mensura* principle, truth can be characterized in terms of best explanatory theories.

[9] See Horwich (1990). Cf. Putnam (1983). p. xiv; (1990), p. 31.

[10] See Putnam (1990), p. 31.

[11] See Walker (1989).

[12] See Putnam (1990), pp. vii–ix. This shows that Hacking's (1983) equation between Peirce and Putnam is problematic.

[13] In this paper, I am primarily interested in the Popperian World 1 (nature) rather than World 2 and 3 (cf. Niiniluoto, 1984, Ch. 9). To deny the independent existence of World 1 leads to ontological idealism.

[14] See Rescher (1982). Rescher's conceptual idealism argues that "nature for us" or "reality-as-we-picture-it" is "mind-invoking" in a conceptual – rather than ontological or historical – sense. But he also makes the more radical conclusion that an altogether mind-independent reality is "an essentially empty idealization" (p. 154), and that "reality-as-we-think-of-it (= our reality) is the only reality we can deal with" (p. 169).

[15] Putnam (1983), p. 211, argues that the concept of total cause involves the non-physical concept of explanation. This does not prove that causal processes are mind-involving.

[16] According to nominalism, an object is green because we apply the predicate 'green' to it. It is more plausible to accept the converse thesis: we can correctly apply the predicate 'green' to objects that have the common quality of greenness (cf. Armstrong, 1978).

[17] See Putnam (1981), p. 53.

[18] For a theory of individuation and identification of physical objects, see Hintikka and Hintikka (1989). In spite of his "semantical Kantianism", Hintikka emphasizes that "constitution" or individuation does not "create inhabitants of any possible world" (Hintikka, 1975, p. 216).

[19] See Niiniluoto (1987a, b).

[20] If the language L is not semantically determinate, i.e., some of the terms of L are vague and do not have unique extensions, language L determines a class of structures – instead of the unique W_L. See Niiniluoto (1987), p. 146.

[21] Cf. Niiniluoto (1991).

[22] In particular, I don't need the assumption that "there is such a thing as *the* correspondence (the One, metaphysically singled out correspondence) between words and things" (Putnam, 1983, p. xi).

BIBLIOGRAPHY

Armstrong, D. M. (1978). *Nominalism and Realism*, Cambridge University Press, Cambridge.

Ellis, B. (1988). 'Internal Realism', *Synthese* **76**, pp. 409–434.

Field, H. (1982). 'Realism and Relativism', *The Journal of Philosophy* **79**, pp. 553–567.

Grim, P. (1991). *The Incomplete Universe: Totality, Knowledge and Truth*. The MIT Press, Cambridge, MA.

Hacking, I. (1983). *Representing and Intervening*, Cambridge University Press, Cambridge.
Hintikka, J. (1975). *The Intentions of Intentionality*, D. Reidel, Dordrecht.
Hintikka, J. and M. B. (1989). 'Toward a General Theory of Individuation and Identification', in *The Logic of Epistemology and the Epistemology of Logic*. Kluwer, Dordrecht, pp. 73–95.
Horwich, P. (1990). *Truth*, Blackwell, Oxford.
Kuhn, T. S. (1990). 'The Road Since Structure', in A. Fine *et al.* (eds.), *PSA 1990*, vol. 2, Philosophy of Science Association, East Lansing, MI, 1991, pp. 3–13.
Kusch, M. (1991). 'Koko totuus totuudesta', *Tiede ja edistys* **16**:4, pp. 284–297.
Niiniluoto, I. (1980). 'Scientific Progress', *Synthese* **45**, pp. 427–464.
Niiniluoto, I. (1984). *Is Science Progressive?*, D. Reidel, Dordrecht.
Niiniluoto, I. (1987a). *Truthlikeness*, D. Reidel, Dordrecht.
Niiniluoto, I. (1987b). 'The Varieties of Realism', in P. Lahti and P. Mittelstaedt (eds.), *Symposium on the Foundations of Modern Physics 1987*, World Scientific, Singapore, pp. 459–483.
Niiniluoto, I. (1991). 'Realism, Relativism, and Constructivism', *Synthese* **89**, pp. 135–162.
Putnam, H. (1978). *Meaning and the Moral Sciences*, Routledge and Kegan Paul, London.
Putnam, H. (1981). *Reason, Truth and History*, Cambridge University Press, Cambridge.
Putnam, H. (1983). *Realism and Reason*. Philosophical Papers vol. 3, Cambridge University Press, Cambridge.
Putnam, H. (1990). *Realism with Human Face*, Harvard University Press, Cambridge, MA.
Rescher, N. (1982). *Conceptual Idealism*, University Press of America, Washington, D.C.
Sellars, W. (1968). *Science and Metaphysics*. Routledge and Kegan Paul, London.
Tuomela, R. (1985). *Science, Action and Reality*. D. Reidel, Dordrecht.
Tuomela, R. (1990). 'Causal Internal Realism', in G. Pasternack (ed.), *Philosophie und Wissenschaften*, Peter Lang, Frankfurt am Main, pp. 165–179.
Unger, P. (1975). *Ignorance: A Case for Scepticism*, Clarendon Press, Oxford.
Walker, R. C. S. (1989). *The Coherence Theory of Truth*, Routledge, London.

QIU RENZONG

HOW TO KNOW WHAT RISES UP IS THE MOON? – ON THE CONCEPT OF REALISM AND THE IRRELEVANCY OF QUANTUM MECHANICS TO THE DEBATE ON REALISM VS. ANTIREALISM

In recent years the debate on philosophical problems in science, particularly in modern physics, has been in the ascendent in China. I would like to refer only to the literature such as [4, 5, 9, 10, 11, 12, 13, 14, 15, 16, 17, 18, 25, 26, 27], but this list is far from complete. Much of the debate involves the problem of realism vs. antirealism in science which might be said to be an important issue in the frontier of contemporary philosophy of science. If this debate is unfolded properly, it is expected that a fruitful result will be brought about with benefit to science as well as to philosophy. However, both in China and abroad there are some confusions concerning the concepts of 'real', 'reality' and 'realism'; the different issues or different dimensions of one issue are too tangled up to unravel in the debate on realism vs. antirealism in science with adverse impact upon the debate.

This paper consists of four sections. In the first section, I try to clarify the concept of reality. In the second section I try to formulate some criteria for philosophical argument and elaborate a case for realism, and point out the metaphysical nature of realism. In the third section, I try to argue that what constitutes the real world is the individual particulars that possess their own self-identity, and argue for the irrelevancy of quantum mechanics to the debate on realism vs. antirealism. In the fourth section, I discuss the semantic and epistemological dimensions of the debate, and argue for the logical and practical possibility of rational theory-choice.

WHAT IS IT RISES UP?: CONCEPTS OF REALITY

Although I do not think that philosophy can be reduced to the therapy of linguistic diseases, as L. Wittgenstein claimed, yet his claim is revealing in the case of the debate on realism vs. antirealism in science. Many scientists as well as philosophers, including the great ones, used

R. S. Cohen, R. Hilpinen and Qiu Renzong (eds.), Realism and Anti-Realism in the Philosophy of Science, 55–73.

or use terms such as 'real', 'reality' and 'realism' in a non-rigorous sense. For instance, 'empirical realism' (I. Kant) or 'phenomenal realism' (N. Bohr). It led to conceptual confusion.

What is meant by 'real' or 'reality' depends on the context of discourse or the context in which the question "Is it real?" is raised. One meaning of real or reality is simply 'existing' or 'existence'. The question is like "Does it exist?" To this kind of question we can give a positive answer saying all exist even which are dependent upon our mind. For instance, we always say something exists in our mind. All sensations, perceptions, desires, beliefs, loves and hatreds etc. do exist, but exist in our mind. It is this 'existence' sense when people say 'empirical realism' or 'phenomenal realism', that refers to the theory which claims that the empirical experience or the empirical world, or the phenomena or phenomenal world does exist, they are not merely illusion. However, when philosophers and scientists argue on "Is the external world real?" in the context of the realism vs. antirealism debate, the term 'real' or 'reality' is used in a sense more than in the sense of 'existence'. Precisely, there is a sense of 'independent of human mind' apart from the sense of 'existence'.[1]

P1: "Does A exist?"

P2: "Is A real?"

These are two different problems. We can say that something exists but is not real. P1 only concerns its existence or non-existence, the questioner does not necessarily want to know its relationship with human mind: dependent upon the human mind or not. The asking of P2 started with the time when one discovered the distinction between appearance and reality. The appearance of a thing actually depends upon the human mind, but its reality does not. When one saw the moon in the sky, the moon in water, the moon in a mirror, the moon in a dream . . . then he/she asked: "Is the moon real?" It means to ask whether the moon exists only in the human mind or not. The answer given by Plato was: What is real is not the moon we see but the idea of the moon: the ideas exist independently of the human mind. When Berkeley put forward his famous thesis, that is "Being is what is perceived", actually he was talking about, or his thesis implies, the question "What is real?" but not the question "What exists?", or "Does the being depend upon human perception or not?" In modern science the question was usually put in the form: "Is a theoretical entity only the creation of our mind, or has it its own existence independent of the human mind?"

I just said that even some great philosophers did not use the terms 'real' or 'reality' in a strict sense, and those philosophers include Karl Popper [21]. His 'pluralistic realism' includes the 'World 2', that is the mental or conscious states and events. In arguing for the 'reality' of his 'World 2' he defines 'reality' as such a property that anything can interact causally with 'World 1'. So far as the mental or conscious state and events interact causally with 'World 3', 'World 2' which is composed of these states and events is 'real'. However, his so-called 'World 2' cannot exist independently of the human mind. So strictly speaking, his 'Three Worlds' theory is a doctrine about 'existence', but not about 'reality'. Similarly in discourses of 'empirical realism' or 'phenomenal realism' the terms 'real' or 'reality' are used in the sense of 'existence'.

In his interesting article 'Five Moons Rise up' Zhang Huaxia [27] also did not notice the difference between the concept of existence and the concept of reality when he argued for his scientific realism in which phenomenal world, classic world, quantum world, empirical world and theoretical world exist. He said in one place of his article:

> Classic reality is only one kind of objective reality, and objective reality is no more than the existence independent of human consciousness (p. 45).

Here he pointed out that the concept of 'reality' contains two senses of 'existence' and 'independence of human mind'. But he said in another place:

> Therefore, the secondary quality of a material object is also one kind of the properties of material world. Recognizing it, recognizing the phenomenal reality, is no more than recognizing some objective reality of the world. It is absolutely not 'a compromise with idealism'. On the contrary, denying it amounts to denying the world which is full of sound and colour, and can be seen and touched, the world is thus described as a dismal picture in which only a swarm of atoms dance in the air (p. 44).

He is right, when he criticized Eddington and other physicists. They like to describe the world as a picture in which only a swarm of atoms (or other more elementary particles) dance in the air; I shall elaborate it later. But what is meant by 'reality' here is actually 'existence'. There is no problem that the secondary quality has its objective content; however, it cannot 'exist independently of the human mind'. That is, although electromagnetic waves of different wave length, or vibrations of the air, or a given distance between molecules of material objects, etc. are independent of the human mind, yet if there had been no creature which is equipped with sense organs and their centers, no creature with the ability of seeing, hearing, touching, there would have been no "world full of sound and color, being seen and touched". The author is faced with the same problem when he was talking about the 'phenomenal world' and the 'theoretical world': How can these two worlds exist independently of the human mind? He seems to feel that there might be some problem, so he said, talking about the theoretical world:

> Scientific theory has the form and implication of its own relatively independent existence (p. 50).

It is not clear what is meant by the form and implication of existence. But here he was talking first about the 'existence' of scientific theory; secondly, he used the qualification of 'relatively independent'. We can accept the 'relative independence' of scientific theory. However, the problem is: Can a theory exist independently of the human mind? A theory is not a biological species which can evolve independently of human mind. The evolution of scientific theory has to rely on human ingenuity.

> I maintain that a theory should refer to, and explain, the external, objective world; this should be one of the aims of scientific theory (p. 50).

He is right when he contrasted theory with the external world. Theory is the model of the world, not the world itself. Therefore, the concept of 'theoretical world' is untenable.

Only if we distinguish the concepts of 'reality' and 'existence', can we know that what rises up is the moon itself or the shadow of the moon? The so-called 'empirical world', 'phenomenal world' and 'theoretical world' are actually the shadow of the moon – the experience of the moon, phenomena of the moon, the theory of the moon, but not the moon itself.

HOW TO KNOW WHAT RISE UP IS THE MOON?: A CASE FOR REALISM

The debate on realism vs. antirealism is a philosophical issue. To resolve a philosophical issue we need to employ philosophical argument. The argument I shall employ here is to compare various philosophical explanations[2] to the following proposition (explanandum): I see that the moon rises up. Its generalized form is that: S perceives that p.

At first, we should put forward some criteria according to which we can say that this philosophical explanation is better than that one. Following the example of the discussion of scientific explanation in philosophy of science, we can work out the criteria for judging philosophical explanations as follows:

1. A good philosophical explanation should be simple. The simplicity is either ontological or structural. The ontological simplicity requires that we do not presuppose any unnecessary entity. The pursuit of this simplicity has been an ideal of philosophers since ancient times. Both Thales's 'water' and Heraclitus's 'fire' and Chinese philosophers' 'Tao' and 'Qi'[3] embody the pursuit for the ideal. Structural simplicity requires that there should be no omittable conjunction, disjunction, appendix, or proviso.

2. A good philosophical explanation should possess greater explanatory power. That is, it can explain not only what other explanations can, but also explain what others cannot.

3. A good philosophical explanation should possess greater predictive power. That is, it can predict not only what other explanations can, but also what others cannot.

4. A good philosophical explanation should possess more fertility. That is, it can lead to more interesting problems and suggest new approaches or new research lines.

Now let us to compare various available explanations (*explanans*) for the proposition 'I see that the moon rises up' according to the criteria mentioned above:

E1. "I see that the moon rises up, because there is an entity outside me called 'the moon' from which the reflected light from the sun stimulates my retina, and thereby the image of the moon is produced on the retina." This is a realist causal explanation.

E2. "I see that the moon rises up, because there is a mechanism so far unknown in my genetic structure which causes me to see the moon rising up at a given time." This is an innatist causal explanation.

E3. "I see that the moon rises up, because God or other supernatural beings use divine or magic power to make me see the moon rising up." This is a theological or magical causal explanation.

E4. "Human life is a dream. That I see that the moon rises up is a part of the dream. And who knows whether it is I who see the 'Chang E'[4] or it is 'Chang E' who see me." This is Zhuangzhouist explanation.

E5. "We cannot go beyond our own experience. That I see that the moon rises up is my own visual experience. It is impossible to provide any explanation for the experience beyond it." This is a Humean explanation, or radically empiricist explanation. Actually it declines to provide any explanation.

E4 and E5 are the simplest explanation in ontology and structure, but they explain nothing, they have no explanatory and predictive power at all. Neither of them is a causal explanation. They do not explain why we have this visual experience ('see that the moon rises up'). And they do not predict anything either: We do not know what we would dream or what we would see at the next moment. The only thing we can do is to record this or that experience and observe the order with which they emerge in my mind.

They are confronted with other troubles. If we explain 'I see that the moon rises up' by the dream, then I ask whether the moon is my dream is the same as the moon in others' dreams or not? If they are the same, how to explain it? If not, there are numerous moons in dream, and this runs counter to people's intuition. And nothing explains why I and other people dream. The radically empiricist explanation is faced with the same troubles.

E2 and E3 are similar. Their difference lies in that: E3 ascribes the cause of our experience to some external supernatural being, whereas E2 ascribes it to some mystical internal natural factor. However, nobody on earth understands the regularity of this supernatural being or natural factor: Why it causes this kind of experience, and not another kind? What will it cause at the next moment? etc., and these remain to be explained. So their explanatory power is very limited, and their predictive power is near to zero.

E1 can explain why what I see is the moon but nothing else. Because the image emerged on my retina was caused by the light from the sun, reflected by the surface of the entity called the moon. The astronomical knowledge can be incorporated into E1, and help it to form counterfactual conditionals such as: "If we had met last time in the middle of months of the lunar calender, and there had been no clouds in the sky, we would have seen the full moon as we see it tonight." And we can predict that we will see the full moon in the middle of months of the lunar calender, see nothing at the beginning or end of months, and see the crescent in between. E1 can lead to many interesting problems such as how about the surface of the moon, its other side, the temperature of its surface, is there any air, is there any trace of life, can human beings survive on it, how to climb up to it, etc.

It might be objected that we can make E2 and E3 possess the same explanatory and predictive power as E1 does. For example, God created the moon and made us see it, he arranged the configuration of the sun, the earth and the moon so nicely that we can predict when there will be full moon or crescent. Or some mystical factor in our genetic structure acts as God. Although E2 and E3 now possess all that E1 does, the cost is that they become more complex than E1 both in ontology and in structure, because they need an extra external supernatural being or an internal mystical factor, and an additional explanation of why they want us to see the moon.

It follows from the argument that E1 is better than any other available explanations. When I see that the moon rises up, the best explanation of it is that: What rises up is the moon. The moon in the first sentence is our visual experience, the moon in the second sentence is the entity which exists independently of our experience.

The realist explanation for our experience entails an ontology or metaphysics. In other words, realism is inescapably metaphysical: the physical objects exist independently of our mind, and the external world is composed of these objects. But when we say that it is the object of the external world which caused our experience, it does not entail that our experience and theory built on the experience must constitute true knowledge of the object. Is there any external world outside experience? Does/can experience provide true knowledge of the world? They are not the same problem, and should have different solutions. F. Engels [6] pointed out that "Is our thinking capable of cognition of the real world? Are we able in our ideas and notions of the real world to produce a correct reflection of reality?" are 'another side' of 'the great basic question of philosophy'. Our contemporary philosophers [1, 3] point out it too.

When some philosophers talk about semantic realism or epistemological realism (for example [16]), they actually deal with another problem. In other words, they are talking about the problem "Are human beings able to obtain true knowledge of the external world?" or "Do our thoughts (mental representations) or sentences (linguistic representations) correspond to the external world?", but not the problem "Does a world exist independent of human mind?" Of course, maintaining or denying that human beings can reach true knowledge, maintaining or denying that a theory can be true, may lead to realism or non/antirealism. But they are different problems logically. As M. Denitt [3] point out, realism is a doctrine which claims the existence of physical entities independent of the mental, and this definition is entirely unsemantic, making no mention of representation or truth. Realism is a philosophical doctrine on the nature of the world, not a philosophical doctrine on the nature of language or thought. He pointed out that people often refuted realism with the following reasoning:

1. If the realist's independent reality exists, then our thoughts/theories must mirror, picture, or represent that reality.
2. Our thoughts/theories cannot mirror, picture, or represent the realist's independent reality,
3. So, the realist's independent reality does not exist [3].

But 1 is false. So 3 cannot be derived from 2. 1 is untenable because it confuses the realist problem with the semantic/epistemological problem. If it is clear that the crucial difference between realism and anti-realism lies in "is it accepted that the external world exists independently of human mind?", then a version of realism such as H. Putnam's 'internal realism' [22] is actually a hidden version of anti-realism, but does not "continue to hold fast to a realist position" [8]. Because he claimed that:

> 'Objects' do not exist independently of conceptual schemes. We cut the world up into objects when we introduce one or another scheme of description." (p. 52).

As pointed out by C. Brown [1], both Kant and Putnam hold that the world we know and talk about is empirically real, but hold also that it is mind-dependent.

WHAT IS THE MOON?: ON THE REAL WORLD OF INDIVIDUALS WITH SELF-IDENTITY AND THE IRRELEVANCY OF QUANTUM MECHANICS TO THE DEBATE ON REALISM VS. ANTIREALISM

It might be argued that the argument developed in the second section only holds in the case of macro-objects, not in the case of micro-objects, because the latter can not be perceived directly by our senses and our mind, and some of them were discovered after scientists had predicted them before. However, it will not spoil the argument in our second section: We can transform the question "What is the better or best explanation of 'I see that the moon rises up' " into the question "What is the better or best explanation of all empirical data about electrons available to scientists?"

On the basis of the argument discussed above, the better or best explanation of "I see that the moon rises up" is that there is an entity which exists independently of me called the moon which caused the emergence of my visual experience.[5] Analogically, there are other various entities that exist independently of me which caused and will cause the emergence of my other various experiences. The external world is composed by these independently existing entities. These entities can be grouped into different classes according to the similarity of their properties, but each element in the class is a single particular individual, from elementary particle to galaxy, but the most delicate individual is the person.

Each individual has its own self-identity which makes it discernible by other individuals and enables us to identify it as distinguished from other individuals. Self-identity means that an object is the same identical with itself (Hume), or means that things are what they are (Russell). Each object has its own 'life' or 'lifetime'. Throughout its lifetime an individual object remains the same and maintains itself through the passage of various outside influences. It is the self-identity that makes it into an individual. The concept of 'self-identity' is closely related with the concepts of 'existence' and 'reality'. A thing is said to exist, that is to say it has self-identity. So it might be said that the existence of a thing presupposes its self-identity. If a thing has no self-identity, how we say it exists? There are two alternatives for a thing without self-identity: It does not exist or it is another thing. If we say a thing without self-identity exists, we cannot distinguish it from other things or we identity it with another thing. A thing without self-identity cannot be said to exist, and thereby cannot be said to be real.

It might be objected that the concept of self-identity does not hold in the case of micro-particles, because they are indiscernible: All particles of the same class have the same physical properties. However, self-identity cannot be reduced to physical properties. The physical properties of two electrons are the same, but their own histories (though very short), and the context in which they exist during their lifetime, are not the same. Just like two homozygote twins, whose genes are identical, but the environment of their development and growth and their personal history cannot be identical. They still have their own self-identity.

On the basis of the concept of individuals having self-identity, philosophers should not, like some physicists, reduce all things in the universe to the sum of some 'elementary' particles. A table has its own self-identity, this self-identity is irreducible. This is the reason why I said that Zhang [27] is right when he criticized Eddington's attempt to reduce a table to the sum of atoms. Objects at different levels [of reality] have their own irreducible self-identities, and the external world is composed of multi-levelled individuals.

The Honourary President of the Sino-British Summer School of Philosophy, Peter Strawson's thesis on objective individualistic ontology [23] may be helpful to our discussion:

> We think of the world as containing particular things some of which are independent of ourselves; we think of the world's history as made up of particular episodes in which we may or may not have a part; and we think of these particular things and events as included in the topics of our common discourse, as things about which we can talk to each other. These are remarks about the way we think of the world, about our conceptual scheme. A more recognizably philosophical, though not clearer, way of expressing them would be to say that our ontology comprises objective particulars (*Ibid.* p. 15).

In [24] he reemphasized his individualistic ontology, and although he accepted that it is reasonable for physicists to claim that the ultimate material objects are atoms, electrons, and further, quarks and fields in some basic sense, but not in the sense of philosophers' primary concern. What mostly concerns philosophers is the conceptual scheme, or general framework of thinking about reality and the world, and even scientists have to start their inquiry within a framework.

There are some elements of individualistic philosophy in Mao Zedong's philosophy [9]. He emphasized that universals only exist in particulars. It means that universals do not exist independently like Plato's 'ideal world' or Popper's 'World 3'. The world is full of particulars in which universals exist. He pointed out:

> These are two cognitive processes: one is from particular to general, the other is from general to particular [19].

That is to say, it is the particular which is the starting-point and end of human cognition, but not the universal. Why? The ontological reason is that the world is composed of numerous particular individuals, and what we have to deal with are all particular individuals. Building this conceptual scheme or thinking framework is very important to the inquiry in all fields and particularly important to the fight against dogmatism and conservatism in today's China.

An individual object cannot be reduced to its constituent parts, nor to its properties or qualities either. The properties of an individual object depend upon the context in which it is situated, and the perception of these properties by the human mind depends at least partly upon the condition or state of the perceiver. This holds for the secondary qualities as well as the primary [15, 18]. To say "The table is red" or "The table has the quality 'red colour' " needs two conditions: the light is adequate and the speaker's system of sense organs-brain as well as her/his psychology is normal. However, an individual object such as a table cannot be reduced to its property or quality such as 'red'. A material object has the property of extension, but its length depends upon the choice of a reference frame: its length contracts in proportion to the speed of motion of reference frame. Under these two conditions it cannot follow that the object does not exist independently of our mind from the fact that 'red' and 'length' will be changed with a choice of reference frame.

From the view of individualistic realism, the problems facing quantum mechanics are irrelevant to the debate on realism vs. antirealism, i.e. irrelevant to the problem "Does the moon exist when nobody see it?"

The concept of relevancy has not been sufficiently discussed in philosophical literature. L. J. Cohen[6] pointed out:

> We say that a proposition R is relevant to the question Q if and only if R gives us a reason to accept or reject a proposition A as an answer to the question Q.

However, neither the indeterminacy, nor EPR correlation in quantum mechanics itself gives us a reason to accept or reject the realist or antirealist answer to the question "Does the external world exist independently of our mind?". Only a conjunction of this property or relationship with other propositions can give us a reason to do so.

1. *Indeterminacy*

That the momentum and position of a micro-object cannot be determinately measured at the same time concerns the problem of measuring some properties of micro-objects, but not the problem of the independent existence of a micro-object itself. It is quite often that the state of an object is changed by observation in a psychological or sociological study. Suppose that a subject S is originally in state p, and the psychologist or sociologist E intends to study S in the state p. If the question asked by E is not proper or is not properly asked, S can be shifted into state q; thereby S gives an answer to the question different from the answer if S is in the state p. In some sense we can say that the state q is 'created' by E or 'emerges' with E's suggestion. However, that the state of a subject can be changed by the experimenter does not give us a reason to reject the independent existence of the subject. In the same vein, that there is no indeterminacy in classical physics does not give us reason to reject the phenomenalist or radically empiricist answer to the problem of "Does the external world exist independently?" According to a phenomenalist or radical empiricist we cannot go beyond our sense-data or experience.

Now there seems to be a physicalist account for the indeterminacy: If we want to measure the position of a particle, we have to operate the apparatus to produce a light wave with short wave length and great energy, and thereby change its speed, and as a result, we cannot accurately measure its speed. If we want to measure its speed, we have to produce a light wave with long wave length and it makes us unable to accurately measure its position. Even if the interference of the apparatus in the measurement is such that it changes the state of the measured particle, or even 'creates' its new state or 'make' its 'emergence', it concerns only the state of the measured particle, not the object itself. In the same vein, if the indeterminacy is not ascribed to the interference by the observation, but to the probabilistic property of the particle itself, it also does not give us a reason to accept the realist answer to the problem "Does the external world exist independently?" because it might be the case that all facts including the probabilistic property of particles are facts only within our experience, and do not reach existence outside our experience.

2. *EPR Correlation*

EPR correlation may be the most bizarre phenomenon. There are optional explanations to this bizarre phenomenon including:

(1) There is a distant action between particle A and particle B.

(2) There is an interference by an observer when measuring any of these two particles.

(3) There is some unknown special interaction or correlation between A and B.

Now some Chinese scholars [13, 27] hold that (1) is already precluded by modern physics, (2) implies that the "realist presupposition would be no longer correct", so the most plausible explanation is (3) which indicates the non-separability and non-locality of quantum objects. However, even if (2) is the most plausible, recognizing the fact that the property or state of particles is interfered by observation or experiment does not necessarily lead to denying the independent existence of quantum objects. Now it has been confirmed by experiments that there is EPR correlation between particles. However, no matter how bizarre the EPR correlation is, it is a property of particles, and nothing here concerns their very independent existence. In the same vein, even if one day (1) or (3) turns out to be more plausible, it does not entail the realist answer to the question of "Does the external world exist independently?" because it might be the case that all these are only within the limit of our experience, that nothing can go beyond our experience.

The conclusion is: problems of quantum physics show how to more deeply understand various bizarre properties (quite different from those of macro-objects) and the mechanism of interaction between micro-object, experimenter/experimental apparatus, and context in which they encounter, but all these are irrelevant to the debate on realism vs anti-realism. In other words, the problem raised in quantum physics is not "Does the moon exist when nobody see it?" but "Is the moon when somebody see it the same as the moon when nobody see it?" [10] If the answer is "the same", it does not give us reason to reject anti-realism, and if the answer is "not the same", it does not give us reason to reject realism. The assertion that the puzzles in quantum physics are relevant to the debate involves a misunderstanding of the concepts 'reality' and 'realism'.

THE MOON AND ITS MODELS: RATIONALITY OF THEORY CHOICE

Our external world is composed of particular individuals like the moon, these individuals interact each other, and are exerted upon by downward and upward causation between different levels of individuals. On the other hand, human cognition is limited in a given space-time, and circumscribed by our logic, and our linguistic and conceptual framework which is not infallible. So the theory of the moon, or its model, formed in a certain historical period is always incomplete. Scientists and philosophers once were over-optimistic, when they believed that if a theory has been confirmed over a long period of time, it can be confirmed as true. In view of the fact that Newtonian theory, which had been well-confirmed for two hundred years, was refuted at last, and hypotheses and theories in modern science were put forward and then refuted successively with a dizzy speed, many are now inclined to over-pessimism and scepticism: to obtain true knowledge is impossible, or as rival theories can be confirmed by the same experimental data, rational choice between two competing and incompatible theories is impossible.

Both over-optimism and over-pessimism are untenable

In the long run we cannot deny that knowledge of the world (including nature, society and human beings) themselves possessed by modern people is much richer, and much more correct than the knowledge possessed by our remote ancestors and by ancient people in the age of Plato and Confucius. Modern science and technology put tremendous power in the hands of modern people, and has brought much more welfare to them. On the other hand, the power is formidable, and created many serious problems facing human beings who are now working to deal with them. This is an intuition shared by common sense. Any logical or philosophical argument cannot reject or eliminate this intuition. However, the question of whether a scientific theory can be said to be true or when it can be said to be true is not a question that can be answered instantly. The life of many theories is full of haps and mishaps. Theory grows and develops with human ingenuity. It is very difficult for us to answer the question "Is it true or false?" at the beginning of its life. Just as when we see a movie, when a character appears on the scene, kids ask their parents: "Is he a good man or bad man?" It is difficult for parents to answer the question, and have to say: "You will find the answer at the end of the movie." Scientific theory is the same: We don't know who is the winner and who is the loser until the game ends. However, during the scientific game, we can judge which theory is better according to some well-received epistemic values. That allows a rational choice between competing theories. These epistemic values include:

V1. Corroboration or support by empirical evidences and the degree of corroboration or support. If a theory (T2) is corroborated by available empirical evidences better than another (T1), or T2's degree of corroboration is higher than T1's, then we have reason to say that T2 is a better theory than T1.

V2. Explanatory power. If T2 can explain all explained by T1, and explain what cannot be explained by T1, then we have reason to say that T2 is better than T1.

V3. Predictive power. If T2 can predict all predicted by T1, and predict what cannot be predicted by T1, then we have reason to say that T2 is better than T1. The predictive power of a theory can be measured by the novelty of the prediction, and the novelty of a prediction can be measured by the distance between the prediction and background knowledge.

V4. Fertility. This means that a theory can lead to interesting scientific problems and/or suggest new approaches and new research lines. If T2 can lead to more interesting problems than T1, or/and it suggests a new approach and a new line of research, but T1 cannot, then we have reason to say that T2 is better than T1.

V5. Simplicity. If T2 is simpler than T1 ontologically and structurally, we have reason to say that T2 is better than T1.

V6. Coherence within a theory. If T2 is more coherent within it than T1, with fewer self-contradictory propositions or self-contradictory consequences in it than T1, we have reason to say that T2 is better then T1.

V7. Compatibility with other well-received theories. If T2 is more compatible with these theories than T1, then we have reason to say that T2 is better than T1.

V8. Compatibility with well-received ontology. If T2 is more compatible with the ontology than T1, then we have reason to say that T2 is better than T1.

In the different disciplines or different fields, different weight can be given to these different epistemic values. For example, V5 may be given more weight in theoretical physics, but V1 be given more weight in applied sciences. And scientists with different value systems may give different weights to different epistemic values.

Some philosophers, like Putnam [22], maintain that because two incompatible theories can get support from the same empirical evidence, they can be said to be equally good, or they have equal claims to truth

(p. 73) So they both merit acceptance. The choice made between them seems to be arbitrary. But even in the case of Newtonian physics cited by him, the theory of action at a distance between particles and the theory of action through fields are not equivalent. They cannot have same value with the epistemic values listed above. Let it be that they are equivalent with V1, but the former may have a higher value with V2,3,4. whereas the latter may have higher value with V5. Perhaps, the choice between these two theories depends upon the relative importance scientists put upon the different epistemic values. Scientists may rationally have different opinions on how much weight should be given to these epistemic values, and make different choices that can be equally justified. Be that as it may, we have no reason to say that any of them have same epistemic values, or they are equally good, or they have equal claims to truth [20].

That a scientific theory is given higher marks for these epistemic values does not entail it must be true. Just as a Chinese proverb says: "We cannot judge if a man is a hero or not according to his success or failure". The success of a scientific theory is not a sufficient condition for its being true. However, the success of a theory does give scientists reason to (sometimes involuntarily) believe or (voluntarily) accept what the theory describes, and represents what is going on in the real world, or what the theoretical entity refers to is the real thing in the external world. Using the metaphor of the moon, we have reason to believe or accept that the moon the theory describes represents the real moon in the sky. Here I say: "give scientists reason to believe or accept" that the theory is not only better than others, but also is true. Because after all there is a logical gap between success and truth (or hero) which seems still to be unbridgable in the near future. Highly probably, a very successful theory turns out to be false. However, that scientists then believed or accepted it as true was not irrational. Even if a theory is successful and gets high marks on epistemic values, it cannot be perfect. Somebody is very sensitive to its shortcomings, and gives it low marks. He must have reason to do so, e.g. he may give different weight to epistemic values than others. If a person gave a theory low marks with no reason, even if the theory turned out to be false eventually, his judgement may not be said to be rational.

On the other hand, the failure of a person cannot entail that he is not a hero. I do not want to repeat many examples in the history of science that have already been elaborated by some philosophers of science to show how a failing theory recovered its vitality and staged a comeback after many years failure. There is an unbridgeable logical gap between failure and falsity too. However, a theory that is failing for a long time cannot have a claim to truth.

If we want to know what rises up is the moon, and how many moons rise up, it can be done only through comparison, competition, discussion and debate between different, rival, or even incommensurable models of a moon and their modification, change, and alternation. All models of the moon, all scientific theories, are unavoidably embedded in a certain logical, linguistic and conceptual framework, and none of these framework are infallible, so none of these models or theories have a right to pretend that it is the only truth reflecting objective reality.

Institute of Philosophy,
Chinese Academy of Social Science,
Beijing, China.

NOTES

[1] When I prepared the first draft of this paper for the National Symposium on Realism vs. Anti-Realism in Science [26] in December 1991 I had not read the article written by M. Denitt [3]. During my visit to the United States in July 1992, I had chance to read it and found his view is similar with mine.

[2] In August 1992 while I was chairing the Third Session of the Sino-British School of Philosophy, Peter Carruthers presented me a copy of his new book where he developed an argument similar to mine but more elaborated.

[3] Chinese Taoists and Confucianists in ancient and medieval times held that the 'Tao' or 'Qi' are the first principle of all things in the universe.

[4] In Chinese legend there is a beauty name 'Chang E' who lives in the moon. Sometimes the moon is called 'Chang E'. The great ancient Chinese philosopher Zhuang Zhou once argued that it is impossible to distinguish real life from the dream, and he said when I dream a butterfly it is impossible to know whether it I dream it or it dreams me.

[5] Of course, this is a simplified formulation. Actually the emergence of visual experience is a complex process which involves the integration of the neural system, especially the brain's, and interference by past experience or background knowledge. However, all these actions cannot provide a reason to reject the cause independent of mind.

[6] J. Cohen presented his definition of 'relevance' at a symposium of the Third Session of Sino-British Summer School of Philosophy held in Tianjin, China.

BIBIOGRAPHY

1. Brown, C. (1988). 'Internal Realism? Transcendental Idealism?' in P. French *et al.* (eds.), *Realism and Antirealism*, University of Minnesota Press, pp. 145–155.
2. Carruthers, P. (1992). *Human Knowledge and Human Nature*, Oxford University Press.
3. Denitt, M. (1988). 'Rorty's Mirrorless World', in P. French *et al.* (eds.), *op cit.*, pp. 157–177.
4. Dong, G-B. (1988). *The Puzzle of EPR Correlation*, Shaanxi Science & Technology Press, Xian.
5. Dong, G-B. (1989). 'Experiments of EPR and the Debate between Realism and Positivism', *Studies in Dialectics of Nature*, no. 4, pp. 31–37.
6. Engels, F. (1988). *Ludwig Feuerback and the Outcome of Classical German Philosophy*, English translation, 1947, Lawrence & Wishart, London, pp. 24–28.
7. Fales, E. (1988). 'How to Be a Metaphysical Realist', in P. French *et al.* (eds.), *op cit.*, pp. 253–274.
8. Gu, S. (1990). 'Remarks on the View of Truth in Scientific Realism', *Journal of Dialectics of Nature*, no. 3, pp. 9–16.
9. Gu, Z-X. (1990). 'On the Three Levels of Epistemic Subjectivity', *Chinese Social Sciences*, no. 6, pp. 85–96.
10. Guan, H. (1991). 'On the Problem of "Does the Moon Exists When It Is Not Seen?" ' *Studies in Dialectics of Nature*, no. 3, pp. 39–45.
11. He, Z-X. (1990). 'Comments on "The Theory of Property and the Theory of Reality in Quantum Mechanics" ', *Philosophical Research*, no. 2, pp. 69–83.
12. He, Z-X. (1990). 'Can Modern Physics Provide Scientific Foundation for the Theory of Epistemic Subjectivity?', *Chinese Social Sciences*, no. 2.
13. He, Z-X. (1991). 'EPR Paradox and Its Philosophical Problems', *Studies in Dialectics of Nature*', no. 3, pp. 28–38.
14. Hu, X-H. (1991). 'Comments on the Discussion of Modern Physics and Epistemic Subjectivity', *Trends in Philosophy*, no. 6, pp. 8–11.
15. Hu, X-H. (1992). 'The Quantum Theory and Realism', *Philosophical Research*, no. 10, pp. 52–53.
16. Laudan, L. (1977). *Progress and Its Problems – Towards a Theory of Scientific Growth*, University of California Press.
17. Liu, S-Z. (1991). 'Modern Physics and Epistemic Subjectivity', *Chinese Social Sciences*, no. 2.
18. Luo, J-C. (1987). 'Objective Realism', *Chinese Social Sciences*, no. 2, pp. 21–36.
19. Mao, Z-D. (1937). 'On Contradiction', *Selected Works of Mao Zedong*, 1967, Peoples Press, Beijing, pp. 274–312.
20. McMichael, A. (1988). 'Creative Ontology and Absolute Truth', in P. French *et al.* (eds.), *op cit.*, pp. 51–74.
21. Popper, K. (1985). *Popper Selections*, ed. by D. Miller, Princeton University Press, Princeton NC, pp. 58–77.
22. Putnam, H. (1981). *Realism, Truth and History*, Cambridge University Press, Cambridge, MASS.

23. Strawson, P. (1957). *Individuals: An Essay in Descriptive Metaphysics*, Metheun, London.
24. Strawson, P. (1992) 'The Theory of Property and the Theory of Reality in Quantum Mechanics', *Philosophical Research*, no. 11, pp. 62–67.
25. Wang, Y-B. (1989). 'The Theory of Property and the Theory of Reality in Quantum Mechanics', *Philosophical Research*, no. 11, pp. 62–67.
26. Xu, L. (1992). 'A Report on the National Symposium on Scientific Realism', *Trends in Philosophy*, no. 5, pp. 1–4.
27. Zhang, H-X. (1991). 'Five Moons Rise up', *Philosophical Research*, no. 10, pp. 42–51.

DEMETRA SFENDONI-MENTZOU

THE REALITY OF THIRDNESS – A POTENTIAL-PRAGMATIC ACCOUNT OF LAWS OF NATURE

INTRODUCTION

The dilemma between realism and anti-realism is focused on the character of reality: what is the real? Is it the singular or the universal, the individual or the species? Anti-realism claims that reality consists of individual, singular, actually existing things, while realism seeks the most essential element of reality in the universal. Thus, when the question arises as to the character of laws of nature, anti-realists can find no place for a real connection between separate cases. If there is nothing other than a long series of independent events, then law merely describes the similarity with which things happened in the past. Realists on the other hand claim that laws of nature express not only a uniformity of behaviour, but an inner string passing through particulars and linking them by a real necessity. It seems, therefore, that the issue, which occupies the minds of many present day philosophers, is still the question of the reality of that which traditionally has been called the *fundamentum universalitatis*. Its history can be traced back to the medieval debate between realism and nominalism and certainly further back to Plato and Aristotle.

A revival of scholastic realism is Armstrong's account of laws of nature.[1] What is interesting about his theory is that the traditional problem of universals, in the sense of the nature of things, becomes that of laws of nature.[2] The central idea propounded by Armstrong's alleged realism is that laws of nature are relations between universals, which link properties with properties and things with things. However, my thesis in this paper will be that a careful reading of Armstrong reveals that some of his most basic assumptions cannot possibly live in harmony with the traditional doctrine of universals. His theory, which starts as a vigorously defended realism, eventually transforms into a disguised nominalism.[3] My pointing out this tension in Armstrong's thought does not diminish my esteem for his admirable work on universals. It is only that I take

R. S. Cohen, R. Hilpinen and Qiu Renzong (eds.), Realism and Anti-Realism in the Philosophy of Science, 75–95.

it as a challenging invitation for a fruitful discussion of the issue in its original scenery: that of Aristotelean-scholastic realism.

To respond to this invitation I will try to profit from the subtle scholastic analyses of some key-concepts and problems connected with realism so as to be able to bring to light Armstrong's concealed leaning toward nominalism. In doing so my hope is that the development of my arguments will provide at the same time the necessary material for my drawing an alternative picture of laws of nature. In my account, which I will call a potential-pragmatic realism, because of its linkage both to Aristotelean-scholastic philosophy and to Peircean thought,[4] Armstrong's belief in the reality of universals will be retained. What will be criticized is his overemphasis on the principle of instantiation on the one hand, and his rejection of the idea of potentiality on the other. By contrast, the burden of my analysis will be on potentiality as a *sine qua non* for a realist theory of laws of nature.[5]

ARMSTRONG'S THEORY OF LAWS OF NATURE

The central idea propounded by Armstrong is this: Laws of nature are relations between universals. The idea that laws of nature link properties with properties and things with things, N(F,G) holds in virtue of a *de re* necessity linking the relations between universals and the uniformity it produces[6] (1983, pp. 85, 86). Furthermore, relations between universals are treated by Armstrong as universals themselves (see *ibid.*, pp. 88ff.), so that he can "assimilate the relation between law and positive instantiation of the law to that of a universal to its instance" (*ibid.*, p. 89). His realism, therefore, which is grounded in a realism about universals (see *ibid.*, p. 7) makes him, as he claims, an opponent of the regularity theory on the one hand, and a defender of the idea of necessity involved in laws of nature on the other.

Armstrong's criticism is focused on the Humean idea that "the law is exhausted by the fact that the observed Fs are Gs, and the unobserved Fs are Gs" (*ibid.*, p. 55). By contrast, he holds that "the law involves an extra thing" the presence of which "serves first to explain why all the observed Fs are Gs, and second, to entail that any unobserved Fs there are will be Gs" (*ibid.*, p. 55). The basis of such necessity can "be found in *what it is to be an F* and *what it is to be a G*" (*ibid.*, p. 77), namely in F-ness and G-ness. If it is a law that all swans are white, then the necessity is grounded in swanhood and whiteness. This is what

Armstrong means in claiming that "the necessitation involved is a law of nature in a relation between universals" (*ibid.*, p. 78).

However, a crucial point in his alleged realism is his insistence on the principle of instantiation, which brings us back to the problem of individuation, much discussed in the Middle Ages: Laws "are abstractions, that is, they cannot exist in independence of other things" (*ibid.*, p. 90). His reason for choosing the principle of instantiation, as Armstrong explains, is to uphold along with the relation between universals the doctrine of naturalism (see *ibid.*, p. 82). And it is only through this combination, he claims, that universals can be brought into the spatio-temporal world. At the same time the idea of powers, dispositions, potentialities[7] is excluded from his scheme.

Even when Armstrong makes some space in his theory for powers, these have nothing to do with potentiality and are used by Armstrong in a pure pragmatic (see 1978b, p. 45) and positivistic sense. Powers are taken as equivalent to "the outcome which things having those properties have the power to produce in various circumstances" (*ibid.*, p. 46). Moreover, it is claimed, "we have access to the properties and relations of things only in so far as the things act upon us, in particular upon our sensory apparatus" (*ibid.*).[8] Therefore, any idea of potentially real power is to be excluded from Armstrong's world-picture, which "contains", as he repeatedly points out, "nothing but particulars and their properties and relations, neither of which can exist in independence of states of affairs" (1983, pp. 83–4). It is thus emphasized that laws are states of affairs and also abstractions from particulars, which instantiate them (see *ibid.*, p. 91; cf. 1983, pp. 113, 121–3).

This is a synoptic picture of Armstrong's theory of laws of nature, which I take to be, in its general lines, a type of 'moderate realism' (Armstrong himself calls it *a posteriori* realism)[9] as opposed both to nominalism in its defense of the reality of universals, and to Platonic realism, in its insistence on the principle of instantiation.

What I must make clear right from the beginning is that in my criticism of Armstrong the basic assumptions of moderate realism will not be questioned, because I share with him the Aristotelean-scholastic belief that universals cannot be up there in heaven but must be brought down to earth (see 1983, p. 85). In effect, I shall not be concerned with the question whether there can be any uninstantiated universals;[10] I shall rather concentrate on the question referring to the character of instantiation and to the nature of the universal *per se*. What exactly does

Armstrong mean by the "existence" of the instantiated law? Is there any space left in his theory for the reality of the universal apart from its presence in its particular instantiations? My claim will be that a careful consideration of Armstrong's principle of instantiation, in the context of scholastic philosophy, will open the door to our tracing his leaning towards nominalism.[11]

Instantiation and "Moderate Realism"

My starting point will be the principle of instantiation which asserts that for "each N-adic universal, U, there exist at least N particulars such that they U". This idea, which is a dominant feature in Armstrong's theory of laws of nature, is the descendent of classical anti-Platonic realism. Indeed, all scholastic philosophers, following Aristotle[12] in his opposition to Plato, believed that the only true substances are single, individual (καϑ' ἕκαστον) concrete objects located in space and time. Therefore, universals cannot have the character of Platonic forms, that is, they cannot be individual substances, existing in a world separate from their particular manifestations.[13] The universal must be found only *in* particular things.[14] This is why scholastic realism, such as that of St. Thomas Acquinas and Duns Scotus is called "moderate realism". We must though notice that in this view the significance of the universal is not diminished. For what offers the grounds for the real unity and intelligibility of individuals is still the universal, which is found both *in rebus* and *in intellectu*,[15] but never in a Platonic transcendental world.

Thus far, I believe, Armstrong is in agreement both with Aristotle and the scholastics. His theory is built on the assumption of the objectivity of universals, "which however cannot *exist* independently of particulars" (1978a, p. XIV). "If universals could exist uninstantiated, then", he claims, following the line of thought of Aristotle and the scholastics, "they would be substances, and would require a relation (a universal) to relate them to their particulars" (1983, p. 165). A universal, therefore, "exists if there was, is or will be particulars having that property or standing in that relation" (1978b, p. 11). It should be noted, however, that this position represents only the first step taken by moderate realism in opposition to Plato. The claim that universals are not *ante res* but *in rebus* is not sufficient to solve the problem. For a most fundamental question is immediately raised: the question concerning the character of the universal *per se*. To appreciate the significance of the issue, a clarification of the meaning of some key-concepts will be requisite.

Some Vital Distinctions

a. Universal as *unum de multis* and *unum in multis*

It is worth noticing that in scholastic philosophy[16] there is a basic distinction in the use of the term universal:[17] On the one hand, the universal means the predicate of many (*unum de multis*) and on the other, that which is *in* many (*unum in multis*). However, the character of universality belongs only to the former. Thus, the term universal is properly used only when one refers to the universal concept or idea in the mind,[18] which has an objective foundation in things (*fundamentum in rebus*). The latter corresponds to what Aristotle called οὐσία[19] and the scholastics, *natura, quidditas, essentia.*

Accordingly, granted that Armstrong's concern is with the ontological basis of laws of nature (1983, p. 3), the use of the term universal is misleading, even for Armstrong himself. As will be indicated, it makes it easier for him to disengage universality from its most fundamental character, that of essence; consequently, it also contributes to his neglecting some vital problems concerning the relation of essence to 'thisness', or in Armstrong's terminology, of universality to particularity.

b. *Esse Essentiae* and *Esse Existentiae*

An analysis of the above topics can lead to false conclusions, if we lose sight of a fundamental distinction in Aristotelian-scholastic tradition between 'being as such' (*esse essentiae*) and 'existent being' (*esse existentiae*). In scholastic philosophy the notion of being (*ens*) enters into the definition of everything,[20] for nothing can be thought of except under this category. In Scotus the *ens communissime suptum*[21] is a metaphysical genus which is ontologically prior to the ten *genera generalissima* and contains with itself the whole universe, both of thought and of physical reality (*esse rationis* and *esse naturae*).[22] This is the "ὄν ᾗὄν" of Aristotelean philosophy, the "*ens secundum suam entitatem*",[23] studied by *philosophia prima.* In this context, the particular, individual thing represents a specifically determined grade of being,[24] since the reality of being does not necessarily entail its existence.[25] Nominalists, by contrast, reject this distinction in their limiting reality in the actually existent world.

This seems to be true also of Armstrong, who treats in several places reality and existence as identical terms. The real particular, he claims, is the existent particular, the real law is the existent law (see 1983, p. 81),[26] which is also characterized as "a thing existing independently of the mind" (*ibid.*, p. 55; cf. pp. 59, 126). The consequences of this seemingly rather innocent confusion will be shown, I hope, in my discussion of the principle of instantiation.

c. *"Essence" and "Existence"*

Before we come to this, it is also worth noting another relevant and equally important distinction, the distinction between "essence" as such (*quidditas*) and "existence", which has been particularly emphasized by St. Thomas,[27] but is also shared, in its general lines by all moderate realists. All created things are compounds of essence and existence, the latter belonging properly to the individuals as such, because existence does not appear to be included in essence.[28] To exist, therefore, is to exist as an individual, particular substance, and not as a universal. It is only Platonic forms that have an independent existence; for moderate realism, and in particular that of Scotus, existence is *haecceitas*, particularity, actuality with determinations of space and time, while essence (Armstrong's universality) considered as such has only a mode of being which is real but not actual.

We must remember that for the scholastics essence is to existence as potentiality is to actuality.[29] This is an analogy that Armstrong totally excludes from his scheme, by rejecting the ideas both of potentiality and essence. Although he follows the scholastics in assuming the involvement of particularity and universality in all individual existent things, he nevertheless rejects the Aristotelean idea of essence, which, as has already been pointed out, in scholastic realism is identical with universality. Furthermore, he rejects the traditional distinction between essential and accidental properties,[30] as well as the category of substance,[31] the latter being an element which is excluded from Aristotle's idea of essence.[32] Universals are, thus, characterized as properties and relations (see 1978a, p. 87, 1983, p. 166). "The properties of a particular", claims Armstrong, "form a single property. The nature of a particular" (1978a, p. 44). Essence is nothing but a conjunction of properties. This is in the line of the doctrines of particularism and naturalism which he is trying unsuccessfully, as is my hope to show, to combine with his theory of universals. According to the former, "there is nothing but particulars having (universal) properties and relations" (1973a, p. 1) and according to the latter, "nothing else exists except the single spatio-temporal world" (1983, p. 82).[33] This fact deprives Armstrong of the possibility of giving a consistent realistic account, and more specifically it deprives him of the possibility of properly investigating the relation between universality and particularity.

THE PRINCIPLE OF INSTANTIATION

Keeping in mind the above remarks, we are now ready for a careful reading of several important aspects of the principle of instantiation. Its claim is that "for each property, P, there exists (not necessarily now) a particular, x, such that Px" (1978b, p. 76),[34] so that "particularity and universality are both involved in all existence" (*ibid.*, p. 1).[35] We must though notice that a mere assumption of this principle is not sufficient for the establishment of moderate realism, for the following questions are immediately raised: (1) what is the relation between essence (universality) and thisness (*haecceitas* or particularity)? (2) how does the essence or common nature become "contracted", or how does it appear as particular in the thing? (3) can common nature of itself be a this (*de se haec*), or, if not, what kind of unity can it have?

a. *The Unity of "Nature"*

Two points deserve our attention here: the issue of the unity of nature and that of its particularized character. (1) As we have already pointed out, all individuals belonging to the same species share a common nature. Socrates is a man in virtue of something real which is called *natura*, namely in virtue of manness, and as Armstrong would put it, this swan (Fa) is white (Ga) in virtue of swanness (F-ness) and whiteness (G-ness). But Socrates's nature as an individual, is numerically different from Plato's nature, as his nature is a part of him.[36] Can we then say that Socrates's nature is as distinct from Plato's nature as it is from a stone's nature? If this is not the case, and if we wish to maintain that Socrates and Plato (or this swan, Fa, and that swan, Fb) have something real in common (F-ness), then this real cannot be a *de se haec* (singular); it cannot be a numerically one, or identical in Armstrong's terminology (see 1978a, p. 87); it must therefore have a unity which is neither numerical nor merely conceptual. This is a specific type of unity which Scotus characterizes as "less than numerical" but none the less real.[37]

In his discussion of the character of what in scholastic philosophy is called essence, Armstrong rejects Scotus's idea of "lesser unity": "I find Scotus's doctrine of 'lesser unity' difficult to understand. It seems to imply that the identity of particulars is different from the identity of essence" (1978a, p. 87). For Armstrong "a property is strictly identical, strictly the same, in all its different instances" (*ibid.*, p. 83). "Identity" he claims, "is just identity" (1978b, p. 112).[38] Thus, "the identical universal is present in each instantiation" (1983, p. 79) and therefore "the law is fully present in each instance" (*ibid.*, p. 102). If this is so, then Armstrong is in trouble; he has to provide an answer to a serious point raised by the scholastics and expressed by Scotus as follows: "If the numerically one nature were fully present, in Dobbin, then he would not be a horse but horseness itself".[39] I am afraid that this is a thorny point which Armstrong would be unable to answer, unless he claimed that the full presence of the universal (or law) turns out in a complete particularization of the universal, if we remember that numerical unity can only be attributed either to Platonic forms or to particular concrete objects.

b. *Particularization of "Nature"*

But this is not the only difficulty that Armstrong is faced with; he must also provide an answer to the question, how the common nature (F-ness), which is not a *de se haec* (singular) becomes, in a sense, a part of the individual (Fa); and consequently, how are laws of nature instantiated? This is a crucial problem and calls for our special attention. At this point Armstrong's answer is very close to that of Scotus. As the subtle Doctor claims, common nature is "contracted" in the individual, by *haecceitas*, so that it becomes, in some respect, identical with it.[40] This is one of the main reasons that Scotus has been accused by C. S. Peirce for his nominalistic attitude: "Even Duns Scotus", claims Peirce, "is too nominalistic, when he says that universals are contracted to the mode of individuality in singulars, meaning, as he does, by singulars ordinary existing things" (8.208).[41]

Indeed, in Scotus's idea of contraction there is an emphasis on the particularization of the universal by *haecceitas*, which must be understood as the *ultima realitas*[42] of the individual and which gives it its actual existence. The particular exists in virtue of its *haecceitas* which makes it this particular thing (Fa) and not something else (Fb). As a result it is contradictory for Socrates's nature to exist in Plato. Each particular thing, he claims, is a compound of quiddity, which is the universal element and haecceity, which is the "*entitas singularis*".[43]

c. *Universality and Particularity*

The idea of the compound of essence and particularity is something shared by all moderate realists, including Armstrong.[44] So they all have to face the problem of defining its character: How is it possible for the individual thing to be a composite of two elements, and at the same time be numerically one? Are particularity and universality completely united, or can they retain their distinct character in this combination? Scotus tried to answer this question by inventing a new term, his famous "formal" distinction,[45] which has given rise to criticisms about Scotus's nominalistic tendencies.[46] What is peculiar about it is the fact that it is very close to logical, but not itself a logical distinction. It is a distinction *a parte rei*, i.e., having an objective ground in reality itself, which however is not physical, since essence and thisness are not two separate things but are "different moments or aspects" of the same particular.

Armstrong's criticism at this point is that although Scotus's solution of "lesser unity" seems to be the most satisfactory one, still it is puzzling, because he talks of "a mere 'formal' distinction". "Though distinct", he claims, "their union is closer than relation" (1978b, p. 3). This is an interesting point witnessing Armstrong's leaning towards nominalism. He is not satisfied with formal distinction, which is already too nominalistic, but he seems to need an even closer relation between universality and particularity. If this were the case, then we would have to assume that the universal becomes particularized in the individual.

Armstrong is not quite clear at this point. He maintains that "particulars and universals do not stand in splendid isolation from each other" (1978a, p. 80). However, he does not appeal to any of the traditional terms to define their relation. The only characterization appearing in his writings is that of a "state of affairs". "Particulars and universals", he claims, "are found only in states of affairs" (*ibid.*, p. 8), the definition being that of "a particular's having a property". What is universal,

F-ness or G-ness, cannot exist but as instantiated or particularized, Fa, Fb, Fc . . . or Ga, Gb, Gc . . . But this cannot be the end of the story, for the most fundamental question for any version of realism has not been answered yet, the question as to the nature of the universal *per se*. Is the universal exhausted in its instantiations, or shall we claim that its reality is wider and much richer in content than that of its manifestations?

d. *The Character of "Nature Per Se"*

If my reading of Armstrong is at all correct, we must assume that he is rejecting the second case, although his theory is based on the assumption of the reality of universals. My contention, thus, is that at this point the scope of his realism is limited to a substantial degree. To be more specific: as already indicated, Armstrong rejects both Scotus's "lesser unity" and "formal" distinction. His alternative to the former is "total" identity and to the latter a much closer relation than that of the "formal" one. This opens the back door to nominalism, which indeed finds a place in Armstrong's world-picture, where "nothing else exists except the single, spatio-temporal, world, the world studied by physics, chemistry, cosmology and so on" (1983, p. 82; cf. p. 126). Laws of nature are not simply relations between universals but "relations between *actually* instantiated universals" (*ibid*., p. 112),[47] and the "law is fully present in each instantiation" (1978a, p. 89),[48] so that we are forced by Armstrong's own analysis to draw the conclusion that law becomes fully particularized, leaving no space in his scheme for universality *per se*.

Furthermore, it is important to notice that the impression of Armstrong's nominalistic inclination is strengthened by his repeatedly characterizing laws of nature as "states of affairs". This is indeed confusing,[49] not only because this characterization cannot fit in the traditional framework, but particularly because it has already been used by Armstrong in his description of the relation pertaining between particularity and universality in the individual thing. In that case a "state of affairs" was defined as "a particular's having a property".

Armstrong tries unsuccessfully, in my opinion, to overcome this difficulty by drawing a distinction between a "thin" and a "thick" particular.[50] A "thick" particular is defined as "a thing taken along with all its properties", while a "thin" particular as "a thing taken in abstraction from all its properties". Furthermore, Armstrong makes clear that in his notion of a "state of affairs" he is using the conception of a particular in the latter sense. But remember, a "state of affairs" is defined as "a particular's possessing a certain property, or two or more particulars standing in a certain relation" (1978a, p. 80). If in this definition we substitute the definition of the "thin" particular for the simple term "particular" the result will be the following: A state of affairs is "a particular in abstraction from all its properties, possessing a property, or two or more particulars in abstraction from all their properties standing in a certain relation". What becomes obvious here is the fact that the relation expressed by "states of affairs" is still a relation between particularity and universality. This unavoidably leads to the conclusion that in Armstrong there is no clear difference between what is taken to be a relation between particularity and universality and what is taken to be a relation between a universality and universality.

My claim is therefore that Armstrong's deep concern not to lose sight of the fact that universals exist only in their instantiations blocks his ability to see that part of the reality of universals, which is also objective but not limited to the mode of actual existence in the spatio-temporal world. This goes hand in hand with his exclusion of potentiality, which, as I hope will be shown, is an inseparable element of universality *per se*.

e. *The Metaphysical Mode of Universal*

For our purposes, it will be helpful to refer to an extremely significant distinction between the physical and the metaphysical aspect of universal made by Scotus[51] in the spirit of Aristotelean metaphysics: although the unity of essence (genus, species) is less than numerical, claims Scotus, and although essence cannot exist apart from the unity of the individual substance, yet it is logically and metaphysically prior to it.[52] Therefore, the essence *per se* cannot be identical with the individualized nature. In other words, although the nature exists objectively only along with its individuating conditions, logically and metaphysically it is prior to these conditions, as absolute quiddity is indifferent to singularity and universality.[53]

We thus arrive at a crucial point concerning Armstrong's instantiated laws. If it is admitted that the universal, which is combined with the particular, ends up in the existence of its instantiation in the spatio-temporal world, then in what respect would this universal be different from its particular instantiations? It must be made clear, therefore, that the universal *in rebus* has not one but two modes of being: the physical and the metaphysical, the latter being the most fundamental one. A more careful consideration of this distinction is likely to throw further light on our analysis of the nature of law.

We have already seen that the logical universal, which is also characterized as *unum de multis*, has an objective basis in things. This is what Scotus calls the physical universal, which is the common nature contracted in the individual substance by *haecceitas*, or what Armstrong means by an instantiated universal (Fa), and respectively an actually existent law. This is where Armstrong's analysis ends, for he does not seem to be interested in tracing any further either the reality of universals or the relation of "contracted" nature both to the universal in the mind and the universal *per se*. Thus he fails to realize what is extremely significant in the context of scholastic philosophy (which is also in line with Aristotelean thought), that the "contracted" nature is only a possibility of a universal, i.e., it is virtually or potentially a universal[54] in the mind, for to be predicable as one thing of many (*unum de multis*), it must be given a numerical unity and this is the work of the intellect.[55] Furthermore, Armstrong fails to see that the instantiated universal (Fa) can only have a limited mode of reality, while common nature, considered *per se*, is unlimited and indeterminate.

We thus come to the basic features of the metaphysical mode of essence or common nature as such. The first thing to remember is that common nature *per se* is not a *res* but a *realitas*, which is neither particular not universal;[56] it exists objectively only with its individualizing conditions, and is conceived by the intellect as universal. It is the mind itself which endows it with its character of universality, for of itself it is indifferent,[57] a mere *natura*: "Horseness is just horseness, neither of itself one nor many, neither universal nor particular".[58] Accordingly, it does not point to any particular individual case and this is exactly what constitutes the indeterminacy of the *ens reale*:[59] "it can be real without being determined to exist in any one thing".[60] We are thus led to the fundamental character of essence or universality *per se*. It can be defined as a mere possibility of existence, just as when it has become existent it is the possibility of a universal.

TOWARDS A POTENTIAL-PRAGMATIC REALISM

If a development along these lines is acceptable, we can draw the conclusion that the involvement of potentiality in the very idea of universality is such that a consistent account of properties of things in terms of universals cannot be given apart from this idea. Furthermore, if it is admitted that the necessity of laws of nature is grounded in the nature of things (F-ness or G-ness), then the characteristics of the metaphysical mode of universality, which have emerged in the preceding analysis, can also apply to laws of nature. We are thus led to a world-picture which combines the Aristotelean emphasis on the idea of potentiality[61] with the Peircean account of laws grounded in the category of Thirdness.[62]

In this picture reality is not restricted to the realm of actual, individual, instantiated things; for there is much more to it, which is real though not completely determined "since not presenting itself in actually objectified form" (6.365), as a concrete existence in space and time. This is the mode of potential being, implying a "capacity for realization", or "some inherent tendency to actuality, which if not thwarted, leads to final completeness of being" (6.365). As C. S. Peirce expresses it, "the *will-be's*, the *actually is's* and the *have beens* are not the sum of the reals. They only cover actuality. There are besides *would-be's* and *can be's* that are real" (8.216). Hence, besides the nominalist's world of actually existing things and properties, all of which come to us through our senses, there is plenty of space for properties of things, interpreted as dispositions, liabilities, tendencies, propensities, as well as for processes taking place in the sub-atomic world, all of which could be taken as instances of a transition from a state of potential to that of actual being. The transformations of nuclei into one another, the mutual annihilation of matter and anti-matter and their conversion into radiant energy, the transformation, in radio-activity, of a neutron in the nucleus into a proton without any external energy being supplied, and finally, the reality of quarks, which are never found as actually existing free particles, can all be seen as manifestations of potentiality in the physical world.

However, as fascinating as the idea of potentiality in the realm of elementary particles is, I will confine myself, for reasons of space restrictions,[63] to a consideration of properties of things, such as hardness, and of dispositions, such as brittleness, fragility, solubility. What is significant about them is that they can be ascribed as real, even if they are not actually tested in the case of properties or even if they are not actually manifested *per se*, in the case of dispositions. This view has many affinities to the theory of "causal powers" propounded by R. Harré and E. H. Madden[64] in those respects that the latter establishes a connection with some Aristotelean and scholastic concepts, such as the reality of natural kinds and the role that the nature of things plays in natural necessity. Finally, a commonly posed idea is the explanation of causality through powerful particulars which leads to a model shifting away from the Humean billiard ball paradigm.

I must say, though, that the very idea of "causal power", as something that can be experienced, is not what I am interested in, my concern being with the categorically unobservable which can offer the right place for the introduction of the idea of potentiality.[65] That is why what is extremely welcome from the theory of "causal powers" is the view that the "concept of power catches what might be called the strong sense of *potentiality* or *potency*", namely, "what would happen as a matter of course, if interfering conditions were absent or taken away" (p. 12).

In this respect, the analysis of qualities, dispositions, tendencies can be given in the form of subjunctive conditionals, or in Peircean terms of 'would-be's'[66] of possible behaviour. For what a subjunctive conditional is about is a possibility or potentiality grounded in the nature of things and referring not to what is contrary to fact, as is the case with Armstrong's counterfactuals,[67] but to the as yet unactualized. In this respect, potentiality is to be understood as that state of capability, disposition, or liability to suffer or produce change in virtue of a thing's nature. To say for instance that the diamond is hard is to assert that it would resist pressure when scratched (8.208). For "*hardness*", as Peirce claims, "is not invented by men, as the word is, but is really and truly in the hard things and is one in them all, as a description of habit, disposition, behaviour" (1.27 n. 1). This is exactly the crucial point, I believe, that offers the real ground for a passage from a realism about properties to a realism about laws of nature.[68] Seen from this point laws can be said to embody in a stronger sense the characteristics of indeterminacy and potentiality, which have already been seen in relation to properties. A further analysis of these features will serve in my relating them to laws of nature and also in drawing my concluding remarks.

Let me start with indeterminateness.[69] As already indicated, nature or essence is indeterminate with respect to particularity. Although it is true that "there is no man unless there is some particular man", nonetheless "There is a real difference between man irrespective of what the other determinations may be, and man with this or that particular series of determinations".[70] The same can be said of laws of nature. The statement, all swans are white, does not point to any particular swan. This is exactly why, not withstanding the fact that there is no Fa unless there is F-ness, law can be real without being determined to be instantiated in any particular case.

Hence, law can be characterized as a "potential aggregate"[71] which is irreducible to any collection of actualities; it only "contains general conditions which permit the determinations of individuals" (6.185). Thus, the content of the statement, "all men are mortal", cannot be exhausted by any multitude of existent things (see, 5.103), either in the past and present or in the future, because it applies to an infinity of future men, including not only concrete but possible men, as well. In other words, "it involves the idea of every possible something",[72] or, as Aristotle would put it, "the idea of that of which there is always something beyond".[73] The proper way, therefore, to describe law is to assert that it has that mode of being which is "in potency of determination", "in equal distance from the physicist who considers it in its concrete determinations and from the logician who considers it as determined to universality".[74]

CONCLUDING REMARKS

In light of the preceding analysis, it can be claimed that, instead of trying to emphasize that aspect of law which appears in its instantiations, it is far more important for realism to appeal to that mode of being of law, which represents a possibility of existence, infinitely rich in content, because of its virtually containing the inexhaustible concrete wealth of its particular instantiations.

This is the core of the difference between Armstrong's instantiated, i.e., existent and actually real law, and that type of law, which is essentially "instantiatable", or potentially real as it points to its instantiations. For the actual manifestations of law are the discrete units which burst the flux of what is essentially a continuum, the very nature of which involves the absence of distinct individuality.[75] And since there is no limit as to the number of its instances, a law of nature can be described both in an Aristotelean and a Peircean fashion as "the potential though not the realized whole",[76] which embraces all phenomena as its possible future manifestations. This view in no way involves a rejection of the principle of instantiation as is understood in Aristotelean-scholastic realism. It only involves a shifting of emphasis from Armstrong's "instantiated" to "instantiatable" law.

What I hope to have shown is, at least, that a consistent attempt at a realist account of laws of nature grounded in universals cannot be given, if trapped in the narrow domain of actually existent law in space-time; what is further needed is an appeal to the infinitely rich reality of that aspect of law, which is in potency of determination, so that the relation between law and its instantiations will be accounted as a relation of *quidditas* to *haecceitas*, or potentiality to actuality.

Department of Philosophy,
Aristotle University of Thessaloniki,
Greece.

NOTES

* In preparing the final version of this essay, commentaries were taken into account. My thanks are due to the participants of the Beijing Philosophy of Science Conference for their helpful comments and suggestions. I am especially grateful to James R. Brown for valuable conversations on the issue and for critical comments on two earlier drafts. I am also indebted to Susan Haack and Andrew D. Irvine for their incisive commentary and useful suggestions on an earlier draft, read at the Dubrovnik Philosophy of Science Conference in April, 1992.

1 D. H. Armstrong (1983).

2 The question concerning laws of nature is quite recent. Plato was concerned with *forms* (ἰδέαι), Aristotle with *essence* or *nature* (εἶδος, οὐσία, φύσις) the scholastics with *universalia*, in the sense of *quidditas* or *natura*. One might say that discussion on laws of nature started with Hume. However, the first serious attempt at a realist account of laws of nature, grounded in Aristotelean-scholastic realism, was made by C. S. Peirce. See for example his, "What is a Law of Nature" in his article under the general title, "Hume on Miracles and Laws of Nature" (see, 1958, pp. 289–321).

3 In her instructive paper (1992) on the relevance of C. S. Peirce's scholastic-realism to contemporary discussions in Philosophy of Science, S. Haack suggests that Armstrong's theory of laws of nature be characterized as "nominalistic aristotelianism" (p. 36). I must say that, although I share the view that each of these characteristics considered separately is true of Armstrong, I would not be happy to use them in conjunction, because they are mutually exclusive.

4 I have elsewhere elaborated the close linkage of Peirce's thought to Aristotelean philosophy. See my (1980) and (1991).

5 In what follows I will be particularly concerned with Armstrong's (1983). However, when necessary I shall also refer to his (1978a) and (1978b), in so far as his theory of laws is grounded in his realism about universals.

[6] The source of inspiration, as Armstrong reports, for the first formulation of this theory, was Plato (*Phaedo* 102–170). However, Armstrong's view is essentially anti-Platonic, as he himself explicitly claims in several places (see, e.g., 1983, pp. 86, 90) and as will also emerge from our analysis.
[7] See, e.g., (1978a) p. 85, (1978b) pp. 5, 157, (1983) pp. 120–121, 135–136, 168. Acceptance of "unrealized possibilities", such as a solid lump of gold with a volume greater than a cubic mile, or a race of white feathered ravens (see, 1983, pp. 17ff.) cannot be equated with acceptance of potentiality, in its classical sense, as will be shown in what follows.
[8] Cf., "The power is constituted the power it is by the *sort* of actualizations it gives rise to in suitable *sorts* of circumstances" (1983, p. 123). As is obvious here, the idea of power is connected with the immediate experienced results.
[9] Armstrong's "*a posteriori* realism", is the view that "just what universals there are in the world, is to be decided *a posteriori* on the basis of total science" (1983, p. 83; cf. 1978a, p. xiii).
[10] As Tooley (1977) has made a substantial and fruitful effort to show. However, an analogous answer as to the potential character of laws will eventually be provided – although through a very different path.
[11] An incisive criticism of the inconsistencies involved in Armstrong's attempted marriage of universals with naturalism, as it appears in his principle of instantiation, is made by A. Irvine in his, "Universals *in rebus* or *in contradictione*?" (unpublished paper, sent to me by the author recently). Irvine's claim is that Armstrong's view seems to fall victim to a traditional difficulty of the "one over many": How can the universal be both completely and not completely present within a certain region of space-time? or how can it have multiple location at all? Irvine, who seems to share with Armstrong the view that universals are indeed located in space-time can see no solution to the problem of instantiation. By contrast, it is my contention that a promising way to get out of the difficulty, while remaining faithful to the basic assumption of the universals *in rebus*, can be found by an appeal to the Aristotelean-scholastic treatment of the universal *per se*.
[12] See, e.g., Arist. *Met.* 1029a 27–28; *Categ.* 2a 11.
[13] See, e.g., Arist. *Met.* 1086b 2ff.
[14] See, e.g., *ibid.*, 1020a 33, 35b 2, b6, 1045a 23–35.
[15] Avicena, *Logica*, fol. 2v. b (ed. Venice, 1508): "*Essentiae vero rerum aut sunt in ipsis rebus, out sunt in intellectu, unde habent tres respectus*".
[16] The contribution of Harris (1959a, 1959b), Boler (1963), Copleston (1953) and Moore (1950) to my understanding of some fundamental scholastic subtleties crucial to the purposes of this paper has been invaluable. More specifically my debt to E. Moore (1950, 1952, 1968) for his pioneering emphasis on the idea of potentiality in C. S. Peirce's scholastic-realism has been acknowledged earlier in my work.
[17] Actually, there is a three-fold distinction of the universal in scholastic philosophy. See, e.g., St. Thomas, *Comm. in Sentt.*, Lib. II, dist. iii, q. 2, art. 2: "*Est triplex universale: quoddam quod est in re, seu natura ipsa, quae est in particularibus . . . est etiam quoddam universale quod est a re acceptum per abstractionem . . . Est etiam quoddam universale ante rem, quod est prius re ipsa*".
[18] See, e.g., Scotus: *Rep. Par.* II. dist. ii, q. 5, n. 12: "*Dico quod universale in actu non est nisi in intellectu, quia non est actu universale, nisi sit unum in multis et de multis*".

[19] See, e.g., Arist. *Met.* 1029b 14–17. It is important however to be indicated that in Aristotle there is no specific word corresponding to the scholastic term '*essentia*'. The meaning of the word 'essence' can be applied to a large number of Aristotelean terms, one of which is οὐσία.

[20] Scotus, *Quaest. de Anima*, q. lxvi: "*sed ens, quod est universalissimum, intrat definitionem omnium* . . . ."

[21] Scotus, *De Rerum Principio*, q. xvi, n. II: "*Et isto mode ens communissime sumptum est genus metaphysicum ad Creatorem et creaturam*".

[22] Scotus, *Quaest. super Elench.*, q. i: "*Ens est duplex, scilicet, naturae et rationis*".

[23] *Ibid.*: "*Logica est de ente communi, sive considerat ens in communi . . . Logica est de hujusmodi intentionibus, quae sunt applicabiles omnibus rebus*".

[24] See Harris (1959b), p. 32.

[25] The above analysis of being is definitely Aristotelean. See, e.g., *Met* 1003a 21–32.

[26] Cf., 1978a, pp. xiii, 10; 1983, pp. 40, 53, 55, 59, 98, 104, 105, 112, 117, 120, 121, 126, 129, 167.

[27] See, St. Thomas, *Summa Theol.* Ia pars. q. 39 art. 2.

[28] "So *esse* denotes simply the act of being, existence being denoted by *ipsum esse*": E. Gilson (1957), p. 38.

[29] See, S. Thomas, *Summa Theol.* Ia pars, q. 39, art. 2. In this respect, Scotus shares with St. Thomas the view that matter stands to form as potency to act. He also defines form in Aristotelean fashion as the actuality of matter. However, matter for Scotus is not only potentiality; it also has an actuality of its own apart from its mere potential existence: "*Forma communicat materiae suam actualitatem et suam actum essendi et suam operationem*" (Scotus, *De Rerum Principio*, q. ix, n. 53). See Harris (1959b), pp. 3, 4, 91; Copleston (1953), pp. 31, 51–52.

[30] See, for example, Arist. *Met.* 1030b 12ff.

[31] See, *ibid.*, 1028a 13ff.

[32] See, Arist. *Met.* 1088a 22.

[33] More of this below.

[34] Cf., 1978b, p. 156; 1983, pp. 82, 91, 97, 165.

[35] Cf., "particularity and universality combined in states of affairs" (1983, p. 122).

[36] Scotus, *Metaph.*, vii, q. 13, n. 21: "*humanitas, quae est in Socrate, non est humanitas, quae est in Platone*. . . ."

[37] Scotus, *Rep. Par.* II, dist. ii, q. 5, n. 11: "*Dico ad quaestionem concedendo . . . quod est unitas extra animam minor quam numeralis, ut specifica* . . . ."

[38] Cf., 1978a, pp. 75, 77, 108, 115; 1978b, pp. 66, 95, 112; 1983, p. 83.

[39] See, Boler (1963), p. 49.

[40] Scotus, *Op. Ox.* II, dist. xlii, q. 4, n. 7: "*universale in sinqulari non est aliud quam singulare*". See, Boler, p. 63.

[41] References to this form are to the *C. S. Peirce* (*1939–58*). 8.208, for instance, refers to volume 8, paragraph 208. See also, Scotus "inclined toward nominalism" (1.560); defended a "halting realism" (6.175); was separated from nominalism "by a hair" (8.11).

[42] Scotus, *Op. Ox.* II, dist, iii, q. 2, n. 15: "*ergo ista entitas non est materia vel forma nec compositum, inquantum quodlibet istorum est natura, sed est ultima realitas entis, quod est materia et quod est forma vel quod est compositum*".

[43] Scotus, *Op. Ox.* II, dist. iii, q. 6, n. 15: "*Sed haec est formaliter entitas singularis, et illa est entitas naturae formaliter*".

[44] See, for example Armstrong (1983), p. 122.
[45] Scotus, *Op. Ox.* II, dist. iii, q. 6, n. 15: "*Sed haec est formaliter entitas singularis, et illa est entitas naturae formaliter*".
[46] See, Harris (1959a), p. 50.
[47] Cf., 1978a, p. 113; 1978b, pp. 130, 131; 1983, pp. 82, 112, 135, 172.
[48] Cf., 1983, pp. 79, 102.
[49] Armstrong seems to realize the difficulty in his making a distinction, although not quite often, between "states of affairs', i.e., tokens, and "sorts of states of affairs", i.e., types (see, 1983, p. 98; cf. p. 172).
[50] See, 1978a, pp. 114ff.
[51] See, Scotus, *In Metaphys.* I, 7, q. 18, n. 6; *Op. Ox.* 2 d. 3, q. 1, n. 7; *Super Praedicam.* q. 2, n. 2. For a discussion of the metaphysical mode of universality, see, Boler (1963), pp. 56ff. See also Claudin Engel-Tiercelin (1992), pp. 59ff.
[52] Scotus, *Op. Ox.* II, dist. iii, q. 1: "*Licet nunquam sit realiter sine aliquo istorum, non tamen est de se aliquod istorum, sed est prius naturaliter omnibus istis*".
[53] *Ibid.*, q. i, n. 7: "*nec ex se una, nec plures, nec universalis nec particularis*".
[54] Scotus, "*est ergo natura in potentia remota ad determinationem singularitas et ad indeterminationem universalitas*". *In Metaph.* 1, 7, q. 18, n. 8; cf. *Op. Ox.* II, d. 3, q. 1, n. 9. See, E. Gilson (1952), pp. 536ff., Moore (1964), p. 405, and Boler (1963), p. 60.
[55] Scotus, *Quaest. in Lib. Periherm*, Lib. I, q. viii: "*Intellectus enim in sua operatione non dependent ab existentia rei, cum et definitiones et demonstrationes fiant de non existentibus, multo magis non existentibus potest attribuere intentiones secundas, quae omnino ab intellectu causantur*".
[56] Scotus, *Quaest. de Anima*, q. xvii, n. 14: "*. . . quae quantum est de se nec est universalis nec singularis, sed de se est indifferens, et tale est objectum intellectus directum*".
[57] Scotus, *Op. Ox.* II, dist. iii, q. 1: "*Non solum ipsa natura est indifferens ad esse in intellectu et in particulari . . . .*"
[58] Avicenna, *Logica*, fol. 12 r. a: "*in se autem hujus nec est universale nec est singulare. Si enim esset univesale, ita quod animalitas ex hoc, quod est animalitas, est universalis, oporteret nullum animal esse singulare*".
[59] For an illuminating analysis of the idea of indeterminacy see E. Moore (1964), pp. 406ff., Fred Michael (1988), pp. 334–335 and Claudin Engel-Tiercelin (1992) especially pp. 57–70.
[60] See, Boler (1963), p. 50.
[61] See, e.g., Arist., *Met.* 1045b 25–1046a 3, 1049b 9–11; *Phys.* 201a 10–13, 257b 7–8, 225b 34–37, 261a 32–36.
[62] For a fuller discussion of Peirce's three ontological categories, see my (1980), pp. 80ff.
[63] For a more extended treatment of the role of potentiality in Q. M. and Physics of elementary particles, see my (1993).
[64] R. Harré and E. Madden (1975).
[65] I have developed this in some detail in my (1985–88).
[66] In Peirce's ontological scheme the 'would-be' is another expression for potentiality and it is interrelated with generality and law of habit, which is expressed by a conditional proposition. For a fuller discussion of the concepts of 'would-be' and 'habit' see my (1980), pp. 70ff., 107ff.
[67] It is to be noted that all statements of counterfactuals Armstrong refers to are state-

ments of uninstantiated law, which "assert only that if, contrary to fact certain sorts of things existed, then these things would obey a certain law" (1983, p. 126; cf., p. 137). "The statement that, it is a law that Fs are Gs supports the counterfactual that if *a*, which is not in fact an F, were to be an F, then it would also be a G" (1983, p. 46).

[68] For a fuller discussion of this connection see my (1992).

[69] The three-fold relation of indeterminacy-potentiality-probability, and its place in C. S. Peirce's categories of Firstness and Thirdness has been elaborated in my 1980 and 1993.

[70] C. S. Peirce (5.312); cf., Arist. *De Caelo* 278a 7–12: "if we supposed that there were but one circle, none the less to be a circle and to be this circle would be different; the one would be the form the other would be form in matter and would be a particular".

[71] See, C. S. Peirce (6.185–188).

[72] For an illuminating discussion of the issue, see F. Michael (1988), pp. 344–345, and C. Engel-Tiercelin (1992).

[73] Arist. *Physics* 207a 1–2: "οὖ ἀεί τι ἔξω ἔστι". This is how Aristotle characterizes the idea of infinity (ἄπειρον), which is treated by him as a potential reality: "δυνάμει εἶναι τὸ ἄπειρον" (*Physics* 206a 19). The same is true of the Aristotelean infinite time, which is real by continuously coming into being. Therefore, time, as a whole, can never be actually, but only potentially real (see, *ibid.*, 206a 8ff.). This is an extremely enlightening idea that could shed light on the issue of laws of nature and in particular of "instantiatable" laws. However this is a topic that deserves special treatment.

[74] This is how E. Gilson (1952, p. 113), describes the metaphysical mode of universality.

[75] See, C. S. Peirce (4.172): "It is only actuality, the force of existence, which bursts the fluidity of the general and produces a discrete unit . . . the possible is general, and continuity and generality are two names for the same absence of distinction of individuals".

[76] Aristotle, *Physics* 207a 21–3: "το δυνάμει ὅλον, ἐντελεχείᾳ δ'οὔ".

BIBLIOGRAPHY

Aristotle (1933–35). *Metaphysics*, 2 vols. Loeb ed. Translated by H. Tredennich. Harvard University Press.

Aristotle (1929–34). *Physics*, 2 vols. Loeb ed. Translated by P. H. Wicksted and F. M. Conford. Harvard University Press.

Armstrong, David M. (1978a). *Nominalism and Realism* (*Universals and Scientific Realism, vol. 1*). Cambridge University Press.

Armstrong, David M. (1978b). *A Theory of Universals* (*Universals and Scientific Realism, vol. 2*). Cambridge University Press.

Armstrong, David M. (1983). *What is a Law of Nature?*. Cambridge University Press.

Boler, J. F. (1963). *Charles Peirce and Scholastic Realism*. University of Washington Press.

Copleston, F. S. J. (1953). *A History of Philosophy* (*Medieval Philosophy, vol. 2*). Image Books Edition (1962).

Duns Scotus, J. (1851–95). *Joannis Duns Scoti Opera Omnia* 26 vols. Vives, Paris.

Duns Scotus, J. (1908). *Duns Scoti Capitalia Opera*. R. P. D. Marie (ed.). Le Havre.

Engel-Tiercelin Claudin (1992). 'Vagueness and the Unity of C. S. Peirce's Realism', *Transactions of the C. S. Peirce Society* **28**, pp. 51–82.

Gilson, Etienne (1952). *Jean Duns Scot.* Vrin.

Gilson, Etienne (1957). *The Christian Philosophy of St. Thomas Aquinas.* Translated by L. K. Shook. Victor Gollancz.

Haack, Susan (1992). ' "Extreme Scholastic Realism": Its Relevance to Philosophy of Science Today', *Transactions* **28**, pp. 19–50.

Harré, R. and Madden, E. (1975). *Causal Powers.* Blackwell.

Harris, C. R. S. (1927a). *Duns Scotus* (*The Place of Duns Scotus in Medieval Thought, vol. 1*). The Humanities Press (2nd ed. 1959).

Harris, C. R. S. (1927b). *Duns Scotus* (*The Philosophical Doctrines of Duns Scotus, vol. 2*). The Humanities Press (2nd ed. 1959).

Irvine, A. D. (1992). 'Universals in Rebus or in Contradictione?'. Unpublished.

Michael, Fred (1988). 'Two Forms of Scholastic Realism in Peirce's Philosophy', *Transactions* **24**, pp. 317–348.

Moore, E. C. (1950). *Metaphysics and Pragmatism in the Philosophy of C. S. Peirce* (Doct. Thesis). University of Michigan Press.

Moore, E. C. (1952). 'The Scholastic Realism of C. S. Peirce', *Philosophy and Phenomenological Research* **12**, pp. 406–417.

Moore, E. C. (1964). 'The Influence of Duns Scotus on Peirce', in E. Moore and R. Robin (eds.), *Studies in the Philosophy of C. S. Peirce.* The University of Massachusetts Press.

Moore, E. C. (1968). 'On the World as General', *Transactions* **4**, pp. 90–99.

Peirce, Charles S. (1931–58). *Collected Papers of Charles Sanders Peirce* (vols. 1–8). C. Hartshorne, P. Weiss and A. Burks (eds.). Belknapp Press of the Harvard University Press.

Peirce, Charles S. (1958). *Selected Writings* (*Values in a Universe of Chance*). Philip P. Wiener (ed.). Dover 1966.

Sfendoni-Mentzou, Demetra (1980). *Probability and Chance in C. S. Peirce's Philosophy* (Doct. Thes. in Greek). University of Thessaloniki.

Sfendoni-Mentzou, Demetra (1985–88). 'A Realist Explanation of Causality Through "Causal Powers" ', *Scien. Ann. School of Phil. of Thess.*, pp. 245–255.

Sfendoni-Mentzou, Demetra (1991). 'Towards a Potential-Pragmatic Account of C. S. Peirce's Theory of Truth', *Transactions of the C. S. Peirce Society* **27**, pp. 27–77.

Sfendoni-Mentzou, Demetra (1992). 'Is there a Logic of Scientific Discovery? A Pragmatic Realist Account of Rationality in Physical Theory', in M. Assimakopoulos *et al.* (eds.), *Historical Types of Rationality.* National Technical University of Athens, pp. 239–250.

Sfendoni-Mentzou, Demetra (1993). 'The Role of Potentiality in C. S. Peirce's Tychism and in Contemporary Discussions in Q.M. and Micro-Physics', in E. Moore (ed.), *Charles S. Peirce and the Philosophy of Science: Papers from the 1989 Harvard Conference.* The University of Alabama Press.

Tooley, M. (1977). 'The Nature of Laws', *Canadian Jour. of Phil.* **7**.

ARCHIE J. BAHM

TENTATIVE REALISM

'Realism' is an epistemological theory claiming that some objects of knowledge exist whether or not they are apparent. In everyday practice, every person is a realist. [Little observation is needed to conclude that all other animals are realists also. The biological evolution of humans was possible only because their physiological organisms were adapted to sensing and responding to real environmental beings and processes.]

Human beings are naturally naive realists. Doubts about accepting apparently real objects as "really real" occur when apparent error appears. Error appears when something appears to be so and not so at the same time. A popular example is a glass of water with a pencil inserted. Visually the pencil appears as broken at the water surface but unbroken to a descending finger. Apparent contradiction often results in uncertainty about whether to trust what is apparently real. Repeated experience with apparent errors tends to generate some distrust about what appears.

Some persons, and theorists, have jumped to conclusions that no apparent reality should be trusted. Some consequent anti-realistic theories include Subjective Idealism, Phenomenalism, Romanticism, Existentialism, Scepticism and Agnosticism. Some persons, and theorists, have ignored error as a subjective deficiency and have continued to be realists, some naively uncritical and some formulating theories about the nature of the real. Speculations about the nature of the real have generated several varieties of metaphysics, some advocating existence as constantly changing and some as unchangeable, some as consisting of uncuttable particles and some as a cosmic whole, some as completely determined (even predetermined) and some as resulting from pure chance, some as structure by eternal forms (Platonism) and some as evolving evolutionarily (Spencer), some as inspired by an *elan vital* and some as designed by an external creator. The history of the physical sciences has produced several different conceptions, including preferring an infinitely powerful Big Bang to an Omnipotent Creator. None of these has provided a final solution to originating doubts.

I propose that we can, as we must, trust what appears as real as long as it continues to appear real. People survive by adapting to their envi-

R. S. Cohen, R. Hilpinen and Qiu Renzong (eds.), Realism and Anti-Realism in the Philosophy of Science, 97–101.

ronments by developing both habits of trusting regularly recurring realistic appearances and habits of being ready to change and adjust to whatever appears as really new. [Intelligence is ability to adapt both to changing apparent realities and to unchanging apparent realities.] When something new appears, common sense response formulates some idea about it and some action to test its nature. If the idea and test seem satisfactory, then, when such novelty recurs, the same idea and test will be reused. As long as such idea and test continue to satisfy, the idea will remain as acceptable knowledge.

But if such idea later becomes unsatisfactory and involves some other error, habitual confidence in it ceases and some new idea and test are undertaken. Pragmatists have summarized the theory that those ideas which work are true, and survive, and those ideas which fail to work are false, and perish. Involved in this view is that an attitude of tentativity not only is desirable but should be recognized as a normal condition of reliable knowing. The word 'tentativity' means "holding on", both "holding on tightly as long as an idea works" and "holding on loosely enough to be willing to let go whenever the idea fails to work".

I propose that 'Tentative Realism' is the best solution to the problem of the nature of reliable knowledge and as a basic assumption of scientific methods. Some will prefer the term 'Pragmatic Realism' or simply 'Pragmatism', although the meaning of this word has been misused so often that misunderstandings may prevent its acceptance.

Is knowledge realistic or subjectivistic? It is both. All knowledge is mental. No non-mental knowledge exists. All objects of knowing are mental because they appear in awareness as objects. Some apparent objects are merely apparent, as those in dreams, day dreams, and idealized objects of desire. Some apparent objects appear as real, i.e., as existing when not apparent. Although, since any actually existing real thing cannot enter into a mind and its appearance, as long as what appears appears as real, and as long as a person trusts one's appearances, the idea that the apparently real object is actually real 'works', and thus realism is thereby assumed. Problems that are involved in such assumption are many.

Is knowledge of apparently real objects certain or uncertain? As naive realists, we are uncritical of appearances and thus remain without feeling uncertain about them. But when appearances contradict, rational doubt begets distrust of appearance. Persons, and scientists, then try to design tests for deciding whether one or the other or both of the appearances seem trustworthy. Tentative Realism accepts the fact that knowledge, including scientific knowledge, normally involves both some certainty (or lack of uncertainty) and some uncertainty. Hence its inclusion of the term "tentative" in its title.

Is knowledge in each mind private or is it communicable? An hypothesis or conclusion is usually regarded as scientific only if it is communicable (to one or some specified number of other persons). The requirement assumes that all minds (or at least the minds of the scientists involved) are alike. But actually, minds are both alike and different, both in biological inheritance, physiological development, and experiential conditioning after birth. Not only personal experiences but cultural influences condition minds in somewhat different ways that make the assumption dubious. Again, although this question may be settled by some thinkers in terms of previously-accepted theories (biological, physiological, psychological, sociological, and metaphysical), the fact that minds are both alike in some ways and different in others needs to be recognized in any acceptable theory. Tentative realism recognizes this fact and advocates that the kind and amount of sameness involved in communication of ideas is itself subject to experimental testing, and that an attitude of tentativity be sustained regarding this kind of realism (the existence of other minds as real is assumed) also. Recurrence of agreement about realistic ideas tends to be regarded as some assurance that the knowledge is reliable.

Is existence unchanging or changing? That is, does apparent regular recurrence of an apparently real object as unchanging imply that it is actually unchanging? Although each person, and each scientist, must rely on one's own experiences, evidences from other sources demonstrate that at least some apparently real objects do change in unapparent ways. Such evidence, and resulting general theories that the existing universe is entirely unchanging or constantly changing or both changing and unchanging, has implications for theory of knowledge. Tentative Realism advocates that the existing universe is both unchanging in some ways and is constantly changing in other ways. This theory implies that the principle of tentativity should be employed both with regard to theories about the existing universe and to implications of such tentativity for each more specific, and, when possible, for each particular, inference regarding the reliability of accepting an apparently real object as actually real.

Do apparent numbers exist only in minds or do numbers exist independently of minds? Many scientists claim that "if it can't be measured, it doesn't exist". Such scientists seem to assert that numbers required for measurement of supposedly real things are also real, and some, at least, that they are more real than the things numbered. Logical and mathematical realisms have their origins in Platonism. The forms of things, and numbers that measure such forms, are eternal in nature, and thus their use implies, and assures us, that, except for any incapacities of minds for knowing them, our knowledge of real things can be certain. Tentative realism regards numbers as apparent objects existing only in minds, and that any inference regarding their existence as real is mistaken. The fact that apparently real things are enumerable does not imply that numbers are somehow embodied in them. Whether and to what extent minds are alike regarding abilities to understand mathematics is itself something that can be tested experimentally, as is often done by teachers of mathematics. But the capacity for observing that two apparently real fingers and two more apparently real fingers always appear as four apparently real fingers seems to be normally common. And generalizations and abstractions, all occurring in minds, normally generate the same inferences regarding numbers. Any logical realist must also assume that minds are alike in capacity to apprehend numbers as apparently real and to generate the same inferences regarding numbers. That numbers are apparent objects is intuited as self-evident. That numbers are apparently real objects involves inference. When, and to the extent that inferences that numbers are actually real 'work', and serve other subjective purposes, must be granted by Tentative Realists. But again, all such 'working' is regarded as tentative.

Is a scientific hypothesis verified by remembered data or by future data or by both? Empiricist theory claims that the truth of an idea is appropriately verified by tracing its origins in previous sensations. But pragmatists claim that such previous sensations no longer exist and so cannot serve to verify, and that verification of an hypothesis can occur only by testing it in future experiences in which the data can be intuited (i.e., as apparently real corroboration). Since some past experiences of apparently real objects can be remembered (and the reliability of memory habits can also be tested pragmatically), and used for purposes of corroboration, Tentative Realism advocates use of both past and future (and present, for any actual assent to verification occurs in some present) apparently real data available for verification. When scientists have agreed upon some universal principles as reliable, then they too may be employed in inferential verifications.

Does science exist only as concern about true theories or also in the results of applying theories in practice? Some philosophies of science claim that science is concern for understanding. Any applications and their consequences are beyond science. Other philosophies of science claim that the purpose of science is to solve practical problems and to improve human welfare and that thus scientific theories are not fully verified until human benefits become obvious. Tentative Realism observes that some scientific efforts are primarily theoretical and that some are primarily practical in intent. Practice, as observing apparently real uniformities, often is a source of hypotheses and their verification. But also, when conclusions become regarded as reliable, they tend to generate applications in practice. Both seem normal and necessary to an adequate theory of science.

If we review the history of science, we can observe that very many of the hypotheses agreed upon as verified by the scientists of earlier periods in history later have been proved false. Such review provides evidence that an attitude of tentativity is needed and that the Tentative Realistic claim that such an attitude is essential to the nature of science adequately conceived is verified. [If we include the whole history of science in our conception of the nature of science, we are reminded that it consists of more false than true hypotheses.]

Department of Philosophy,
University of New Mexico,
U.S.A.

F. WALLNER AND M. F. PESCHL

COGNITIVE SCIENCE – AN EXPERIMENT IN CONSTRUCTIVE REALISM; CONSTRUCTIVE REALISM – AN EXPERIMENT IN COGNITIVE SCIENCE

1. DEVELOPMENT AND EVOLUTION OF CONSTRUCTIVE REALISM

There are various origins for the development of the ideas of Constructive Realism – they are coming from philosophical as well as (natural) scientific disciplines. CR, thus, does not stand in a certain philosophical tradition. Its uniqueness and strength are founded in the fact that its theses are the result of a process of establishing relations between very different sources and information. Of course there a multitude of points of reference can be found; they acted, however, rather as a *stimulating* instance for the development of Constrictive Realism. The view of CR is influenced by:

(a) the ideas of Ludwig Wittgenstein which played an important role in the development of the function of Constructive Realism in respect to the sciences;

(b) the second point was the experience with *interdisciplinary* cooperation and research in the (natural) sciences – by discussing and investigating the process of science we learned about the implicit and in most cases unspoken needs of the scientists and sciences of our time.

Another experience, which was very important for the development of Constructive Realism, was the debacle of philosophy of science in respect to its normative claims. On the one hand it seems to be clear that the reflections being made in philosophy of science are deficient, if they are only describing the processes of science; on the other hand the last decades of discussion in philosophy of science have shown that it is getting more and more unclear, how normative claims can be legitimized in the context of the sciences. This uncertainty of philosophy of science led to the following development: scientists not only felt themselves not understood and not in charge of the philosophy of science, but also they turned to metaphysical philosophy or irrational ideas.

R. S. Cohen, R. Hilpinen and Qiu Renzong (eds.), Realism and Anti-Realism in the Philosophy of Science, 103–116.

On the one hand we have to see this in the context of the development of natural sciences in recent decades, and on the other hand modern epistemology did not react in an adequate manner to this development. Modern technologies, especially the use of modern *computer technology* in the process of research, accelerated the trend to instrumentalism (which already can be found in the basic structure of (natural) sciences); i.e., without having given up the claim of (definite and objective) knowledge this claim has been increasingly circumvented and has been replaced by contexts of functioning relations. This implies a lot of needs for explanations which remain, however, in most cases unspoken, because they do not have a direct influence on the development of research. Because the natural sciences minimize the claim of recognizing the world, the area of irrational explorations is continuously increasing.

In many cases this has the consequence that irrational strategies of explanations (i.e., strategic methods are not rationally legitimized) make use of results or elements of natural sciences. This is completely harmless if these explanations are only for entertainment. It can have unpleasant or even dangerous consequences if such explanations are used as normative claims. In the view of such a dichotomy – between contexts which can be dominated and irrational explanations – the claims of occidental science to lead to scientific knowledge are given up. It could be, of course, that this already has been in the structure of science for a long time – Heidegger pointed out that occidental science would be in principle wrong. In contrast to that the Constructive Realism tries to save this kind of knowledge by a new interpretation of its procedure, i.e., scientific knowledge is understood as a specific form of results of actions.

At first we want to discuss alternative approaches in modern epistemology. An increasing alienation can be found between epistemology and philosophy of science on the one hand and natural sciences on the other hand. Naturalistic epistemology took into account this process; the most famous and most successful is represented by the evolutionary epistemology. Naturalistic epistemologies take into consideration the knowledge of modern natural sciences. Hence, they are explicitly doing something which already can be implicitly found in Kant's philosophy. In most cases, however, it remains unclear in modern naturalistic epistemologies which argumentative role disciplinary results are playing in their epistemological concepts.

We are assuming *cognitive science* to be a continuation of naturalistic epistemologies. Compared to naturalistic epistemologies cognitive science has an *interdisciplinary* structure – the problem is, however, that there does not exist a theory yet making clear the methods of cooperation of different disciplines. Thus it is not possible to rationally execute interdisciplinary cooperation. As an implication of this lack of theories, cognitive science is in a strange ambivalence: the use of computers offers to a very large extent the possibility of hiding the fact that there are methodological difficulties. Using computers often hides the possibility of controlling the relevance and plausibility of data, whereas in the manual manipulation of data the inadequacy of the used method can be noticed (which gets clear in the discussion with other disciplines). The application of computers makes complex theories explicit and thus ignores or at least is not primarily interested in the critique of the applied method. The orthodox AI approach, for instance, represents such an example where the question of functioning has a higher priority than the question of understanding and basic knowledge.

Having this in mind, cognitive science is often in danger of being dominated by an instrumentalistic approach – the higher priority is making perfect the processes of cognition, the basic problems of cognition are completely omitted. Another aim of Constructive Realism is the development of a methodology of cognitive science as an *interdisciplinary* cooperation between various disciplines taking part in the discussion of very basic problems concerning, for instance, the questions of what is knowledge, what is language, what is knowledge representation, etc. – an answer for these questions is very important, because, if we want to investigate the process of science, we have to have at least an idea of what we understand, for instance, by knowledge or how this knowledge is represented in the disciplines, etc. We will look for new paradigms in cognitive science as well as in philosophy of science – the Constructive Realism approach represents the basis for the investigations and considerations in this dynamic development.

2. STRUCTURE AND METHODS OF CONSTRUCTIVE REALISM

Looking at the structure of CR we can see that it is standing in a twofold position: on the one hand it is positioned between methodology and ontology; on the other hand it can be found between naturalistic epistemology and theory of action. What does this mean in detail? We are investigating methodological rules as instructions for acting and thus criticize methodologies of single disciplines in the context of theories of acting. CR develops ontological concepts from the result of naturalistic epistemologies. It also reveals the implicit ontologies of results from single disciplines which arise as placebo effect of certain methodological strategies.

2.1. *The Concept of "Strangification"*

Compared to classical metaphysical theories or to theories of philosophy of science of our century, CR refers neither implicitly nor explicitly to a meta level. The meta level has been replaced by another methodological action – it could be compared to a methodological trick: we are referring to this process as "*strangification*". This does not mean only one strategy, but rather a *set of strategies* having one thing in common: they are transferring one (logical) system of propositions from their original context into another context – this system is judged then out of this context. This process can not be planned in advance, because there is no metatheoretical standardization – it rather represents a game with different contexts. Changing the context many times enables us to get new insights, perspectives and views in the structure of the system of propositions. The important point is that, if we are looking at these contexts where the systems of propositions get to be absurd, we are noticing the implicit assumptions and considerations of this system – i.e., we are investigating its tacit knowledge.[1]

One could think, for instance, of orthodox Artificial Intelligence which is strangificated in the context of the humanities; in this moment and in this context the *insufficiency* of the (in most cases unreflectedly) applied methods of computer science (i.e., mathematical logic, etc.) for modeling or even better for understanding cognitive processes gets clear. Think, for instance, of the models in cognitive psychology which make in most cases use of the symbolic processing paradigm of Artificial Intelligence (e.g., J. R. Anderson).[2] The consequences which can be drawn from such a strangification are two-fold: in one case the scientists ignore the results of this strangification (because they pretend that these results are "not relevant or adequate") or they are consequently trying to look for an alternative approach which is, of course, more uncomfortable than the first case.

We are differentiating three types of strangification: linguistic, ontological and pragmatical strangification, excluding contexts in which a system of statements gets absurd represents the first type of strangification. Thus, this method enables the revealing of (implicit) assumptions as well as showing up the domain of application of a system of statements without having to fall back upon metatheoretical standardization instances. Talking in meta levels is not prohibited; it can be seen rather as an example for the strategies of strangification.

Applying a system (a set) of methods of one discipline to a very different discipline represents the *ontological type of strangification.* We are thinking, for instance, of the application of hermeneutical procedures in the natural sciences. Gestalt perception in biology, for instance, forms the boundary of the application of quantitative methods.

We are also thinking of the introduction of quantitative methods in the humanities. The aim is not, however – as normally assumed – to make the humanities more exact or to give them the methods to "correctly" quantify these phenomena that can be quantified; it is quite the reverse: the application of quantitative methods in the humanities makes clear the phenomena which can not be quantified – it enables a better understanding for structures which are unquantifiable. As an example we could think of psychology (if we think of it as a discipline of the humanities) and its (absurd) methods and trials to quantify intelligent behavior. What can be shown is the naiveté of natural scientists and their methods – these methods are too simple and too naive to quantify or even estimate such a complex phenomenon as human intelligence; it is, however, not only the method, but also the scientists and their assumptions which have to be analysed. That is the third kind of strangification in which the social and organizational context of the scientists is observed – we are calling it *pragmatical strangification.*

Thus, CR has a strict interdisciplinary structure – it is, however, not ontological, but rather methodologically motivated. We have to give up the claim of traditional European epistemologies of complete insight; the methodologies of CR have another aim: on the one hand they represent the basis for the construction of knowledge, on the other hand for the indirect insights into these constructions. As an implication of this we do not need a metatheoretical legitimation for the methodologies of CR.

2.2. *Two Types of Reality*

In German there exist two words for 'reality': we will refer to *Realität* with 'reality' and to *Wirklichkeit* with environment(w). Stimulated by

H. Maturana we are distinguishing (in contrast to evolutionary epistemology) between two types of reality: we understand by "Wirklichkeit" (environment(w)) the world we are living with, i.e., the world which is *presupposed* to our perceptions and to our processes of life. "Realität" (reality) means our *cognitive world* being the result of a process of construction. This differentiation can be derived from different contexts.[3] We could compare it to, as well as differentiate it from, Kant's differentiation between "Ding-an-sich" (thing-in-it-self) and the things' appearances. The dilemma of Kant's approach, which claims the unification of the "Ding-an-sich" and transcendental ego, was the reason for introducing this differentiation. It avoids the differentiation (being presupposed implicitly as well as explicitly by Kant) between absolute and human knowledge, as such a differentiation represents a principal misunderstanding of the concept of knowledge according to CR. It rather introduces the differentiation between the world we are living with as *environment(w)* and the constructed world of our cognition as *reality.*

By making use of the strategies of strangification CR offers to get a better understanding of reality; the environment(w) can not be understood. We can gain control, however, over environment(w) by our constructions of reality. In respect to our knowledge we can understand only what we have constructed.

The differentiation between reality and environment(w) does not serve the relativization of our knowledge; it rather avoids the fact that we are surrendering ourselves to the success of our constructs in the environment(w). As an implication we are suggesting a proposal for the organization of the process of research: we should differentiate between the instrumentalistic aspect in science and the aspect of construction of reality. Instrumentalism gives itself up to the criterion of "working well", the latter claims to impart knowledge.

It was one of the biggest mistakes in European history to admit to an *empirical success* of constructed knowledge a *judicial instance* over our knowledge. The alternative for such a hypostatization of empiricism can be found not only in ideological and arbitrary assumptions. We rather must not lost sight of the constructive character of empirical instances. This implies that empirical control of theoretical constructs represents nothing but the comparison of two (different) constructs (of our cognitive reality). Thereby we do not give up principally the relation to environment(w); the relation to environment(w) is rather *indirect.* If we are comparing two different constructs (of reality), we are "leaving" reality; we are testing to what extent reality endures beyond the correlations of our constructions. Stated another way we could formulate the following: By making use of empirical control we are applying a strangification "transforming reality to environment(w)". Having in mind this context of *indirect control*, empirical testing seems to be very useful.

3. KNOWLEDGE REPRESENTATION AND CONSTRUCTIVE REALISM

This differentiation into two domains of reality has very interesting and important implications for a central problem in Constructive Realism as well as in cognitive science: the problem of *knowledge representation*. We are understanding knowledge as the result of cognitive processes in natural (or artificial) cognitive systems – hence, knowledge can also be seen as the result of complex *information processing*. As we will see in the course of this section, we have to use this metaphor of cognition being information processing in a rather careful manner for the following reasons.

3.1. *Symbolic Knowledge Representation*

In most cases information processing and knowledge representation is reduced to *linguistic* knowledge, i.e., at least three steps of mapping are assumed:

(a) there is a mapping from environment(w) to the nervous system, i.e., stimuli coming from the environment are transformed into patterns of neural activations.
(b) these neural activities are mapped to explicit linguistic structures. This domain is assumed to be what we have called "cognitive reality" above – this implies that this domain is restricted to explicit natural language.
(c) these natural language symbols are mapped to a symbol system of a formal language; this system offers the possibility of processing "knowledge" on a *syntactic* level. This implies that knowledge processing is reduced to the manipulation of syntactic structures representing only the very "surface" of our language.

Hence, the semantic as well as the pragmatic aspect of knowledge are reduced to the level of *syntactics*. Such a reduction is well known (and was doomed to failure) in philosophy of science: early R. Carnap is one of the most prominent exponents of this approach.[4] In this book *Logische Syntax der Sprache*, Carnap tries to reduce the semantic aspect of scientific language to syntactics. He is not even aware of the pragmatic aspect at this time; a few years later Carnap himself recognized the shortcomings of such an unjustified reduction. We can learn from this development in philosophy of science that if we are interested in the investigation of structures of knowledge (or of reality) we have to go beyond this syntactic border and seek for *alternative mechanisms* of constructing knowledge. Here we are going the long way of giving up classical AI.

If we are looking at the orthodox approach of Artificial Intelligence or traditional cognitive science we can observe this unjustified dominance of language in their representation mechanisms: (natural) linguistic structures are transformed into very complex systems of rules, frames, schemata, semantic networks, etc. – *these symbol systems obtain their meaning only in the observer's or interpreter's cognitive domain*, i.e., a symbol is represented in a meaningless bit pattern having a meaning only if it is interpreted by the user. These computer symbols are *lacking* the *direct relation to the environment(w)* as they are the abstract result of processes taking place in the (cognitive) reality. This approach of manipulating (meaningless) symbols is – from the perspective of computer science – very convenient; if we are interested in very basic problems concerning basic questions of what is knowledge, what is language, what is cognition, etc., such an approach is *not adequate* and we have to look out for alternative forms of presenting knowledge in a way which is not as restricted as in this symbolic approach (for a more detailed discussion, see M. Peschl).[5]

3.2. *Neural Computing – an Alternative Form of Understanding Knowledge*

Cognition can not be reduced to a process of manipulating symbols – we rather have to see its *neural* foundation in a cognitive system's nervous system. This implies that we have to understand the metaphor of information processing in a more *general* way: cognitive information processing also includes basic processes such as physically interacting with an environment or physical interactions within the nervous system – these processes are the foundation for language, formal manipulations, etc. Knowledge is not understood in such a restricted manner of linguistically explicit units any more. It is the result of a process of self-organization, adaptation and *construction.* Hence, we prefer the notion of *knowledge construction* to the notion of knowledge representation. The Parallel Distributed Processing paradigm, which is often called connectionism, neural computing, or PDP, offers such mechanisms for presenting knowledge in an alternative way and makes it possible to achieve a deeper understanding of the process of *construction* of knowledge as Constructive Realism outlined. Rumelhart and McClelland[6] represent the traditional exponents of the newer form of connectionism which is a dynamically emerging field in computer science as well as in cognitive science. In short, the aim is to simulate neurons which are assumed to be the functional "atoms" of our brain and cognition. These neurons, which are called *units*, represent a very abstract model of natural neurons. They are forming a huge network which is highly interconnected; each of these units is working in parallel and computes its own activation by summing up the product of the activations and weights of the other units. The output acts as an input for all the other units being connected to this unit.

What about knowledge representation in such networks? We cannot find a linguistic representation in the form of explicit symbols any more. We rather find a set of weights being responsible for the behavior of the whole network. These weights are changing by applying a learning rule, i.e., by a *locally* slight change of the value of the weights the *global* behavior of the whole network is changing as these changes are taking place in each of these weights. Simply spoken, PDP networks are "learning" to associate patterns of activities in a subsymbolic way, i.e., in a way which is not restricted to symbols any more. The PDP paradigm is comparable as well as different from the structure of knowledge construction. This special relation enables us to improve the insights in the structure of knowledge construction by strangification. The domain of acting is strangificated into the domain of connectionist structures.

3.3. *The Constructivist Approach to Knowledge*

The point in CR is not to assume a given body of knowledge which is mapped to the computational structures (for instance, symbols or PDP networks); we are trying to make transparent the fundamental processes of constructing knowledge. From a more abstract point of view we have to look at the Parallel Distributed Processing paradigm as a very complex form of pattern processing. It very much depends on the *computational environment* in which such a PDP network is applied: if the input and output are encoded symbolically (for instance in Sejnowski's NETtalk)[7] the same problems will arise as in the approach of orthodox symbol manipulation; the only advantage is the way in which these symbols are processed: by propagating patterns of activations and by interpreting these patterns as a code for symbols. Such an approach does not help us in solving the problems of reduction of knowledge representation to linguistically explicit structures.

The alternative approach being suggested by M. Peschl[8] tries to model these processes in a *bottom up* manner. This means that the suggested cognitive model is linked through sensors and effectors to its environment(w). By learning in the highly recurrent connectionist network an *internal cognitive reality* is built up. The structure of this way of understanding knowledge of the cognitive system's environment depends on the following items:

(a) the cognitive system's current state of activities, which itself is determined by
(b) the ontogenesis of the cognitive system (i.e., by its very private history)
(c) by the structure (and architecture) of the nervous system
(d) by the structure of the sensors and effectors.

These four factors determine the construction of knowledge, i.e., the process of how knowledge is constructed, by integrating the stimuli from environment(w) and the internal stimuli to the system's cognitive reality. Hence, a closed feedback loop is established between the cognitive system, the cognitive reality and environment(w). The relation between these can be studied by simulating such a cognitive system having this epistemological foundation of Constructive Realism (i.e., differentiation between cognitive reality and environment(w), etc.).[9]

4. COGNITIVE SCIENCE AND CONSTRUCTIVE REALISM – A MUTUALLY INTERDEPENDENT RELATION

In this section we are going to discuss the relations between cognitive science and Constructive Realism in detail. If we are looking at traditional approaches in cognitive science we can see that the role of philosophy and epistemology is quite unclear. This uncertainty is rectified by Constructive Realism. CR, however, does not understand itself as the philosophy of cognitive science; rather it understands itself as an instance of services which are offered to the (natural) sciences (this is standing in relation to L. Wittgenstein's concepts of philosophy). There is, however, a difference between Wittgenstein's concept and the concepts of CR: L. Wittgenstein's philosophy "leaves everything as it is", whereas CR claims to support the self reflection of sciences and thus is capable of changing them. CR instructs the sciences to newly rethink central aspects of method and the claims which are made for objects – by applying this procedure many self-misunderstandings of science could be revealed.

In short, CR offers the cognitive sciences a new ontology by redefining the role of ontology itself. It designs concepts of the world for the processual descriptions of cognition and knowledge in cognitive science – the most general concept is the differentiation between reality and environment(w) as it has been presented and discussed above. Similar things could be said about the constitutive role of the interdisciplinarity of cognitive science. It is the point of CR to show that interdisciplinarity is possible only if we give up the traditional ontological concepts and ideas (Wallner 1989).[10]

Why is cognitive science interrelated with Constructive Realism? To answer this question we have to present briefly our approach to cognitive science: we are understanding cognitive science as the continuation of traditional epistemology in the context of the modern knowledge (of natural science) and in the context of the new technologies of our century. This means that it is our aim to *integrate* traditional approaches of epistemology with results and approaches from computer science and neuroscience. As has been mentioned above, this requires an *interdisciplinary* approach to the phenomena being investigated in cognitive science.

If we are looking at traditional approaches, however, we can see that the investigations are rather multidisciplinary than interdisciplinary (cf., Stillings 1987, Adams 1985, Posner 1989, Osherson 1990, etc.),[11] i.e., there are results coming from different disciplines (and their methods and paradigms). These results are presented to the other participating disciplines and (in the best case) compared to each other. There are, however, *no* consequences drawn from this process of comparing. Only the *consequent integration* of results enables fruitful interdisciplinarity. The *concept of strangification* in Constructive Realism could possibly offer such a strategy for consequent integration. Results from neuroscience, for instance, are strangificated into the context of epistemology or neurocomputing. One thing (among many others) we can learn from this strangification is that relations can be found between *epistemology, computer science* (especially Parallel Distributed Processing, McClelland 1986)[12] and *neuroscience*. Such strangification is only possible because of the similar structures of these disciplines.

As has been mentioned, the differentiation between (cognitive) reality and environment(w) plays an important role because we are understanding (all kinds of) knowledge as the product of a process of construction. This construction is triggered by environment(w) and is the result of a learning process in the (cognitive) reality. On this basis we are investigating the basic questions of what is knowledge, what is knowledge representation, cognition, language, etc., in the interdisciplinary context of the disciplines of epistemology, computer science and neuroscience.[13] Each of them plays a specific role:

(a) neuroscience represents the *empirical* part, i.e., the empirical results represent a constraint for epistemology and computer science. On the other hand we have to be aware of the problems of methodology and "empirical input" – they could be investigated by epistemology;

(b) epistemology has a threefold function:
   (1) to structure the outline of the research programs
   (2) to make interdisciplinary interventions by strangifications
   (3) to show insights into the structure of cognitive reality.

(c) computer science's aim is to combine these two disciplines by making use of *simulation* techniques – as mentioned, the Parallel Distributed Processing approach (PDP, connectionism) plays an important role in our understanding of the role of computer science in cognitive science. This means that computer science tries to find models coming up to the claims of neuroscience as well as of epistemology – the PDP approach is capable of meeting these requirements and represents the foundation of our research and considerations. Computer science generates (from an observer's point of view) cognitive behavior by applying the simulation methods of the Parallel Distributed Processing paradigm. By learning in a PDP network we are building up an artificial (cognitive) reality being triggered by the (artificial) environment(w). Thus we are capable of studying structures of knowledge, how they evolve, and how they are constructed; we could even learn something about the constructive relation between reality and environment(w).

Thus we can find a *twofold* relation between cognitive science and Constructive Realism; on the one hand CR provides the methodology of interdisciplinary cooperation between the disciplines – as has been mentioned, the strategy of strangification and the differentiation between reality and environment(w) are the most important concepts being applied in our approach to cognitive science. On the other hand cognitive science represents a very interesting source for CR for two reasons:

(i) generally speaking cognitive science investigates the phenomenon of knowledge – thus it provides the Constructive Realism with information being of great interest, because it is one of the aims of CR to get deeper insights into the structures of knowledge (i.e., in the structure of reality);

(ii) the methods being suggested by CR can be tested in the interdisciplinary cooperation of our approach to cognitive science. As a consequence of this testing process the methods of CR could be improved, adapted, etc., because we are trying to develop the concepts of CR not only by speculative ideas, but also by considering the problems of the actual process of research and thus need feedback from the sciences.

5. CONCLUDING REMARKS

CR is a concept of science and research necessary in order to rescue the idea of our occidental science in the face of a completely changed domain of social structures. While the aim of occidental science according to the original ideas was the democratization of human conditions, the history of our universities often shows that strong tendencies counteracted this structure. If we do not establish the already discussed difference between environment(w) and reality, the structure of science tempts us to accept an undemocratic hierarchy, because the structure of the world must be mirrored in the organization and structure of science. If we assume research groups constructing alternative constructions of reality are not only possible, but rather desired, it is because they enrich our insights into reality. Therefore reorganization of the university cannot solve the rather fundamental problems of university structures unless we are willing to rethink and reflect the ontological presuppositions of science.

Institut für Wissenschaftstheorie
und Wissenschaftsforschung,
Universität Wien, Austria.

NOTES

[1] Polanyi, M. (1966). *The Tacit Dimension*, Doubleday & Company, Inc., Garden City, New York.
[2] Anderson, J. R. (1985). *Cognitive Psychology and its Implications*, New York.
[3] E.g., Wallner, F., Thesen zu Maturana. Eine Aufforderung zum Widerspruch, Research Report, Dept. of Epistemology & Cognitive Science, Univ. of Vienna, IWTF-ECS-89-2.
[4] Carnap, R. (1936). *Logische Syntax der Sprache*, Wien.
[5] Peschl, M. F. (1989). 'An Alternative Approach to Modelling Cognition', in Proceedings of Man & Machine Conference published by John von Neumann Society for Computing Sciences, Hungary, pp. 46–56 and also *ibid.* (1990). 'Cognition and Neural Computing – an Interdisciplinary Approach', in Proceedings of International Joint Conference on Neuronal Networks, Washington (IJCNN '90), Lawrence Erlbaum Associates, Publishers, pp. 110–113.
[6] Rumelhart, D. E. and McClelland, J. R. (1986). *Parallel Distributed Processing, Explorations in the Microstructure of Cognition*, Volume I: Foundations, MIT Press, Cambridge Massachusetts. *Ibid.* (1989). 'The Architecture of Mind: A Connectionist Approach', in Posner M. I. (ed.), *The Foundations of Cognitive Science*, MIT Press, Cambridge, Massachusetts, pp. 133–159. McClelland, J. L. and Rumelhart, D. E. (1988). *Parallel Distributed Processing, Explorations in the Microstructure of Cognition*, Volume II: *Psychological and Biological Models*, MIT Press, Cambridge, Massachusetts.
[7] Sejnowski, R. J. (1981). 'Skeleton Filters in The Brian', in Hinton and Anderson (eds.), *Parallel Models of Associative Memory*, pp. 189–212.
[8] Peschl, M. C. (1990). 'Cognition and Neural Computing – an Interdisciplinary Approach', in Proceedings of International Joint Conference on Neural Networks, Washington (IJCNN '90), Lawrence Erlbaum Associates, Publishers, pp. 110–113. *Ibid.* (1990). 'A Cognitive Model Coming up to Epistemological Claims – Constructivist Aspects to Modeling Cognition', in Proceedings of International Joint Conference on Neural Networks 1990 (IJCNN '90), San Diego, CA, IEEE, Vol. III, pp. III-657–662.
[9] Wallner, F. (1990). 8 Vorlesungen zum Konstruktiven Realismus, Wiener Universitätsverlag.
[10] Wallner, F. (1989). Maturanas möglicher Beitrag zur Epistemologie, Wissenschaftliche Nachrichten, Jänner.
[11] Stillings, N. A., Feinstein, M. H. and Garfield, J. L. *et al.* (1987). *Cognitive Science, An Introduction*, A Bradford Book, the MIT Press, Cambridge, Massachusetts. Adams, R. G. (1985). 'Cognitive Science: A Tripartite Approach', in: *Cognitive Systems* 1-1, pp. 17–29. Posner, M. J. (ed.) (1989). *Foundations of Cognitive Science*, MIT Press, Cambridge, Massachusetts. Oshersons, Daniel N. (ed.) (1990). *An Invitation to Cognitive Science*, MIT Press, Cambridge, Massachusetts.
[12] McClelland, J. L., Rumelhart, D. E and Hinton, G. E. (1986). 'The Appeal of Parallel Distributed Processing', in: Rumelhart, D. E., *Parallel Distributing*, Vol. I, MIT Press, Cambridge, Massachusetts.
[13] The reduction to these three disciplines has been suggested by M. F. Peschl (1990). Auf dem Weg zu einem neuen Verständnis der Cognitive Science, *Informatik Forum*, Juni.

JAMES ROBERT BROWN

PHENOMENA

PROOFS AND PICTURES

My interest is *phenomena*, but I'll begin with something apparently quite different – mathematical diagrams. Consider first, the following diagram which accompanies the proof of the Pythagorean theorem in Euclid's *Elements*. We can stare at it for days and still not see that the theorem is true; we need a proof. The common view of diagrams in mathematics is this: they provide a heuristic aide, a help to the imagination when following a proof. Diagrams cannot justify and are not to be confused with real proofs, which are formulated in words and symbols; at most they play a psychological role.

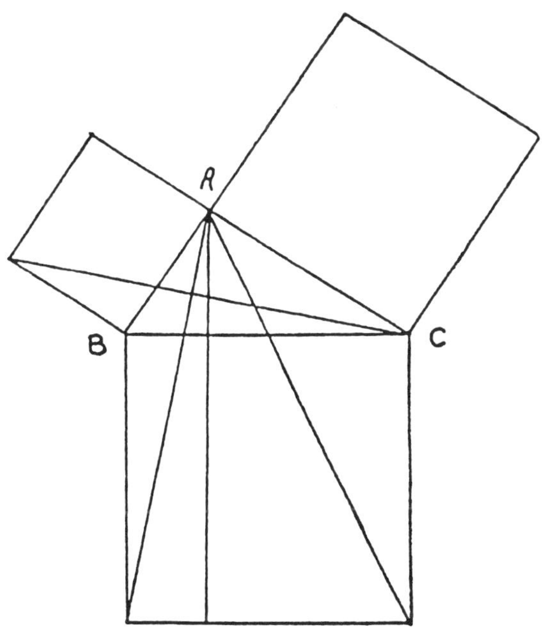

Fig. 1. Diagram that accompanies the proof of the Pythagorean theorem.

R. S. Cohen, R. Hilpinen and Qiu Renzong (eds.), Realism and Anti-Realism in the Philosophy of Science, 117–129.

For examples such as this one, the standard account seems right. But there are a few rare and remarkable examples where something quite different is going on. The following theorem is from number theory; and it has a standard proof (by mathematical induction) which uses no diagrams at all. But it can actually be proven with a diagram. (Take a moment to study the proof, to see how it works.)

Theorem: $1 + 2 + 3 + \ldots + n = n^2/2 + n/2$

Proof:

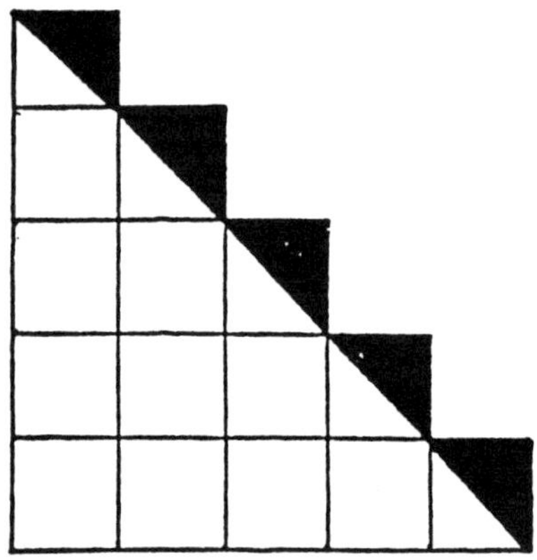

Fig. 2. A diagram as a proof.

Of course, there is lots of interpreting going on to make this a proof. For example, we must consider the individual unit squares as numbers and we must bring some geometric background information to the effect that a square with sides of length n has area n^2. But these sorts of interpretive assumptions are not less innocuous than those made in a typical verbal/symbolic proof.

The moral I think we should draw from examples like this is simple but profound: *we can prove things with pictures.* In spite of the fact that the diagram seems to be a special case ($n = 5$), still we can see all generality in it. And the proof does not work by suggesting the 'real' proof, since in the diagram there is nothing which corresponds to the passage from n to n + 1 which is the key step in any proof by mathematical induction.

Those who hesitate to accept the picture as a proof might think that the picture merely indicates the existence of a 'real' proof, a standard proof by mathematical induction. Perhaps they even wish to appeal to the well known distinction between discovery and justification: the picture is part of the discovery process while true justification comes only with the verbal/symbolic proof. But consider: would a picture of an equilateral triangle make us think there is a proof that all triangles are equilateral? No. Yet the above picture makes us believe – rationally believe – that there is an inductive proof of the theorem. The picture is evidence for the existence of a 'real' proof and the 'real' proof is evidence for the theorem. But we have transitivity here; so the picture is evidence for the theorem, after all.

Let us put this connection between theories and pictures in the background for now and turn to our main concern, phenomena.

DATA VS. PHENOMENA

My point of departure is a notable recent analysis of phenomena by James Bogen and James Woodward who make a "distinction between phenomena and data" (1988, p. 305). The former are constructed[1] out of the latter.

> Data, which play the role of evidence for the existence of phenomena, for the most part can be straightforwardly observed. However, data typically cannot be predicted or systematically explained by theory. By contrast, well-developed scientific theories do predict and explain facts about phenomena. Phenomena are detected through the use of data, but in most cases are not observable in any interesting sense of that term (1988, p. 305).
>
> Data are . . . idiosyncratic to particular experimental contexts, and typically cannot occur outside of those contexts . . . Phenomena, by contrast, are not idiosyncratic to specific experimental contexts. We expect phenomena to have stable, repeatable characteristics which will be detectable by means of a variety of different procedures, which may yield quite different kinds of data (1988, p. 317).

There are several important features and consequences of this view. Among the more important are these: explanation is not a relation between theories and observable facts; nor is prediction; theories are not tested by comparing them with experience; and observation – whether theory-laden or not – is "much less central to understanding science than many have supposed" (1988, p. 305).

Typical of Bogen-Woodward phenomena are weak neutral currents. The associated data are bubble chamber photographs. The relevant theory which is supported by all of this is the Weinberg-Salam theory of weak interactions. It is supported, not by the data, but by the phenomena of weak neutral currents. The existence of the phenomena is in turn supported by the data, the photographs. According to Bogen and Woodward, the data are far too messy to serve as evidence for any theory; the phenomena play a crucial and irreducible intermediate role in the process of scientific inference.[2]

A cursory glance at the social sciences suggests that the phenomena/data distinction is important here, too, perhaps even more so. Social scientists seem to do (at least) two quite distinct things. One is to establish phenomena: e.g., that there is widespread child abuse, that x% of the population is homosexual, that suicide rates in some cultures are higher than in others, that x% of women are physically abused by their male companions, that there is an x% unemployment rate, etc. These are often extremely difficult to ascertain as we might imagine, especially when questions of sexuality are involved. The second job is to explain these phenomena. And it is indeed phenomena that social theory attempts to explain. Economists try to tell us why we have high unemployment (a downturn in the economy), not why Joe Blow is out of work (perhaps he was an incompetent worker); and Durkheim told us why Protestant societies have higher suicide rates than Catholic ones (they are socially less cohesive); he doesn't tell us why Joe Blow killed himself (perhaps he was depressed after losing his job).

A number of examples, even if only briefly described, should help to explain and reinforce the data/phenomena distinction.

EXAMPLES

High energy physics abounds with examples. We are all quite used to having information from this field presented to us twice over, first as data in the form of a photograph, then as phenomena in the form of an artist's drawing. Here is a typical example (Figure 3).

What high energy physics does is explain and predict the drawing on the right; not the photo on the left. The chicken scratches on the left are far too variable, idiosyncratic, and downright messy for any theory to deal with. Theories in high energy physics only try to cope with the phenomena as represented in the artist's drawing.

The so-called mechanical equivalent of heat was established by James Joule in the middle of the 19th Century. That is, "That the quantity of heat produced by friction of bodies, whether solid or liquid, is always

Fig. 3. The reaction $K^- + p \rightarrow \Lambda + \pi^0$. The photo is a datum; the drawing is a phenomenon.

proportional to the quantity of force expended". In a large number of repeated experiments involving a paddle wheel contraption that heated a quantity of water when the paddles were driven by falling weights, Joule established

> That the quantity of heat capable of increasing the temperature of a pound of water (weighed in vacuo, and taken at between 55° and 60°) by 1°F, requires for its evolution the expenditure of a mechanical force represented by the fall of 772 lbs through the space of one foot (Joule 1850, p. 82).

This was not the result of any simple observation, but the culmination and processing of an enormous amount of data.

> The method of experimenting was simply as follows: The temperature of the frictional apparatus having been ascertained and the weights wound up . . . the roller was refixed to the axis. The precise height of the weights above the ground having then been determined by means of the graduated slips of wood . . . the roller was set at liberty and allowed to revolve until the weights reached the flagged floor of the laboratory, after accomplishing a fall of about 63 inches. The roller was then removed to the stand, the weights wound up again, and the friction renewed. After this had been repeated twenty times, the experiment was concluded with another observation of the temperature of the apparatus. The mean temperature of the laboratory was determined by observations made at the commencement, middle and termination of each experiment (Joule, 1850, p. 66).

It was out of the enormous amount of data that he had that the phenomena of the mechanical equivalent of heat was brilliantly and painstakingly constructed by Joule.

The Periodic Table of the chemical elements provides another illustration. The Periodic Table is a classification scheme of the elements in accordance with these properties. It is a paradigm of the construction of phenomena out of data. The phenomena are the entries in the table – the chemical elements and their properties: atomic weights, atomic numbers, chemical similarities, etc.

There is no algorithm for making phenomena out of data – it is a fallible process. Dimitri Mendeleev ordered the elements according to their increasing atomic weights. But he noticed that atoms with similar chemical properties recurred periodically at fairly regular intervals. By lumping together those which are chemically similar he created a classification of the elements known as the Periodic Table.

Though brilliantly conceived, Mendeleev's taxonomy was somewhat problematic. In the case of a few elements, ordering them by increasing weight was at odds with ordering them in accord with their chemical properties. And the discovery of isotopes (which have different weights but are chemically identical) made matters even worse. This was the background for Henry Mosely's work, begun in 1913.

The characteristic frequencies associated with each of the elements is due, according to Bohr's theory of the atom, to electrons in orbit around the nucleus falling to lower orbits. When they fall from one energy level, or shell, to a lower one they emit a photon of the appropriate energy, or frequency.

Mosely fired cathode rays at several of the heavier elements and recorded the x-ray frequencies produced. He focused on a particular series known as the K_α-lines in a large number of elements. What he discovered is that as the atomic number increases by 1 (i.e., as $Z \rightarrow Z + 1$), the quantity $(4/3 \times \nu(Z) \times R)^{1/2}$ also increases by 1. This led to the following formula for the frequencies of K_α-series for the element with atomic number Z: $\nu(Z) = (Z - 1)^2 \times (1/1^2 - 1/2^2) \times R$. (Where R is the Rydberg constant, known independently, and $1/1^2 - 1/2^2$ is associated with the first and second energy levels.)

Mosely's classification and Mendeleev's coincide except in a few cases. For example, potassium preceded argon in Mendeleev's table, but Mosely reversed them. This resulted in Mosely's Periodic Table being in full agreement with both the recurring regularities of the chemical properties and with the increasing atomic numbers. There is no internal tension as there was in Mendeleev's taxonomy.

The history of the Periodic Table illustrates all sorts of interesting things about phenomena. In it we see, of course, the construction of precise elements out of the hodge-podge of chemical data. And we also see that the Table is not explanatory; it is just a taxonomy. But there is one more feature of the Table which strongly supports the data/phenomena distinction. When the table was being constructed there were 'gaps' in it; that is, nothing had ever been observed which corresponded to certain places in the table (e.g., germanium, Ge). Any theory (such as Bohr's) that attempted to explain the features of the Table would be required to account for every place in the Table, including the gaps. (Or explain why the gaps had to exist, as quantum mechanics does in the case of the very heavy elements – they are unstable.)

I've been saying 'gaps in the table'. Strictly, this is wrong: the table is complete; the gaps are in the data. This means that the entries in the Periodic Table cannot be identified with what is actually observed, with data – since there is none (or was none at the time) – but must instead be thought of as phenomena.

PHENOMENA AND NATURAL KINDS

The world is full of data, but there are relatively few phenomena. My suggestion is rather simple: *phenomena are abstract entities which correspond to visualizable natural kinds*. When scientists construct the phenomena out of a great mass of data what they are doing is singling out what they take to be genuine natural kinds. In Plato's gruesome metaphor, they are trying to cut nature at its joints. To this I would only add: at nature's visualizable joints.

The shift in the ordering structure of the Periodic Table, from atomic weights to atomic numbers, shows the complexity and ingenuity that is sometimes involved in constructing phenomena out of data. But it also shows the importance of natural kinds and their essential properties in scientific thinking. Mosely expressed it well when he summed up his experimental work:

> We have here a proof that there is in the atom a fundamental quantity, which increases by regular steps as we pass from one element to the next. This quantity can only be the charge on the central nucleus, of the existence of which we already have definite proof (Quoted in Trigg 1975, p. 32f).

Notice that Mosely is not claiming to have discovered that the nucleus has an electronic charge, any more than he is denying that the elements have an atomic weight. His claim is about which of these existing properties is 'fundamental', or essential (chemically), and which is not.

THOUGHT EXPERIMENTS

Thought experiments deal with phenomena.[3] Obviously, they don't deal with actual experimental data – this much is true by definition. But the fact that they involve picturable processes suggests that we need to keep something observation-like centrally involved.

In Einstein's elevator, to cite one important example, the observer inside cannot tell whether she is in a gravitational field or accelerating. A beam of light passing through would bend downward if the elevator were accelerating, so, by the principle of equivalence, it would also

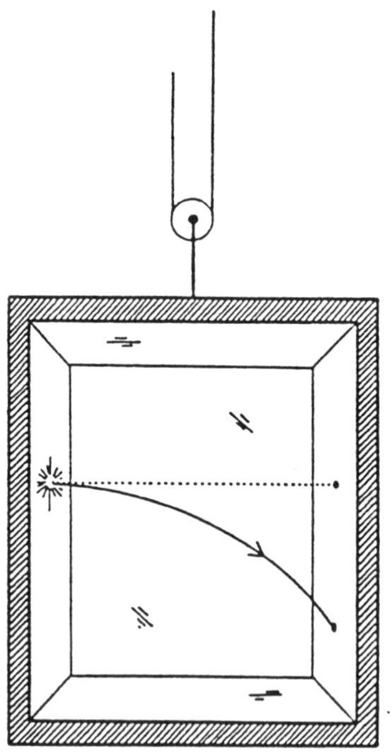

Fig. 4. Light bending in Einstein's elevator.

bend downward in a gravitational field. The conditions required to make such an observation are so extreme that any actual observer would have to be a puddle on the floor of the elevator. The observation in this thought experiment is of phenomena, not data.

Newton's bucket thought experiment provides an instructive example in a different way. The thought experiment asks us to imagine the different stages of a bucket partly filled with water as it is released and allowed to 'rotate'. The water and bucket would be initially at rest with respect to one another, and the water surface would be flat. Next they would be in relative motion. In the third stage they would again be at rest with respect to one another, but this time the surface of the water would be concave.

Why the difference between stages one and three? Newton's explanation is simply this: In the first stage the water and bucket are at rest with respect to absolute space, and in the third they are rotating with respect to absolute space.

After Leibniz, Newton's most forceful critics were Berkeley and Mach.

Fig. 5. Three stages in Newton's bucket thought experiment.

Did they deny that absolute space was the best explanation for the observed difference? Not really; instead, they denied the observed difference itself. They denied that in a universe without distant masses (the fixed stars) the water would climb the walls of the bucket. Clearly, Berkeley's and Mach's fight with Newton is not a dispute over empirical data; it's not even a fight over rival explanations of what is given in the thought experiment – it is a fight over the phenomena.

What we can see from these two thought experiments is that phenomena must be playing a role in scientific inference, a role which is distinct from data. Though phenomena are picturable, they exist at a high level of abstraction.

In passing, a word about the theory-ladenness of observation. No one nowadays believes in raw data; observations are always conceptualized. (This is undoubtedly one of the great results of modern philosophy of science, due to Hanson, Kuhn, Feyerabend, Sellars, and many others.) Isn't 'phenomena' just another name for this, that is, for theory-laden data?[4] In many cases the distinction between phenomena and conceptualized or theory-laden data will seem artificial. (High energy physicists like to say that they can just see the subnuclear process right in the bubble chamber photo.) But there are clear cases which cannot be treated as theory-laden observations. The elements of the Periodic Table are phenomena, and no doubt some of them, e.g., Fe (iron), might successfully be treated as observable in some theory-laden way. But there are others, e.g., Ge (germanium), which (at the time of Mendeleev) were simply not seen at all. Similarly, the phenomena of a thought-experiment, e.g., the light bending in Einstein's elevator are not actually seen at all either. So these examples of phenomena cannot be reduced to actual observable data, theory-laden or not.

PHENOMENA AND INFERENCE

How is it possible that a great and grand theory can seem to be justified by only a tiny bit of sketchy visualized information? Recall the mathematical example from the very beginning of this paper. A lesson about inference can be learned from that rare form of mathematical proof. The claim made there was that the diagram is a perfectly good proof. One can see complete generality in the picture, even though it only illustrates the theorem for $n = 5$. The diagram does not implicitly 'suggest' a 'rigorous' verbal or symbolic proof. The regular proof of this theorem is by mathematical induction, but the diagram does not correspond to an inductive proof at all (where the key element is the passage from n to n + 1).

One of the morals to be drawn from this example is of great importance for the philosophy of mathematics, especially concerning the nature of proof. But the moral I want to draw here is just this: We can in special cases correctly infer theories from pictures, that is, from visualizable situations.

The great inductive leap is really from data to phenomena; once we have the phenomena, the further inference to theory often can be actually rather small. This is because of the following feature of natural kinds.

> *Any natural kind has an essential property (or set of properties) that makes it the kind that it is. If any member of a kind has essential property ϕ then every member of the kind has ϕ.*

If any sample of water, e.g., has chemical composition H_2O, then they all do. Notice our reluctance to make a similar inference about, say, the colour of ravens. We balk at: If any raven is black, then they all are. We hesitate because of our belief that colour is not an essential property of ravens. We do believe that all ravens are black, of course, but this belief is based on the observation of an enormous number of ravens. By contrast, our beliefs about, say, the mass of intermediate vector bosons are based on only a handful of scratchy bubble chamber photographs. So, either our physics colleagues have absurdly low standards when compared to bird watchers or something remarkably different is going on in each case. Clearly, it is the latter. There is a profound difference between the two cases and it has to do, I suggest, with phenomena as natural kinds. In particular, if any vector boson has mass m then they have mass m. Natural kind inference is quite different than enumerative induction, the principle used in inferring the colour of all ravens.[5]

While I have invoked natural kinds and their essential properties to account for some of the aspects of phenomena, my commitment to natural kinds is not too deep. Perhaps *patterns* would be a better notion.[6] Instead of seeing phenomena as constructed out of data, I should take patterns to be so constructed. First, patterns would avoid the controversial metaphysics of essences, etc.; second, patterns are obviously abstract and hence clearly different from observable data; and third, inferences from patterns are quite unproblematic and so lend themselves to quick conclusions of the sort we see in the mathematics and thought experiment examples given above. For now, at any rate, I prefer to remain agnostic and leave this an open question, and for the sake of consistency I'll stick with talk of natural kinds.

Of course, the question arises whether we really have a natural kind (or an essential property of a natural kind) on our hands or not. Is mass really an essential property, and colour not? It seems like an *a priori* assumption, and to some extent I dare say that it is. But the view that the colour of ravens is not an essential property while the micro-structure of water is, is at least in part based on very broad experience and the past success of various classes of theories that we hold. Theories based on micro-structure have been enormously successful while those based on colours have not. So the construction of phenomena out of data is based on more than the data itself. It is theory-laden, but it needn't be laden with the theory that it will subsequently be used to test.

This raises an interesting an important point that I can only mention here: the construction of pseudo-phenomena. Many scientific works are replete with drawings. E. O. Wilson's *Sociobiology*, for example, has almost no photographs, but has several beautiful drawings of animals in various activities. One of these shows two dinosaurs fighting. Needless to say, this was seen by no paleontologist. It is not a datum, but a phenomenon. But is it a real phenomenon? I will leave to others the assessment of sociobiology. But I will point out that the theory – like any other – is in the business of explaining phenomena, not data. Perhaps it even does this brilliantly, which is why many find it persuasive. But there is a lot of room to ask pointed questions about the construction of such phenomena (are they pseudo-phenomena?) and the role that values may have played.

Let me now, by way of conclusion, summarize the main points. Phenomena are to be distinguished from data, the stuff of observation and experience. They are relatively abstract, but have a strongly visual character. They are constructed out of data, but not just any construction will do. Phenomena are natural kinds that we can picture. They show up in thought experiments and they play an indispensable role in scientific inference mediating between data and theory. So let's attend to them.[7]

Department of Philosophy,
University of Toronto,
Canada.

NOTES

[1] This term is unfortunately loaded. Sociologists of science often use 'construction' to mean 'social construction', the very opposite of an independently existing object or fact, as for instance, Pickering does when he speaks of 'constructing quarks.' I'm using the term in a more innocuous sense perfectly compatible with describing the objective truth. For example, when a mathematician 'constructs' a function, she is not creating it anew, but merely (though perhaps very cleverly) characterizing it in terms of other already given mathematical objects.
[2] For further discussions, rich with examples, see Woodward 1989, Kaiser 1991, and Kaiser this volume.
[3] For more on thought experiments see my *The Laboratory of the Mind: Thought Experiments in the Natural Sciences.*
[4] This challenge came from Simon Blackburn during the conference.
[5] For more on natural kind reasoning see Harper (1989). Generally, this is unexplored territory and deserves a great deal more attention.
[6] Thanks to Mary Tiles for making this point. I'm grateful for helpful discussions on a number of other topics in this paper.
[7] The themes of this paper will be treated in greater depth in my forthcoming book, *Smoke and Mirrors: How Science Reflects Reality.*

BIBLIOGRAPHY

Bogen, J. and Woodward J. (1988). 'To Save the Phenomena', *Philosophical Review.*

Brown, J. R. (1989). *The Rational and the Social*, Routledge, London and New York.

Brown, J. R. (1991). *The Laboratory of the Mind: Thought Experiments in the Natural Sciences*, Routledge, London and New York.

Harper, W. (1989). 'Consilience and Natural Kind Reasoning', in J. Brown and J. Mittelstras (eds.), *An Intimate Relation: Studies in the History and Philosophy of Science* [*Boston Studies in the Philosophy of Science*, v. 116], Kluwer, Dordrecht.

Joule, J. (1850). 'On the Mechanical Equivalent of Heat', *Philosophical Transactions of the Royal Society.*

Kaiser, M. (1991). 'From Rocks to Graphs – The Shaping of Phenomena', *Synthese*, 111–133.

Kaiser, M. (this volume). 'Empirical versus Theoretical Progress in Science'.

Trig, G. (1975). *Landmark Experiments in 20th Century Physics*, Crane, Russell & Co., New York.

Wilson, E. O. (1975). *Sociobiology: The New Synthesis.*

Woodward, J. (1989). 'Data and Phenomena', *Synthese*, 393–472.

ALLAN FRANKLIN

THERE ARE NO ANTIREALISTS IN THE LABORATORY

One of the continuing polemics in the philosophy of science has been the battle between the realists and the antirealists. Although there are probably as many variants of these positions as there are adherents of them, I shall adopt, and defend, Bas van Fraassen's characterization of realism, "Science aims to give us, in its theories, a literally true story of what the world is like; and acceptance of a scientific theory involves the belief that it is true" (1980, p. 8). I shall argue that we have good reasons to believe in both the truth of scientific laws and theories and in the reality of the entities involved in those theories.

The battle has flared up since van Fraassen's seductive account (1980) of an antirealist position, that of constructive empiricism.[1] In his own words, "Science aims to give us theories which are empirically adequate; and acceptance of a theory involves as belief only that it is empirically adequate" (1980, p. 12), where empirical adequacy means that what the theory "says about the observable things and events in the world, is true – exactly if it 'saves the phenomena' " (1980, p. 12).[2] I shall discuss this view in some detail later.

My own position, which one might reasonably call 'conjectural' realism, includes both Sellars's view that "to have good reason for holding a theory is *ipso facto* to have good reason for holding that the entities postulated by the theory exist"[3] (1962, p. 97), and the 'entity realism' proposed by Cartwright (1983) and by Hacking (1983). Both Hacking and Cartwright emphasize the manipulability of an entity as a criterion for belief in its existence. "We are completely convinced of the reality of electrons when we regularly set out to build – and often enough succeed in building – new kinds of device that use various well-understood causal properties of electrons to interfere in other more hypothetical parts of nature" (Hacking, 1983, p. 265).[4] Cartwright also stresses causal reasoning as part of her belief in entities. In her discussion of the operation of a cloud chamber she states, ". . . if there are no electrons in the cloud chamber, I do not know why the tracks are there" (Cartwright, 1983, p. 99). In other words, if such entities don't exist then we have no plausible causal story to tell. Both Hacking and

R. S. Cohen, R. Hilpinen and Qiu Renzong (eds.), Realism and Anti-Realism in the Philosophy of Science, 131–148.

Cartwright grant existence to entities such as electrons, but do not grant 'real' status to either laws or theories, which may postulate or apply to such entities.

In contrast to both Cartwright and Hacking, I suggest that we can have good reasons for belief in the laws and theories governing the behavior of the entities, and that several of their illustrations implicitly involve such laws.[5] I shall present an illustration of my own later. I agree with them, however, that we can go beyond Sellars and have good reasons for belief in entities even without such laws.

1. ARE THERE REALLY K MESONS?

Before discussing some of the philosophical issues, I would like to present an example from contemporary physics which illustrates my view. I will argue that this experiment provides good reasons to believe in both the existence of K mesons, as well as in the truth of several laws involved in the operation of the experimental apparatus.

I do not, however, wish to imply by this that experiment always provides good reasons for belief in entities and laws.[6] I do wish to argue that it can, and does, do something I believe Van Fraassen would deny. One must also be careful about which entities the experiment argues for and which are assumed on the basis of prior knowledge. Thus, in the example that follows, the experiment provides reasons for belief in K mesons, but the existence of electrons, protons, and pions and their properties is regarded as unproblematic. One should also note that the status of an entity may change during the course of an experiment. In the case of the experiments designed to demonstrate the existence of weak neutral currents, the existence of the currents was shown only after all the data was taken and an argument constructed that the events seen could not be due to neutron background or to the failure of the apparatus to detect muons[7] (See Galison, 1987 for details). As David Cline, one of the experimenters, remarked toward the end of one of the experiments, "At present I don't see how to make these effects [the neutral currents] go away" (Galison, 1987, p. 235). I note also that in this experiment the existence of neutrinos, neutrons, pions, muons, and electrons was taken as given.

The evidence for the reality of entities may also change over time. Thus, when Gell-Mann gave a seminar on the quark model in its early stages, before any experiments were conducted, he suggested that one might regard them either as useful mathematical devices or as real entities. After experiments on deep inelastic electron scattering were completed there was more reason to believe that quarks were real. Subsequent experiments strengthened that belief.

The experiment I wish to discuss was designed to measure the K^+_{e2} branching ratio, the fraction of all K^+ mesons that decay into a positron and a neutrino (Bowen *et al.*, 1967). The first order of business was to obtain a supply of stopped K^+ mesons. The experimental apparatus is shown in Fig. 1. The group obtained an unseparated positive beam of momentum 530 MeV/c from the Princeton-Pennsylvania Accelerator (PPA). This beam included pions and protons, in addition to the K^+ mesons (kaons) needed. The kaons were identified by their range in matter and by time of flight. The beam telescope consisted of our scintillation counters, C_1, C_2, C_3, and C_4, with 6.7 cm of copper placed before the stopping region, which was counter C_3. A stopped particle was indicated by a coincidence between C_1, C_2, and C_3 with no pulse in C_4 ($C_1C_2C_3\bar{C}_4$). The copper eliminated virtually all of the protons, which have a shorter range in matter than kaons, before they reached C_3. In addition, protons do not decay into positrons.[8] Pions were a more serious problem. There were about 1000 times as many pions as kaons in the beam. Most of the pions, which have a longer range than kaons, passed through the stopping region and counted in C_4, and were eliminated, reducing the ratio of pions to kaons to about 100 to 1. Time of flight provided additional discrimination. Particles of the same momentum (this was a momentum selected beam) but different masses have different velocities and therefore different times of flight. The internal proton beam at the PPA consisted of bunches of protons separated in time by 34 ns. Thus, particles were produced every 34 ns. A signal from the RF (radiofrequency) system of the accelerator signalled the production of particles and could therefore be used to time the beam particles. For the beam transport system used in this experiment the difference in time of flight between pions and kaons was 8 ns, so a narrow coincidence of 3 ns ($C_1C_2C_3\bar{C}_4$ + RF) was used to separate kaons from pions. The background of unwanted pions was reduced to approximately 5%.[9]

Particles from decays at approximately 90° to the incident beam were detected by two scintillation counters, C_5 and C_6, and then passed through a gas Cerenkov counter, which was set to detect high energy positrons. The time between the K^+ stop and the decay particle was recorded for each event.

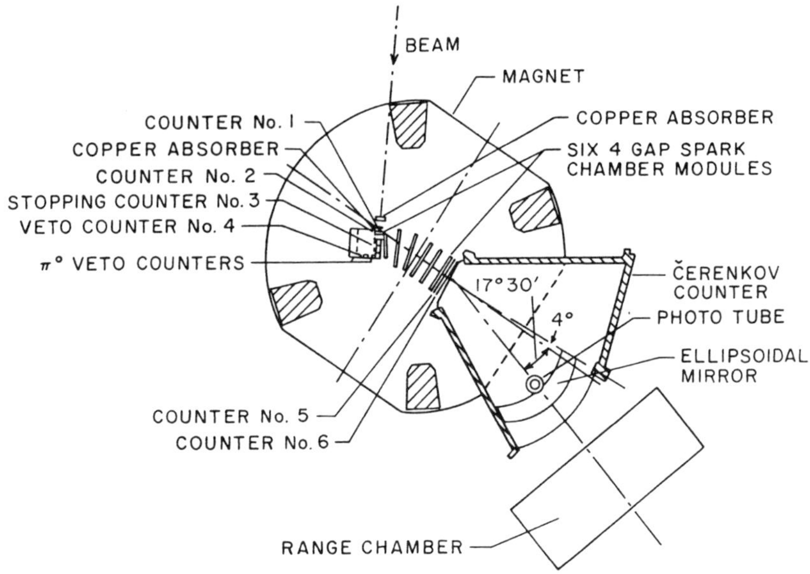

Fig. 1. Plan view of the experimental apparatus used to measure the K^+_{e2} branching ratio. From Bowen *et al.* (1976).

A K^+ decay was identified by a coincidence between a stopping K^+ signal $(C_1C_2C_3\bar{C}_4 + RE)$ and a decay particle pulse (C_5C_6). If the events were really due to K^+ decays the time distribution between the K^+ stop and the decay pulse should match the known K^+ lifetime. This experimental check was performed, the lifetime measured, and the positive results found are shown in Fig. 2. An electronic gate was used to eliminate pion background, as shown in the figure.

Thus, the experimenters determined that the particles had a definite charge, mass, and lifetime[10] which, in addition, agreed with the known properties of the kaon.[11] If there are no K mesons, then we have no plausible explanation of what was being observed. It would seem odd, in such circumstances, to refer to the kaons as merely useful fictions, even if the particles could not be directly observed in van Fraassen's sense of unaided human sense perception. I note that the three properties were sufficient to identify the particle as a kaon.[12]

This procedure also seems to me to be quite similar to identifying a

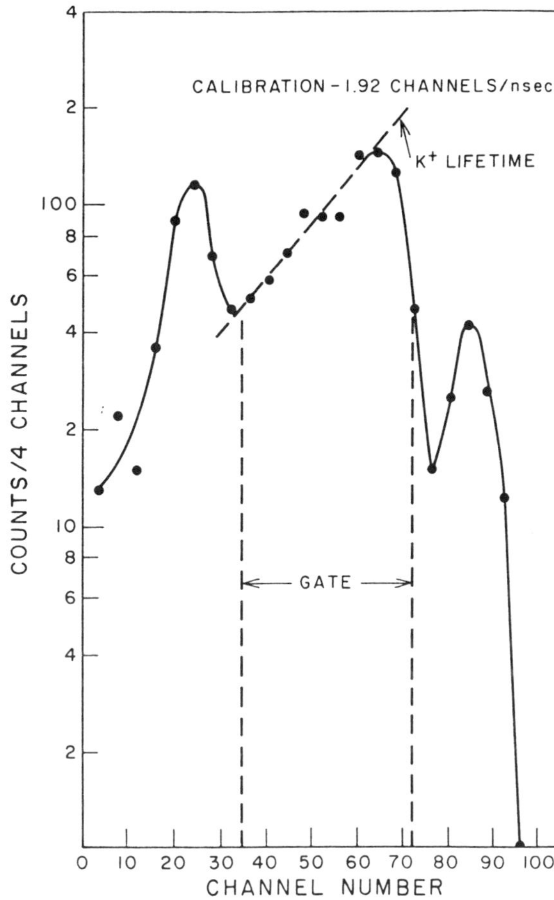

Fig. 2. The decay time spectrum obtained by Bowen *et al.* (1967). The K^+ lifetime is shown.

person by noting that measurements on them of all the properties listed on a driver's license give definite values. Suppose we had an entity who had a definite height, weight, gender, hair color, eye color, date of birth, and home address, which were, in addition, exactly those listed on the driver's license of Bas van Fraassen. Would we not be justified in concluding not only that there is a real Bas van Fraassen, but that the entity we are observing is one and the same person. Yes, the skeptical reader, or an antirealist, might reply, but we can observe the entity directly. Suppose, however, that we had determined all of these quantities without such direct viewing, as well we might have.[13] Would we not still be justified in believing in the real existence of the philosopher? This seems to me to be by far the best explanation of the observations. If there is no Bas van Fraassen then we are faced with a remarkable and bizarre set of coincidences – similarly for K mesons.[14]

It might be argued that there is a difference between the arguments for the existence of K mesons and the existence of Bas van Fraassen. Van Fraassen is an individual, whereas K mesons are a type.[15] I do not believe that this is a valid objection. I could, for example, name my K mesons. Lest this be regarded as totally facetious, let me point out that scientists are, in fact, already referring to individual elementary particles by name.

> 'Here . . . in the center of our Penning trap resides positron (or anti-electron) Priscilla, who has been giving spontaneous and command performances of her quantum jump ballets for the last 3 months'. There can be little double about the identity of Priscilla during this period, since in ultrahigh vacuum she never had the chance to trade places with a passing antimatter twin. The well-defined identity of this elementary particle is something fundamentally new, and deserves to be recognized by being given a name, just as pets are given names of persons (Dehmelt, 1990, p. 539).

A further example of the existence of, at least, a small number of entities, namely xenon atoms is shown in Fig. 3. A group of scientists used a scanning, tunneling electron microscope to manipulate a group of xenon atoms to spell out 'IBM' (The company the scientists worked for is clear). This is an example in which the manipulability of the objects themselves argues for their existence. As Hacking might say, "If you can spell IBM with them, they are real".

In establishing the observed properties of the kaons we have also made use of, and made a commitment to, several physical laws. Thus, the momentum of the particles was fixed by requiring them to travel in a circle of fixed radius in a known magnetic field and then determining the momentum by using $F = ma$, $a = v^2/r$, and $F = qvB$ where F is the force on the particles, q, m, v, and a are its charge, relativistic mass, velocity, and acceleration, r is the radius of curvature, and B is the known magnetic field. From this we find that momentum $= mv = qBr$. We have also made use of the dependence of the range of charged particles on their charge, mass, and velocity. Our use of these laws to establish the prop-

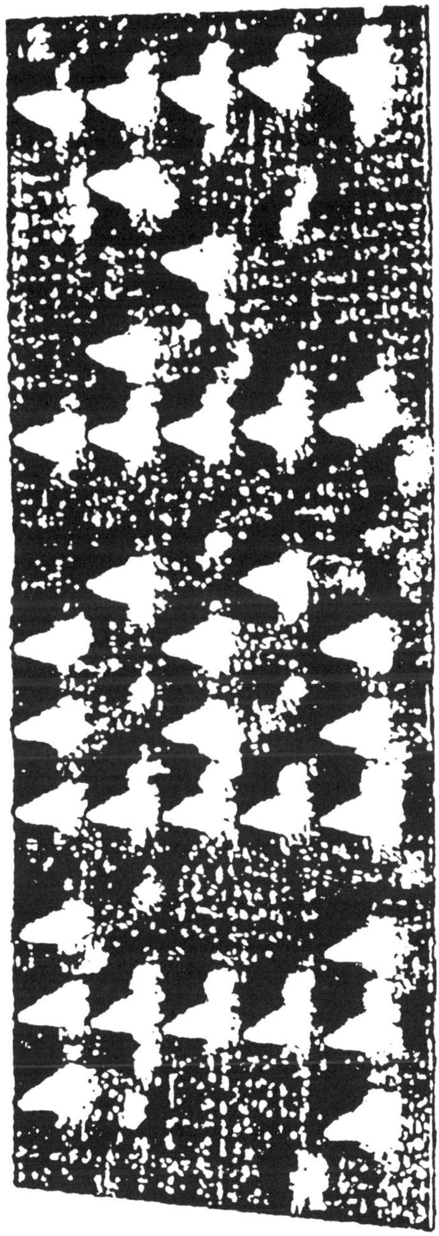

Fig. 3. A magnified picture of xenon atoms spelling out IBM.

erties of kaons seems to me to give them the same epistemic status as the particles and their properties. The successful performance of the measurements gives us reason to believe not only in the kaons, but also in the laws. If the laws weren't valid it is hard to imagine that the measurements would be possible.[16] In contrast to Sellars's view, these laws do not involve K mesons *per se*, but are laws obeyed by all charged particles. Belief in these particular laws does *not* give us good reasons for belief in K mesons.

An example where the support for a law does give us good reason to believe in an entity is the following. The discovery of the Ω^- particle supported the eightfold way, a particle classification scheme, which both predicted and accounted for the existence of certain elementary particles, including the Ω^-. This discovery also gave us good reason for belief in kaons, which are part of the scheme. This will be discussed further below. Similar support would be provided for the existence of elements by the discovery of new elements predicted by Mendeleev's Periodic Table.[17]

2. IS CONSTRUCTIVE EMPIRICISM EMPIRICALLY ADEQUATE?

a. *Is It Empirical?*

How might van Fraassen react to the argument given above for the existence of K mesons?[18] His discussion in *The Scientific Image* (1980, pp. 75–77) states that measurement of a particle's properties, such as the charge on the electron, does not imply that the particle exists. In his view a theory leaves blanks for experiment to fill in and claims that this type of experiment, "shows *how that blank is to be filled in if the theory is to be empirically adequate* (emphasis in original)" (1980, p. 75). I shall discuss in detail later why I believe that this analysis, based on Millikan's experiments, is historically inadequate. I also do not believe that this is a sufficient answer to the argument in favor of the existence of kaons. First, there is no theory of kaons for which the blanks have to be filled in, unless it is the statement that every particle has a mass, a charge, and a lifetime.[19] Second, the experimenters measured three properties of the particle simultaneously, rather than just one, providing more evidence for its existence. It is hard to imagine filling in three blanks simultaneously without believing that there is actually something that has the properties.

Van Fraassen would also regard the K meson or its properties as unobservable because the story I have recounted involves instrumental detection and not unaided human perception. "A calculation of the mass of a particle from the deflection of its trajectory in a known force field is not an observation of that mass" (1980, p. 15). There is a real question as to why van Fraassen privileges unaided sense perception in arguing for the existence of entities and for the validity of measurements.[20] Human perception is notoriously unreliable. It can be influenced by weather conditions (mirages), the state of the body (alcohol, drugs, etc.), stress, and so on. Eyewitness identification in trials has been shown to be far from infallible, and optical illusions do occur. It seems to me that the arguments one gives for the validity of human observations are, in fact, the same arguments one gives for the validity of instrumental observations and measurements and that neither is privileged over the other. (For details of the strategies used to establish the validity of experimental results see Franklin (1986, Chapter 6) and Franklin and Howson (1988). A similar point on instrumental detection has been made by Bogen and Woodward (1988).)

Despite this argument, van Fraassen might very well still deny the existence of the kaon on the grounds that we have only measured or calculated its properties and not observed it directly. It is not clear to me, however, how he would explain the measurements if not by the existence of a particle. As Cartwright might say, "If there are no kaons then we have no explanation of our measurements".[21]

Van Fraassen might respond that, although there are indeed K mesons, constructive empiricism deals only with the attitude one should take toward theories that involve kaons. He might further state that the evidence in support of such theories only gives us reason to believe the theory is empirically adequate and does not give us reason to believe in the existence of kaons, whereas the measurements I have cited do support such existence. I don't believe that such an evidential distinction can be maintained, that between evidence in support of the existence of a particle such as the kaon and evidence that supports a theory of kaons, and therefore their existence.

Consider the eightfold way mentioned above. One found evidence in favor of the existence of the Ω^- particles, i.e., its mass and charge. The existence of such a particle was predicted by the eightfold way and therefore its observation supports, at the very least, the empirical adequacy of the theory. But, the content of the theory is precisely the existence of certain particles and their properties. Therefore, I believe that van Fraassen would have to accept that detection of the Ω^- provided support for the existence of K mesons. I don't believe that having accepted this argument that van Fraassen is then justified in excluding evidence that supports a theory, but does not bear directly on the entities involved in the theory. For example, observation of atomic parity violation supports the Weinberg-Salam unified theory of electroweak interactions. It also supports the existence of the intermediate vector bosons that the theory uses to explain this effect.[22] It would seem then that he would have to agree with Sellars and myself that any evidence that supports a theory gives *ipso facto* good reasons for believing in the entities postulated by that theory. Once van Fraassen has accepted the arguments given in favor of the existence of K mesons, as I believe he should, then he has lost the battle.

A second point concerns the question of what van Fraassen means when he says that acceptance of a theory involves only the claim that what it says about "observable phenomena" (1980, p. 57) is true. He states earlier that "the term 'observable' classifies putative entities (entities that may or may not exist" (1980, p. 15). It seems clear that he cannot regard only entities as observables, for then theories would make very few statements concerning observables. Theories also make statements about measurable quantities and certainly acceptance of a theory must involve, at the very least, the belief that what it says about such measurable quantities is true. If a constructive empiricist does not subscribe to at least this much then the science they believe in is so impoverished as to scarcely merit the name.

To make this clearer, let us consider a kinetic theory which consists of a hard sphere model of gas molecules and Newton's laws. One consequence of this theory is that PV = constant, at constant temperature, where P is the pressure of the gas, and V is its volume. We would certainly regard V as measurable, if not observable. For a reasonably sized container, which has the shape of a rectangular solid, we can determine the volume using only a ruler and the naked eye. What then of pressure? We can certainly detect pressure with our unaided senses. All one needs to do is to dive beneath the surface in a swimming pool. This would seem to make pressure an observable. In order to see whether or not what the theory says about these observables is true, we must not only be able to detect pressure, we must be able to measure it. To do this we will need measuring devices or instruments which not only will go beyond the unaided senses[23] but will also, no doubt, involve a theory of the apparatus.[24] I don't see how a constructive empiricist can avoid allowing quantities like pressure and volume to be observables. Otherwise theories will have virtually no empirical content. Once, however, it is granted that pressure is an observable then other quantities detected only with instruments also become observable, such as the mass, lifetime, and charge, etc., of particles. That being granted, we can then say that we have good reasons, based on observation, for the existence of entities.

b. *Is It Adequate?*

One of van Fraassen's arguments in favor of constructive empiricism is that it provides an accurate description of scientific practice. "However, there is also a positive argument for constructive empiricism – it makes better sense of science, and scientific activity, than realism does and does so without inflationary metaphysics" (1980, p. 73). I wish to dispute this claim.

If scientists are interested only in empirical adequacy then we cannot explain why so much effort is devoted to resolving a contradiction between two theories, each of which is empirically adequate within a certain range of phenomena. Why should scientists have worried about the contradiction between Maxwell's electromagnetic theory and Bohr's atomic theory? Both were, after all, empirically adequate within their own ranges of phenomena. It was the fact that Maxwell's theory predicted that the Bohr atom would be unstable (the Bohr atom should decay in 10^{-9} seconds). Why not accept the empirical adequacy of the theories for their separate phenomena, and let it go at that? This is not what scientists do, or what they did in this case. They work to resolve or eliminate the contradiction. The scientific realist has an explanation of this. The reason scientists act as they do is that when they accept a theory they believe in its truth, and truth cannot be contradictory.

Van Fraassen might respond that Bohr's theory shows that Maxwell's is empirically inadequate, or vice versa. But this will not work. It is true that the two theories are incompatible, but there is nothing in the available data on atomic spectra that contradicts Maxwell's theory. The theory didn't say anything about atomic spectra. It is true that there was no classical electromagnetic explanation of the phenomena, but that might just provide a reason for looking for one. Still, there already existed an empirically adequate theory of atomic spectra, namely Bohr's theory. Why search for another one if your aim is only saving the phenomena?

An antirealist might respond that the removal of the contradiction is necessary in order to construct a unified theory of the phenomena, and that unified theories have, in the past, turned out to be more empirically adequate, i.e., they fit more phenomena than the separate theories they replace. I agree that the search for unified theories is part of scientific practice, and that it has sometimes, although certainly not always, been successful in the past. Recall Einstein's or Heisenberg's futile searches for a unified field theory late in their careers.[25] I do not believe the antirealist position offers any advantage here over the realist view in the explanation of scientific practice. At best, it seems equally good. This judgment of equality depends, of course, on believing that the search for a unified theory is usually, if not always, successful, and that such unification results in theories that are more empirically adequate. I am not convinced that this is always the case. The unifications we remember in the history of science do satisfy this criterion, but we tend to forget those that don't.

My suggestion of equality does not hold for van Fraassen's explanation of the episode of Millikan's measurement of the charge on the electron. As mentioned earlier, he views Millikan's measurement of the charge on the electron as filling in a blank in atomic theory. Although he admits that the experiment was "a test of the theory that there exists this elementary electrical charge, it was not surprising at this time that such tests should bear out theory" (1980, p. 75). He concludes that, "In a case such as this one, *experimentation is the continuation of theory by other means*" [emphasis in original] (1980, p. 77).

While I agree with him that at the time most, but certainly not all, physicists believed that there was an elementary electrical charge, the question was an open one. We can look at the introductions to Millikan's own papers during this period to see what he regarded as the important questions and answers (I note that at this time the question of whether or not there was a fundamental unit of electrical charge and what the value of that unit was were closely related).

In 1911 Millikan reported that it had been found possible

> To present direct and tangible demonstration through the study of the behavior in electrical and gravitational fields of this oil drop, carrying its captured ions, of the correctness of the view advanced many years ago and supported by evidence from many sources

that all electrical charges, however produced, are exact multiples of one definite elementary, electrical charge, or in other words, that an electrical charge instead of being spread uniformly over the charged surface has a definite granular structure, consisting, in fact, of an exact number of specks, or atoms of electricity, all precisely alike, peppered over the surface of the charged body.

To make an exact determination of the value of the elementary electrical charge which is free from all questionable theoretical assumptions and is limited in accuracy only by that attainable in the measurement of the coefficient of viscosity of air (Millikan, 1911, p. 350).

The ordering of these statements seems to indicate Millikan's view of their relative importance.

By 1913 Millikan regarded the question of charge quantization as settled on the basis of his own previously published (1911) work. "The total number of changes which we have observed would be between one and two thousand, and *in not one single instance has there been any change which did not represent the advent upon the drop of one definite invariable quantity of electricity, or a very small multiple of that quantity*" [emphasis in original] (Millikan, 1911, p. 360). His 1913 paper begins, "The experiments herewith reported were undertaken with the view of introducing certain improvements into the oil drop method of determining *e* and *N* and thus obtaining a higher accuracy than had been in the evaluation of these most fundamental constants" (Millikan, 1913, p. 109). He is, in fact, filling in the blank, but only because he regarded the more important question of charge quantization as having already been answered.

During the course of these measurements Millikan was engaged in a controversy with Felix Ehrenhaft concerning the quantization of charge. (For details see Holton [1978] and Franklin [1986, Ch. 5].) During 1912 and 1913 a lull occurred in the controversy and opinion, as well as other experimental results, was generally favorable to Millikan. Ehrenhaft and two of his students, Zerner and Konstantinowsky, returned to the attack in 1914 and 1915 (Ehrenhaft, 1914; Zerner, 1915; and Konstantinowsky, 1915). Millikan answered the criticism, I believe successfully, in a paper entitled "The Existence of a Subelectron?" (Millikan, 1916).

The vast preponderance of evidence and the opinion of the physics community strongly favored Millikan and he was awarded the Nobel Prize in Physics in 1924. The presentation speech by Gullstrand further emphasizes the importance of experimentally establishing charge quantization.

Millikan's aim was to prove that electricity really has the atomic structure, which, on the basis of theoretical evidence, it was supposed to have . . . By a brilliant method of investigation and by extraordinarily exact experimental technique Millikan reached his goal . . . Even leaving out of consideration the fact that Millikan has proved by these researches that electricity consists of equal units, his exact evaluation of the unit has done physics an inestimable service, as it enables us to calculate with a higher degree of exactitude a large number of the most important physical constants (Gullstrand, 1965).

The filling in of the blank of the value of *e*, as van Fraassen has it, while of great importance, is secondary to the issue of charge quantization. Even three years after Millikan had received the Nobel Prize, O. D. Chwolson, a respected physicist, wrote

It (the Millikan-Ehrenhaft dispute) has already lasted 17 years and up to now it cannot be claimed that it has finally been decided in favor of one side or the other, i.e., that all researchers have adopted one or the other of the two possible solutions of this problem. The state of affairs is rather strange (Chwolson, 1927).

Although by that time most physicists would have disagreed with this assessment, it indicates that the issue had not been closed by 1911 as van Fraassen states.

I believe that the realist account that Millikan 'discovered' or gave strong evidence for the existence of a fundamental unit of electricity, and then made a precise measurement of that unit, is a far better description of this episode than van Fraassen's view that Millikan was merely filling in a blank left by theory.

3. CONCLUSION

I believe that the discussion of kaons has shown that we can have good reasons to believe in both the existence of entities and in laws, and that our belief in the laws is a belief that it is true. If my arguments are correct then constructive empiricism is not philosophically justified. I also believe that I have cast doubt on the empirical adequacy of constructive empiricism. It has a vague notion of 'observable', which I believe is far too strict to apply to the practice of science. If one extends the notion as I have suggested then it supports a realist position. Constructive empiricism is, at best, no better than the realist account in discussing the elimination of contradictions between theories or the unification of theories. It also fails to give an adequate account of Millikan's experiment, which is better explained by a realist position.

Nevertheless, showing that constructive empiricism is not satisfactory provides only a small amount of support for the realist view. The elimination of one of many alternative explanations and there are many of them) only slightly strengthens our belief in the remaining alternatives.[26] I have, however, also given positive arguments for the particular version of realism that I support. This includes the discussion of the existence of kaons, the elimination of contradictions, and the historical account of Millikan's experiments.

Supporting a realist position does not, however, mean that I believe in either the absolute truth of the laws or in the 'real' existence of the entities. It means only that I think we have good reasons for believing in the truth of the laws and in the existence of the entities.

Department of Physics,
University of Colorado,
U.S.A.

NOTES

[1] See, for example, the recent book on realism edited by Leplin (1984) and the volume devoted to discussion and criticism of van Fraassen's view edited by Churchland and Hooker (1985).

[2] Van Fraassen admits that 'observable' is a vague predicate but argues that one can still make use of it. I believe, as will be discussed later, that his characterization of observable is, in fact, too strict.

[3] John Stachel has raised the interesting question as to whether or not my view commits me to arguing that 'virtual' particles such as those referred to in the empirically very successful quantum electrodynamics actually exist or whether I regard them as useful visualizations or interpretations of the mathematical formalism. I tend to lean toward the latter view, but I must admit I have not thought about the issue sufficiently. Although I do regard Feynman diagrams, which include virtual particles, as very useful pictures, each line or vertex in such diagrams can be replaced by a mathematical expression, but how one should regard such lines is not clear to me.

[4] Morrison (1990) has argued that manipulability is not sufficient to establish belief in an entity. She discusses particle physics experiments in which particle beams were viewed not only as particles, but also as beams of quarks, the particle constituents, even though the physicists involved had no belief in the existence of quarks. Although I believe that Morrison's argument is correct in this particular case, I do think that manipulability can, and often does, give us good reason to believe in an entity. See, for example, the discussion of the microscope in Hacking (1983). More recently scientists have used the scanning tunneling microscope to spell out IBM with xenon atoms. This seems to me to be very good evidence for such atoms.

[5] In Cartwright's discussion of the electron track in the cloud chamber, for example, she can identify the track as an electron track rather than as a proton track only because

she has made an implicit commitment to the law of ionization for charged particles, and its dependence on the mass and momentum of the particles. The momentum is measured by curvature in a known magnetic field in a way similar to that discussed below for K mesons. A similar point has been made by Morrison (1989). See also my review of Cartweight in *Foundations of Physics* (1984).

[6] I am grateful to Bob Ackermann for raising this point.

[7] I note that it is the argument that is constructed, not the weak neutral currents, as some have stated (see Pickering, 1984).

[8] Although modern theory does predict proton decay into a positron, both the measured proton lifetime and the predicted rate give a decay rate that is too small to be of any significance in this experiment.

[9] The reader may note that the pion also decays into an electron and a neutrino, with a branching ratio approximately seven times that of the kaon. These pion decays could easily be separated from kaon decays by their momentum. The momentum of an electron from pion decay is 69.8 MeV/c, while that for an electron from kaon decay is 246 MeV/c. The experimental apparatus had a momentum resolution of 1.9%, which could easily distinguish between the two decays.

[10] In fact, the experimenters had assumed that the charge on the particle was e. They did, however, measure the range, the radius of curvature in a known magnetic field, and the time of flight of the particles. From these three measurements, the three unknown quantities, the charge, mass, and velocity of the beam particles could have been calculated.

[11] The reader may object that using agreement with known properties of the kaon already assumes that the kaon exists. I believe, however, that establishing that a particle has a definite mass, charge, and lifetime is sufficient to establish its existence. Recall J. J. Thomson's 'discovery' of the electron by showing that cathode rays had a definite charge to mass ratio.

[12] There are of the order of 100 elementary particles, each with different properties (in particular each particle has a unique mass) so that specifying three properties serves to identify the particle.

[13] One could, for example, measure the weight of the entity by using a scale which was located behind an opaque screen, but which had a remote readout visible to the observer, and so on.

[14] I note that this story agrees with Cartweight's causal view, but differs slightly from Hacking's because it is the K mesons themselves that are under investigation.

[15] I am grateful to Zeno Switinjk for pointing this out and also for providing the quotation from Dehmelt, which argues against his own point.

[16] I have used the term 'valid' to describe the laws. I don't think the laws have to be true in order to have a successful measurement, and I don't know how to make sense of the term "approximately true". What I mean by valid is that to within the required experimental accuracy the laws give correct results. This does not mean that I do not feel that we are justified in our belief that the laws are true. Observation of evidence entailed by a theory should, and indeed does, strengthen our belief that the theory is true. It also strengthens our belief that the theory is empirically adequate, but that is beside the point.

[17] For a discussion of this historical episode and a discussion of the evidential value of prediction and accommodation see Howson and Franklin (1991).

[18] For the sake of economy I am attributing the arguments that might hypothetically be offered by a defender of constructive empiricism to its foremost proponent, Bas van Fraassen. He might not, of course, agree that he would offer any of these arguments.

[19] This is not strictly true. The current theory of strongly interacting particles, quantum chromodynamics (QCD), does, in principle, predict the mass of the kaon and other elementary particles, given the masses of the up, down, and strange quarks. Because there are more particles than quarks, one can use the observed masses of some of the particles to calculate the quark masses, and then proceed to calculate the masses of the other particles. These are very difficult and complex calculations and, at present, are accurate to approximately 20%. This theory was not available in 1967, when the experiment was performed.

[20] Van Fraassen actually privileges vision over the other senses. No one would regard hearing something as evidence for the existence of an entity. Recall the old television commercial "Is it live or is it Memorex?" in which a listener cannot distinguish between a recording and a live person. The sense of touch also does not provide reliable evidence for the existence of an entity. Remember the story of the five wise men each touching a different part of an elephant and reaching different conclusions as to its nature.

[21] Cartwright (1983), pp. 87–99, argues that when we have a causal explanation of a measurement or an observation then we are justified in making an inference to the most probable cause, i.e., an entity. I agree. Van Fraassen might ask why we need an explanation at all.

[22] I note that more direct evidence for the existence of these particles was found later.

[23] It may very well be true that any detection device, such as a mercury barometer, will have an output detectable by unaided human senses, but that is not the point here.

[24] For a discussion of how one comes to believe in an experimental result see Franklin (1986), Chapter 6, and Franklin and Howson (1988). For further discussion of the theory-ladenness of measurement see Franklin (1989).

[25] The search for unification also applies to noncontradictory theories. Take, for example, the V-A theory of weak interactions and quantum electrodynamics. These were not contradictory. Each was empirically adequate for its own range of phenomena. The search for unification was successful, leading to the Weinberg-Salam unified theory of electroweak interactions. This unified theory was widely hailed as a major achievement and led to the prediction and observation of new phenomena, weak neutral currents, atomic parity violation, and the observation of intermediate vector bosons, the W and Z° particles. Similarly, Maxwell's theory unified the separate theories of electricity and magnetism and led to the observation of electromagnetic radiation.

[26] For a discussion of this see Franklin and Howson (1988).

BIBLIOGRAPHY

Bogen, J. and Woodward, J. (1988). 'Saving the Phenomena', *The Philosophical Review* **47**, pp. 303–352.

Bowen, D. R. *et al.* (1967). 'Measurement of the K^{+}_{e2} Branching Ratio', *Physical Review* **154**, pp. 1314–1322.

Cartwright, N. (1983). *How the Laws of Physics Lie*, Oxford University Press, Oxford.
Churchland, P. and Hooker, C. (eds.) (1985). *Images of Sciences*, University of Chicago Press, Chicago.
Chwolson, O. D. (1927). *Die Physik, 1914–1926*, G. Kluge (trans.), F. Viewig and Sohn, Braunschweig.
Dehmelt, H. (1990). 'Experiments on the Structure of an Individual Elementary Particle', *Science* **247**, pp. 539–545.
Ehrenhaft, F. (1914). 'Die Quanten der Elektrizitat', *Annalen der Physik* **44**, pp. 657–700.
Fraassen van, B. (1980). *The Scientific Image*, Clarendon Press, Oxford.
Franklin, A. (1986). *The Neglect of Experiment*, Cambridge University Press, Cambridge.
Franklin, A. and Howson, C. (1988). 'It Probably is a Valid Experimental Result: A Bayesian Approach to the Epistemology of Experiment', *Studies in History and Philosophy of Science* **19**, pp. 419–427.
Franklin, A. *et al.* (1989). 'Can a Theory-Laden Observation Test the Theory?' *British Journal for the Philosophy of Science* **40**, pp. 229–231.
Galison, P. (1987). *How Experiments End*, University of Chicago Press, Chicago.
Gullstrand, A. (1965). *Nobel Lectures in Physics, 1922–41*, Amsterdam, pp. 51–53.
Hacking, I. (1983). *Representing and Intervening*, Cambridge University Press.
Holton, G. (1978). 'Subelectrons, Presuppositions, and the Millikan-Ehrenhaft Dispute', *Historical Studies in the Physical Sciences* **9**, pp. 161–224.
Howson, C. and Franklin, A. (1991). 'Maher, Mendeleev, and Bayesianism', *Philosophy of Science* **58**, 574–585.
Konstantinowsky, D. (1915). 'Elektrische Ladungen und Brownsche Bewegung sehr kleiner Metallteilchen im Gase', *Annalen der Physik* **46**, pp. 261–297.
Leplin, J. (ed.) (1984). *Scientific Realism*, University of California Press, Berkeley.
Millikan, R. A. (1911). 'The Isolation of an Ion, A Precision Measurement of Its Charge, and the Correction of Stoke's Law', *Physical Review* **32**, pp. 349–397.
Millikan, R. A. (1913). 'On the Elementary Electrical Charge and the Avogadro Constant', *Physical Review* **2**, pp. 109–143.
Millikan, R. A. (1916). 'The Existence of a Subelectron?' *Physical Review* **8**, pp. 595–625.
Morrison, M. (1990). 'Theory, Intervention, and Realism', *Synthese* **82**, pp. 1–22.
Sellars, W. (1962). *Science, Perception, and Reality*, Humanities Press, New York, p. 97.
Zerner, F. (1915). 'Zur Kritik des Elementarquantums for Elektrizität', *Physikallische Zeitschrift* **16**, pp. 10–13.

KOSTAS GAVROGLU

CAN THEORIES OF CHEMISTRY PROVIDE AN ARGUMENT AGAINST REALISM?

Right after the formulation of quantum mechanics by Heisenberg, Born, Jordan and Schrodinger, most chemists became aware of the amazing explanatory power of the new quantum mechanics, yet it was difficult to see how this newly developing explanatory framework would be assimilated into the chemists' culture. Many chemists were apprehensive that such an assimilation might bring lasting, and not altogether welcome, changes to their culture. But for some, it was a risk worth taking. Neville Sidgwick in his influential book *The Electronic Theory of Valency* written just after the dramatic developments of 1926, would have no inhibitions about letting the new quantum mechanics invade the realm of chemistry. He expressed an unreserved enthusiasm about the new quantum mechanics. Faced with the full development of the new mechanics by Heisenberg and Schrodinger, but not, yet, with an application of the theory to a chemical problem, Sidgwick in the very first lines of the preface to his book attempted to clarify the methodological stumbling block that he sensed would be in the way of his fellow chemists

> In developing a theory of valency there are two courses open to the chemist. He may use symbols with no definite physical connotation to express the reactivity of the atoms in a molecule, and may leave it to the subsequent progress of science to discover what realities these symbols represent: or he may adopt the concepts of atomic physics – electrons, nuclei, and orbits – and try to explain the chemical facts in terms of these. But if he takes the latter course, as is done in this book, he must accept the physical conclusions in full, and must not assign to these entities properties which the physicists have found them not to possess. He must not use the terminology of physics unless he is prepared to recognize its laws. I have endeavoured to conform to this principle, and not to lay myself open to the reproach of an eminent physicist, that 'when chemists talk about electrons they use a different language from the physicists'. I have been careful to avoid as far as possible the introduction of any physical hypotheses which are not already sanctioned by those who are best qualified to judge of them.[1]

Exactly fifty years later, at a Conference in 1977 commemorating half a century of Valence theory, Charles Alfred Coulson gave the closing talk. He was a mathematician by training and a writer of what became the standard text-book on valence. Here is a passage from his talk.

R. S. Cohen, R. Hilpinen and Qiu Renzong (eds.), Realism and Anti-Realism in the Philosophy of Science, 149–170.

Valence Theory is about bonds: what they are and how many there are from each atom . . . So fifty years of valence theory really means fifty years of changing ideas about a chemical bond. The first third of the period . . . was necessarily concerned with identifying the electronic nature of the bond, and in escaping from the thought forms of the physicist . . . Has the chemical bond now done its job? Have we grown to that degree of knowledge and that power of calculation that we do not need it? . . . This is a tantalizing question. Chemistry is concerned to explain, to give us insight and a sense of understanding. Its concepts operate at an appropriate depth and are designed for the kind of explanation required and given. If the level of enquiry deepens, as a result of our better understanding, then some of our older concepts no longer keep their relevance . . . Whether "something bigger" that should replace the chemical bond, will come to us or not is a subject for a Symposium bearing for its title: the changing role of chemical theory.[2]

Coulson could not have expressed any better the often misunderstood relationsip between chemistry and physics. Borrowing and using concepts first proposed by physicists, and, subsequently, "escaping from the thought forms of the physicists" had really been the dominant trend, not only in the first years of quantum chemistry, but also in much of the history of physical and structural chemistry in the latter part of the 19th century as well.

And though chemistry was striving to free itself from the "thought forms of the physicist", the theoretical issues raised by this escape have been, on the whole, ignored by philosophers of science. I am not, of course, saying that there have been no philosophical discussions of the problems of chemistry, since questions about the phlogiston and the Daltonian atoms have been very systematically discussed. But the neglect of the philosophical problems arising out of the particularities of chemistry after the development of microphysics since the middle of the 19th century is also a fact.

In all the discussions about scientific realism, there is at least one undeclared assumption. If pressed, the various arguments, claims and conclusions can be illustrated by examples in physics. And though it is acknowledged that biology and the social sciences pose problems of their own, there is a rather peculiar silence about chemistry. Nothing is said about chemistry because there is the implicit assumption that chemistry is reducible to physics and, hence, physics is the paradigmatic science, the science that really tells us about Nature. It is claimed, not unjustifiably, that the specific role of mathematics in physics renders the problems of scientific realism to be clearly delineated. Furthermore, it is said that physics deals with the fundamental entities of the world and there are no intrinsic limitations as to how deeply it might probe. Whether it studies the planets, billiard balls, atoms, nuclei, electrons, quarks or superstrings, it is still physics and the change of scale does not oblige the change, as it were, of the discipline itself – as it would be the case in biology and chemistry. Although, on the whole, I agree with this view, I have two reservations. The first, which I mention without further discussion, is that it is not too clear to me that when we are discussing quarks or superstrings or the big-bang we are doing theoretical, rather than mathematical physics. My second reservation is that the view which confines the study of realism predominantly to the problems of physics is just a convenient mode of discourse and it neglects the theoretical particularity of chemistry. And what is much more important, it supposes an absolute reductionism of chemistry to physics. If nothing else, these undeclared assumptions deprive philosophy of science of a vast area where issues about the ontological status of theoretical entities and the criteria for empirical adequacy for the acceptance of a theory have been passionately debated. Ever since the end of the 19th century chemists have been debating whether their science may not be the "science of bodies which do not exist". Or whether the unsettling discovery of radium implied that "in relation to the ponderable, we seem to be creating a chemistry of phantoms." The history of chemistry is also a history of the attempts of the chemists to establish its autonomy with respect to physics. Otherwise the chemist would be at a loss about the underlying ontology, and would never be sure whether chemistry should be doing the describing and physics the explaining, The chemists have passionately debated these issues, and the myth of the reflective physicist and the more pragmatic chemist is, if anything, historically untenable.

Since most of the problems debated in realism refer to quantum mechanical entities or, to be more precise, to entities predominantly used in quantum mechanics, I would like to present an account of the historical becoming of quantum chemistry in such a way as to highlight some of its philosophical problems.

Quantum chemistry has been invariably regarded as another instance of the success story of quantum mechanics. Textbooks, leaflets, encyclopedias and the vast majority of scholarly articles discuss the successful application of quantum mechanics to chemical problems. Its birth and early development in the late twenties and early thirties was during the period when physics was reasserted as the unquestionable paradigm for a "good and proper" scientific discipline. Certainly the community of physicists was under the spell of promises to be delivered by Dirac's reductionist program, expressed as a theoretically correct, but practically meaningless dictum

> The general theory of quantum mechanics in now almost complete . . . The underlying physical laws necessary for the mathematical theory of a large part of physics and the whole of chemistry are thus completely known, and the difficulty is only [that] the exact application of these laws leads to equations much too complicated to be soluble.[3]

There was, thus, no more fun in chemistry, its laws had been deciphered and what was left was a matter of devising shrewd methods for carrying out the calculations.

The point I want to argue is that the beginnings and the establishment of quantum chemistry involved a series of issues which transcend the question of the application of quantum mechanics to chemical problems. Quantum chemistry developed an autonomous language with respect to physics and what appeared to be disputes over methods were, in fact, discussions concerning the collective decision of the chemical community about methodological priorities and ontological commitments. The outstanding issue to be settled in the community turned out to be the character of theory for chemistry and, therefore, a reappraisal of the praxis of the chemists. Is it the case that, especially after the advent of quantum mechanics, all the different theoretical and cultural factors precipitated in such a way as to determine and impose a common way of doing chemistry? I am inclined to think that, as a rule, disputes and disagreements were as much about getting the correct solution to a problem as they were part of a rhetoric about how to go about solving similar kinds of problems. In many instances the scientific papers had a strong rhetoric 'propagandizing' various changes in the chemists' culture. In other words. during the 1930s the discussions and disputes among chemists were to a large extent about the new legitimizing procedures and consensual activities to be incorporated within the chemists' culture.

Drawing up a program to examine the nature of the chemical bond presupposed a particular attitude on how to construct a theory in chemistry, on how much one 'borrows' from physics and what the methodological status of empirical observations for theory building is. There were basically two different research traditions. Heitler and London insisted on an approach which, while not as reductionist as Dirac's pronouncement of 1929, followed this path of orthodoxy. Pauling and Mulliken had a strong inkling to semi-empirical methods whose only criterion for acceptability was their practical success. To suppose that the question of a stronger command over the mathematical details is the sole differentiating criterion between the two styles is quite misleading. The difference could only be understood in terms of two different cultures for doing (quantum) chemistry. It is a matter of explicating the theoretical and methodological coherence of the proposed schemata, and realizing that they constitute two diverging programs. At times it appeared to be a confluence of different styles of research and, at the same time there was an uneasy feeling that not all differences could be reconciled.

THE MYSTERIOUS BOND

Among the first successful applications of quantum mechanics was the calculation of the energy levels of the hydrogen atom and of the hydrogen molecular ion. By the beginning of 1927 many physicists and chemists talked about the possibilities of quantum mechanics to deal with actual *chemical* problems. The simplest, but deeply intriguing, chemical problem was the formation of the hydrogen molecule. The 'mechanism' responsible for a such a formation – the homopolar bond – was quite puzzling, since it joined two electrically neutral atoms to form a molecule. Earlier on, there had been quite a few suggestions to explain the homopolar bond, but it was only since 1916 and within the framework of the old quantum theory, that some remarkably insightful proposals by Gilbert Newton Lewis provided a rather simple rule to deal with such a puzzling bond. Lewis had proposed that chemical bonding – both the ionic type as well as the homopolar type – could be explained in terms of shared electron pairs. It was a semi-empirical 'theory'. He had started from the observation that almost all the molecules had an even number of electrons. By requiring the molecule to have eight electrons in the outer orbits of each atom he argued that two neutral atoms could only be joined together by the sharing of pairs of electrons between them.

In 1923, after the publication of Lewis's book *Valence*, the Faraday Society organised a meeting at the University of Cambridge entitled the "Electronic theory of valency". J. J. Thomson was the Chairman of the meeting and in his opening statement he referred to the tasks awaiting the chemists. He felt that the problem of the chemical bond dominated chemistry and that the problem itself found its most suggestive mode of expression in terms of electrons.[4] Lewis gave the introductory talk and though he thought that the aim of the meeting was to secure a better mutual understanding of divergent points of view, the expressions he used in his presentation failed to convince the audience of his tolerance about the kind of pluralism he advocated. He asserted that the "cardinal phenomenon of all chemistry" was the formation of electron pairs. For Lewis this was an actual pairing and not a convenient mode of explanation. He could not offer an explanation of the mechanism itself, but he had no doubt that sooner or later quantum theory would provide an explanation of electron pairing.

HEITLER AND LONDON IN ZURICH

During their short stay in Zurich, Fritz London together with Walter Heitler managed to solve one of the outstanding problems of chemistry by using the wave mechanical methods of Schrodinger. They showed that the mysterious chemical binding of two *neutral* hydrogen atoms to form a hydrogen molecule could *only* be undertood in terms of the principles of the new quantum mechanics.

Undoubtedly the simultaneous presence of both Heitler and London in Zurich was one of those unplanned happy coincidences. Walter Heitler was born in Karlsruhe in 1904. His father was a professor of engineering. His interest in physical chemistry grew while he attended lectures on the subject at the Technische Hochschule and through these lectures in physical chemistry he came into contact with quantum theory. Wishing to work in theoretical physics, he first went to Berlin, but found the atmosphere not too hospitable especially since a student was left to himself to choose a problem and write a thesis. Only after its completion would the 'great men' examine it. After a year in Berlin he went to Munich and completed his doctoral thesis with Karl Herzberg on concentrated solutions. After completing his thesis, Sommerfeld helped him to secure funding from the International Education Board, and he went to Copenhagen to work with J. Bjerrum on a problem about ions in solutions. He was not particularly happy in Copenhagen. Determined to work in quantum mechanics, he convinced Bjerrum, the Education Board and Schrodinger to spend the second half of the period for which he received funding in Zurich.

Fritz London was born in 1900 and his father was a Professor of Mathematics at the University of Bonn. To graduate from the University of Munich in 1921 he wrote a thesis in philosophy under the supervision of one of the best known phenomenologists, Alexander Pfander. In his thesis he discussed questions of deductive systems. After teaching at the Gymansium, he went to Sommerfeld's institute at the University of Munich and started working in physics. His first paper in physics, in 1925, was on the broadening of the spectral lines. In 1926 he was offered an assistantship from E. E. Ewald at the Technical College in Stuttgart. There he started work in quantum mechanics publishing papers in transformation theory and on the quantum mechanical interpretation of Weyl's unifying electromagnetism and gravitation. London, too, went to Zurich on a Fellowship from the International Education Board.

When they met in Zurich, Heitler and London decided to calculate the 'van der Waals' forces between two hydrogen atoms considering the problem to be "just a small 'by the way' problem". Nothing indicates that London and Heitler were either given the problem of the hydrogen molecule by Schrodinger or that they had detailed discussions with the latter while they were proceeding with their calculations. The acknowledgements in their paper were quite diplomatic: They thank Schrodinger for his hospitality and "for the kind and supportive interest with which he has guided our work". Linus Pauling, who was also in Zurich during the same time as Heitler and London, notes that neither he nor Heitler and London discussed their work with Schrodinger, who, nonetheless, did know what they were all working on as witnessed by Robert Mulliken's reminiscences of his visit to Zurich in 1927. Schrodinger had told Mulliken that there were two persons working in his Institute and who had some results "which he thought would interest me very much; he then introduced me to Heitler and London whose paper on the chemical bond in hydrogen was published not long after".[5] It may be worth noting that Ewald remembered that the question of the homopolar bond was in London's mind before going to Zurich, and Pauling recalled discussions with Heitler about bonding when they were both in Munich in 1926.

BINDING FORCES

Heitler and London's initial aim was to calculate the interaction of the charges of two atoms. They were not particularly encouraged by their first results, since the attraction due to the 'Coulomb integral' was too small to account for the homopolar bond between two hydrogen atoms. But they were puzzled by the presence of the 'exchange integral' whose physical significance was not evident at all. Heisenberg's work on the quantum mechanical resonance phenomenon which had already been published was not of particular help to Heitler and London, since the exchange was part of the resonance of two electrons one of which was in the ground state while the other was excited and both were in the same atom. Heitler remembered that they were stuck and "we did not know what it meant and did not know what to do with it".

> Then one day was a very disagreeable day in Zurich; [there was the] Fohn. It's a very hot south wind, and it takes people different ways. Some are very cross . . . and some people just fall asleep. . . . I had slept till very late in the morning, found I couldn't do any work at all . . . went to sleep again in the afternoon. When I woke up at five o'clock I had clearly – I still remember it as if it were yesterday – the picture before me of the two wave functions of two hydrogen molecules joined together with a plus and minus and with the exchange in it. So I was very excited, and I got up and thought it out. As soon as I was clear that the exchange did play a role, I called London up; and he came as quickly as possible. Meanwhile I had already started developing a sort of perturbation theory. We worked together until rather late at night, and then by that time most of the paper was clear. . . . Well, I am not quite sure if we knew it in the same evening, but at least it was not later than the following day that we knew we had the formation of the hydrogen molecule in our hands. And we also knew that there was a second mode of interaction which meant repulsion between two hydrogen atoms – also new at the time – new to the chemists too. Well the rest was then rather quick work and very easy, except, of course, that we had to struggle with the proper formulation of the Pauli principle, which was not at that time available, and also the connection with spin . . . There was a great deal of discussion about the Pauli principle and how it could be interpreted.[6]

The paper was sent for publication on June 30, 1927. In the manuscript, the introduction and the discussion of the results are in London's handwriting. The calculations are in Heitler's handwriting. Heitler wrote the part on the Pauli principle and molecular forces as well and London the conclusion. Heitler and London started their calculations by considering the two hydrogen atoms coming slowly close to each other. Electron 1 belonged to atom a and electron 2 to atom b or electron 2 belonged to atom a and 1 to b. Because the electrons were identical, the total wave function of the system was the linear combination of the wave functions of the two cases.

$$\psi = c_1\psi_a(1)\psi_b(2) + c_2\psi_a(2)\psi_b(1)$$

The problem now was to calculate the coefficients c_1 and c_2. This they did by minimizing the energy

$$E = \frac{\int \psi^* H \psi \, d\tau}{\int \psi^2 \, d\tau}$$

They found two values for the energy

$$E_1 = 2E_0 + \frac{C + A}{1 + S_{12}}; \quad E_2 = 2E_0 + \frac{C - A}{1 - S_{12}}$$

The integrals C (Coulomb integral) and A (exchange integral) had negative values, but A was larger than C. E_1 implied $c_1/c_2 = 1$ and E_2 implied $c_1/c_2 = -1$. Hence the wave function of the system could now be written as

$$\psi_I = \psi_a(1)\psi_b(2) + \psi_a(2)\psi_b(1)$$

$$\psi_{II} = \psi_a(1)\psi_b(2) - \psi_a(2)\psi_b(1)$$

Up to this point, the spin of the electrons was not taken into consideration. The symmetry properties required by the Pauli exclusion principle, were satisfied only by ψ_I. It was the case when the electrons had anti-parallel spins. But ψ_I corresponded to E_1. E_1 was less than $2E_0$, the sum of the energies of the two separate hydrogen atoms, and, hence, it signified attraction. ψ_{II}, which, when spin was taken into consideration was a symmetric combination, corresponded to E_2. But E_2 was greater than $2E_0$, and it implied repulsion. The "mechanism" responsible for the bonding between the two neutral hydrogen atoms was the pairing of the electrons which became possible only when the relative orientations of the spins of the electrons were antiparallel. To form an electron pair it did not suffice to have only energetically available electrons, but the electrons had to have the right spin orientations. The homopolar bonding turned out to be a pure quantum effect, since its explanation depended wholly on the electron spin which had no classical analogue. As Heitler and London noted in their paper, such a result could only be described very artificially in classical terms.[7] They found the bond energy to be 72.3 kcals and the internuclear distance 0.86 Angstroms to be compared with the experimental values of 109.4 and 0.74 respectively.

They soon realised that the proposed exchange mechanism obliged them to be confronted with a fundamentally new phenomenon. They had to answer questions posed by experimental physicists and chemists, like "What is really exchanged? Are the two electrons really exchanged? Is there any sense in asking what the frequency of exchange is?"

> It became gradually clear to me that it has to be taken as a fundamentally new phenomenon that has no proper analogy in older physics. But I think the only honest answer today is that the exchange is something typical for quantum mechanics, and should not be interpreted – or [one] should not try to interpret it – in terms of classical physics.[8]

Both London and Heitler in all their early writings repeatedly stressed this 'non-visualizability' of the exchange mechanism. It is one aspect of their work which in the name of didactic expediency has been consistently misrepresented.

Though it appeared that the treatment of the homopolar bond of the hydrogen molecule was an 'extension' of the methods successfully used for the hydrogen molecular ion, there was a difference between the two cases that led to quite radical implications. It was the role of the elusive Pauli principle. In the case of the hydrogen molecule ion, its solution was a successful application of the Schrodinger equation where the only forces determining the potential are electromagnetic. A similar approach to the problem of the hydrogen molecule leads to a mathematically well defined, but physically meaningless solution where the attractive forces could not be accounted for. There was a need for an additional constraint, so that the solution would become physically meaningful. At least part of the theoretical significance of the original work of Heitler and London was that this additional constraint was not in the form of any further assumptions about the forces involved. Invoking the Pauli exclusion principle as a further constraint led to a quite amazing metamorphosis of the physical content of the mathematical solutions. These solutions became physically meaningful and their interpretation in terms of the Pauli principle brought about the new possibilities provided by the electromagnetic interaction.

London, in his subsequent publications, proceeded to a formulation of the Pauli principle for cases with more than two electrons and which was to become more convenient for his later work in group theory: the wave function can at most contain arguments symmetric in pairs; those electron pairs on which the wave function depends symmetrically have antiparallel spin. He considered spin to be the constitutive characteristic of quantum chemistry. And since two electrons with antiparallel spin are not identical, the Pauli principle did not apply to them and one could, thus, *legitimately* choose the symmetric solution, With the Pauli principle it became possible to comprehend 'valence' saturation, and as it will be argued in the future work of both Heitler and London, spin would become one of the most significant indicators of valence behaviour and, in the words of Van Vleck, would forever be "at the heart of chemistry".[9]

CHEMISTS AS PHYSICISTS?

Among the meetings where questions related to chemical bonding and valence were exhaustively discussed there was the "Symposium on atomic structure and valence" organized by the American Chemical Society in 1928 at St. Louis.

G. L. Clark's opening remarks in the 1928 meeting of the American Chemical Society noted some of the difficulties associated with atomic physics, but ascribed them to the failure of the chemists to test "their well-founded conceptions with the facts of physical experimentation, and that far too few physicists inquired critically into the facts of chemical combination". He thought that physicists and chemists were "firmly entrenched, each in his own domain, a certain long-range firing of static cubical atoms against infinitesimal solar atoms has ensued, with few casualties and few peace conferences.[10]

Clark was not alone in attempting to specify the newly acquired consciousness about this strange relationship between the physicists and the chemists. Worth Rodebush, one of the first to receive a doctorate in 1917 from the newly established Department of Chemistry at Berkeley under the chairmanship of Lewis, went a step further than Clark. He asserted that the divergent paths of physicists and chemists were being drawn together after the advent of quantum theory and especially after Bohr's original papers. But in this process the physicist seemed to have yielded more ground than the chemist and the former appeared to have learned more from the latter than was the case with the chemists. Rodebush gracefully remarked that it was to the credit of the physicist that he could now calculate the energy of formation of the hydrogen molecule by using the Schrodinger equation. But the outstanding tasks for a theory of valence was to predict the existence and absence of various compounds, and the nature of valence which can be expressed by a series of small whole numbers leading to the law of multiple proportions. The "brilliant theories" of Lewis accounted for the features of valence "in a remarkably satisfactory manner, at least from the chemist's point of view".[11] London's group theoretical treatment of valence was considered as an important piece of work even though it did not provide answers to all the queries of the chemist as, for example, the differences in degree of stability between chemical compounds.

Perhaps the most cogent manifestation of what would become the characteristic approach of the American chemists was Harry Fry's contribution in this Symposium. He attempted to articulate what he called the pragmatic outlook. He started by posing a single question. What would be the kind of modifications to the structural formulas so as to conform with the current concepts of electronic valency? Such a question, he suggested, should by no means lead to a confusion of the fundamental purpose of a structural formula which is to present the number, the kind and the arrangement of atoms in a molecule as well as to correlate the manifold chemical reactions displayed by the molecule.

> The opinion is now growing that the structural formula of the organic chemist is not the canvas on which the cubist artist should impose his drawings which he alone can interpret . . . On the grounds that practical results are the sole test of truth, such simple system of electronic valence notation may be termed 'pragmatic'.[12]

'Chemical pragmatism' resisted the attempts to embody in the structural formulas what Fry considered to be metaphysical hypotheses: Questions related to the constitution of the atom and the disposition of its valence electrons. It was the actual chemical behaviour of molecules that was the primary concern of the pragmatic chemist, rather than the imposition of an electronic system of notation on these formulas which was further complicated by the metaphysical speculations involving the unsolved problems about the constitution of the atom. Fry had to admit the obvious fact that as the chemists will know more about the constitution of the atom, they would be able to explain more fully the chemical properties. He warned, though, that premises lying outside the territory of experiences of sensations are bound to lead to contradictory conclusions quoting Kant and becoming, surely, the only chemist to use Kant's ideas in order to convince other chemists at a Conference about an issue in chemistry!

CONCEPTIONS RELATED TO THEORY BUILDING IN CHEMISTRY

In April 1926, Linus Pauling, supported by a Guggenheim Fellowship, arrived in Munich where he planned to work at the Institute of Theoretical Physics. He was twenty five years old and had received his doctorate from the California Institute of Technology working with Roscoe Dickinson on the structure of molybdenite. When he met Sommerfeld, the latter suggested that Pauling work on the electron spin. Pauling did not follow this advice, since his main interest was in chemical problems, especially of atoms with many electrons. He had a solid background in the old quantum mechanics and had exceptionally good training in theoretical physics and mathematics for a person whose main interests were in chemistry. His first paper based on quantum mechanics had appeared in 1926 and it contained a calculation of the values of the Sommerfeld fine-structure screening constants.

While in Zurich he spent most of his time attempting to treat the interaction of two helium atoms, trying to get good approximate values to some integrals that come in the Heitler-London analysis of simple molecular systems. Despite discussions with both Heitler and London, they did not invite him to participate in their work. And, as he said in his interview, the group theoretical methods were not particularly appealing to him.

Almost everything in the series of Pauling's papers starting in 1931 and titled *The Nature of the Chemical Bond* are included in his book of the same title. There are, however, some details of significance. In the opening paragraph of the first paper in the series, Pauling stated his assessment of the situation concerning work on the chemical bond as well as the method he would follow.

> During the last four years the problem of the nature of the chemical bond has been attacked by *theoretical physicists*, especially Heitler and London, by the application of quantum mechanics. This work has led to an approximate theoretical calculation of the energy of formation and of other properties of simple molecules . . . and has also provided a *formal justification* of the rules set up in 1916 by G. N. Lewis for his electron bond. In [this] paper it will be shown that many more results of *chemical significance* can be obtained from the quantum mechanical equations, permitting the formulation of an *extensive and powerful set of rules* for the electron-pair bond *supplementing* those of Lewis.[13]

Texts of this sort are, in a way, pace setting texts; they are rhetorical texts contributing to the formation of the chemists' culture, to the way chemists view others and themselves. It was the theoretical physicists who applied quantum mechanics to a chemical problem, but at the same time Pauling considered his own work as an extension of their program. He declared his applications will provide "many more" results which can be obtained in the *form of rules* supplementing other rules – of Lewis, in fact, who had formulated them much earlier than the advent of wave mechanics! But since Lewis' cardinal rule – that of electron pairing – had been given *formal justification*, one can formulate *new* rules *supplementing* Lewis's rules! Interestingly, concerning the question of the relationship of the various alternatives to the Lewis schema, Heitler and London thought that their work replaced that of Lewis, whereas Pauling and Mulliken considered theirs as supplementing it.

The rules which are formulated later in the paper were provided with a kind of quantum mechanical justification and they were by no means rules derived from first principles. Pauling's papers were mathematically sophisticated and from the calculations he had published it was evident that he was at home with the details of quantum theory. And, furthermore, to be able to present, in his book, such a coherent and convincing argumentation about the nature of the chemical bond with little mathematics is not a tribute to Pauling's ability to polarize his work, but is indicative of his method. In this manner Pauling was able to inaugurate the language of quantum chemistry which could be used by chemists in a practical manner.

To account for chemical bonding, Pauling's schema made use of the attraction expressed in the quantum mechanical resonance between two 'more basic' structures. But the ontological status of these 'more basic' structures was rather problematic. In 1944 George Willard Wheland, who was a student of Pauling's and one of the strongest propagandists of the theory of resonance, published his *The Theory of Resonance and its Application to Organic Chemistry*. Appropriately, the book was dedicated to Pauling. Wheland's view was that

> resonance is a man-made concept in a more fundamental sense than most other physical theories. It does not correspond to any intrinsic property if the molecule itself, but instead it is only a mathematical device, deliberately invented by the physicist or chemist for his own convenience.[14]

At the time, Pauling did not seem to disagree with such an assessment. But when a later edition of Wheland's book appeared in 1955, a lively correspondence ensued between the two about the actual character of resonance theory. Wheland thought that resonance was not an intrinsic property of a molecule, but instead

> something deliberately added by the chemist or physicist who is talking about the molecule. In anthropomorphic terms, I might say that the molecule does not know about the resonance in the same sense in which it knowns about its weight, energy, shape and other properties that have what I would call real physical significance.[15]

Pauling disagreed. "I feel that in your book you have done an injustice to resonance theory by overemphasizing its man-made character."[16] Their correspondence continued and neither appeared to be convinced by the other. What Pauling greatly emphasized was not the arbitrariness of the concept of resonance, but its immense usefulness and convenience which "make the disadvantage of the element of arbitrariness of little significance".[17] This according to Pauling became the constitutive criterion for theory building in chemistry. It was the way, as he had noted, to particularize Bridgman's operationalism in chemistry. In fact, Pauling felt more at ease with the Schrodinger approach than with matrix mechanics and did not worry about questions of interpretation of quantum mechanics. "I tend not to be interested in the more abstruse aspects of quantum mechanics. I take a sort of Bridgmanian attitude toward them".[18]

In his analysis of resonance, Pauling expressed in the most explicit manner his views about theory building in chemistry. He asserted that the theory of resonance was a chemical theory, and, in this respect, it had very little in common with the valence-bond method of making approximate quantum mechanical calculations of molecular wave functions and properties. Such a theory was "obtained largely by induction from the results of chemical experiments".[19] The development of the theory of molecular structure and the nature of the chemical bond, Pauling asserted in his Nobel speech in 1954, "is in considerable part empirical – based upon the facts of chemistry – but with the interpretation of these facts greatly influenced by quantum mechanical principles and concepts".[20]

Both the discussions with Wheland and a vicious attack against his theory by chemists in the Soviet Union[21] prompted Pauling to include a discussion of the character of theory in chemistry in the third edition of his book in 1960. The theory of resonance was not simply a theory embodying exact quantum mechanical calculations. Its great extension has been "almost entirely empirical, with only the valuable and effective guidance of fundamental quantum mechanical principles". Pauling emphasized that the theory of resonance in chemistry was an essentially qualitative theory, which, "like the classical structure theory, depends for its successful application largely upon a chemical feeling that is developed through practice".[22] Pauling himself has repeatedly stressed the rather empirical character of his theory of resonance.

> My work on the nature of the chemical bond and its application to the structure of molecules and crystals has been largely empirical, but for the most part guided by quantum mechanical principles. I might even contend that there are four ways of discussing the nature of the chemical bond: the Hund-Mulliken way, the Heitler-London way, the Slater-Pauling way, and the Pauling semi-empirical way.[23]

Though the method of molecular orbitals was first introduced by Hund, it was Mulliken who provided both the most thorough treatment of the different kinds of molecules, and the theoretical and methodological justifications for legitimizing the molecular-orbital approach. Mulliken was born in 1897 and received his doctorate from the University of Chicago in 1921 working with D. W. Harkins on isotope separation, especially of mercury. He had worked at the University of Chicago and at Harvard as a National Research Fellow and by 1926 he was an assistant professor at New York University. When he did his foundational work on the method of molecular orbitals he had become associate professor at the University of Chicago and had spent some months travelling in Europe, before his extended stay there in 1930 as a Guggenheim Fellow.[24]

In a remarkable paper titled 'On the method of molecular orbitals' published in 1935, Mulliken expressed his views on what he considered to be the most characteristic and differentiating aspects of his theory. The Heitler-London method

> follows the ideology of chemistry and treats every molecule, so far as possible, as composed of definite atoms . . . It has had the notable success as a qualitative conceptual scheme for interpreting and explaining empirical rules of valence and in semi-quantitative, mostly semiempirical calculations of energies of formation.[25]

The method of molecular orbitals departs from "chemical ideology . . . and treats each molecule, so far as possible, as a unit". This seemingly terminological – or shall we say procedural? – difference highlights the more theoretical issues involved in the study of molecular physics.

> It is the writer's belief that, of the various possible methods, the present one may be the best adapted to the construction of an exploratory *conceptual scheme* within whose framework may be fitted both chemical data and data on electron levels form electron

spectra. A procedure adapted to a broad survey and interpretation of observed relation is here aimed at, rather that (at first) one for quantitative calculation, which logically would follow later.[26]

Mulliken had realized that one of the reasons for the poor quantitative agreement using the molecular orbital approach was because of the inability of this theory to include the details of the interactions between the electrons. But even though their quantitative inclusion would make a theoretical calculation from first principles an impossible job, "their qualitative inclusion has always formed a vital part of the method of molecular orbitals used as a *conceptual scheme* for the interpretation of empirical data on electronic states of molecules".[27] Such considerations, in fact, led to the qualitative explanation of the paramagnetism of oxygen – one of the main weaknesses of the valence bond approach.

The approach where theoretical developments had their foundations in the experimental results and where they were more or less generalizations of these results, was the standpoint from which Mulliken built up his own approach to the question of chemical bonding. For him "quantum mechanics, following experiment, demands the existence of stationary states of energy, for molecules as well as for atoms" (my emphasis).[28] Coming to his critique of Heitler and London he remarked that their "very valuable method . . . should be carefully distinguished from [their] valence theory" (emphasis in the original).[29] The complicated calculations were indeed a reason for not preferring the Heitler-London method, but more important is the fact that this methods failed to provide a detailed insight into the nature of the changes which took place in the electron orbits when atoms came together. It was quite surprising that a large part of the Mulliken's critique was based on a wrong interpretation of the "exchange energy" which he thought was "connected with the exchange or jumping back and forth of the electron between the two nuclei".[30] Both Heitler and London in their subsequent papers strongly emphasized the impossibility to give a visualizable interpretation of the exchange energy.

London and Heitler's theory is enticingly simple but, in opinion of the author, really does not hit the nail on the head . . . The presence of unpaired electrons and their spins, and their pairing in molecules, undoubtedly act usually as convenient indicators of valence and of the formation of valence bonds respectively, but even then in the author's opinion, they conceal something which is more fundamental . . . what is fundamental is that we have in the 1s orbit in hydrogen an orbit whose character permits it to become considerably more firmly bound when the hydrogen atom unites with another hydrogen atom.[31]

Summarizing his overall approach, Mulliken placed himself in the antipodes of Heitler and London

> The concept of a discrete, whole number property of atoms called valence, is less fundamental from the point of view of quantum theory than a continuous conception of chemical binding . . . The fact that valence electrons almost always occur in pairs in saturated molecules appears to have after all no fundamental connection with the existence of chemical binding. It can be adequately explained on the basis of the fact that, because of the Pauli principle and the properties of the electron spin, each type of molecular orbit can be occupied by just two electrons.[32]

Many a time such discussions have been portrayed as discussions among the more sophisticated protagonists which did not really touch the rank and file. I do not think so. The settling of these issues deeply affected laboratory practices and research agendas and were never snubbed by the chemists as a whole. The issues were suggestive of the various ways available to map the undefined and undelineated middle ground that some called physical chemistry and some molecular physics. In a manner analogous to the situation in spectroscopy, or even thermodynamics, the question could be asked as to whose domain quantum chemistry was. Was it an activity for physicists or chemists? Or was it a subject for physical chemistry or atomic physics? How would the boundaries be drawn, what would be the methodological priorities, the ontological commitments and, above all, what would be the character and extent of the practitioners' allegiances to physics and chemistry? Most importantly, the question as to the character of theory in quantum chemistry dominated the minds of many chemists. The problem of the chemical bond contributed more than any other problem, so that quantum chemistry could articulate its own autonomous language with respect to both physics and chemistry, chart its own theoretical agenda and formulate its own theoretical framework. The beginnings and the establishment of quantum chemistry involved a series of issues which transcend the question of the application of quantum mechanics to chemical problems. The outstanding issue to be settled in the community turned out to be the character of theory for chemistry and this is something which could provide a useful insight to the problems raised in the discussions concerning realism. I have tried to present a reading of the developments in quantum chemistry so that at least some of the problems relevant to the discussions of realism could be made a little clearer. And, perhaps, working in the history of science, I have the luxury of posing the problems by avoiding their philosophical discussion!

I wish to thank Bob Cohen, Theodor Benfey and Ana Simoes for their valuable comments. Part of this work has been done during my stay as the Edelstein International Fellow for 1992–1993 at the Beckman Center for the History of Chemistry. I wish to thank Arnold Thackray for his hospitality and for the many enlightening conversations.

Department of Physics,
National Technical University,
Athens Greece.

NOTES

[1] Sidgwick (1927), iii.
[2] Coulson (1970), p. 259 and p. 287.
[3] Dirac (1929), p. 714.
[4] Thomson (1924), p. 450.
[5] Mulliken (1965), p. S7.
[6] Archive for the History of Quantum Physics (A.H.Q.P.) Interview with Walter Heitler.
[7] One of the main drawbacks of such an approach is the non-existence of the contribution of the ionic terms. Though they do not lead to any appreciable differences in the case of the hydrogen molecule, their *a priori* absence in similar kinds of calculations had been systematically examined by Van Vleck and Sherman (1935).
[8] A.H.Q.P. Interview of Walter Heitler.
[9] Van Vleck (1970), p. 240.
[10] Clark (1928), p. 362.
[11] Rodebush (1928), p. 513.
[12] Fry (1928), pp. 558–559.
[13] Pauling (1931c), p. 1367.
[14] Wheland (1944), p. 31.
[15] G. W. Wheland to L. Pauling January 20, 1956.
[16] Pauling to Wheland January 26, 1956. Wheland papers at the University of Chicago Library.
[17] Pauling (1960), p. 95.
[18] A.H.Q.P. Interview of Linus Pauling, p. 9.
[19] Pauling (1939), p. 219.
[20] Pauling (1955), p. 92.
[21] Kursanov *et al.* (1952); Tatevskii, Shakhparanov (1952); Moyer Hunsberger (1954).
[22] Pauling (1960), 220.
[23] Pauling, Private communication.
[24] Mullikan (1989).
[25] Mullikan (1935), p. 376.
[26] *Ibid.*, p. 375.
[27] *Ibid.*, p. 378.

[28] Mulliken (1931), p. 350.
[29] *Ibid.*, p. 353.
[30] *Ibid.*, p. 355.
[31] *Ibid.*, pp. 359–360.
[32] *Ibid.*, pp. 386–387.

BIBLIOGRAPHY

Clark, G. L. (1928). 'Introductory Remarks in the Symposium on Atomic Structure and Valence', *Chemical Reviews* **5**, pp. 361–364.

Coulson, C. A. (1970). 'Recent Developments in Valence Theory – Symposium: Fifty Years of Valence', *Pure and Applied Chemistry* **24**, pp. 257–287.

Dirac, P. A. M. (1929). 'Quantum Mechanics of Many-electron Systems', *Proceedings of the Royal Society* **A123**, pp. 714–733.

Fry, H. S. (1928). 'A Pragmatic System of Notation for Electronic Valence Conceptions in Chemical Formulas', *Chemical Reviews* **5**, pp. 557–568.

Gavroglu, K. (1995). *Fritz London (1900–1954) A Scientific Biography*, Cambridge University Press.

Gavroglu, K. and Simoes, A. (1994). 'The Americans, the Germans and the Beginnings of Quantum Chemistry', *Historical Studies in the Physical Sciences* **25**, 47–110.

Heitler, W. (1955). 'The Theory of Chemical Bond', *Arkiv fur Fysik* **10**, pp. 145–156.

Heitler, W. (1967). 'Quantum Chemistry: The Early Period', *Inter. Jour. of Quantum Chemistry* **1**, pp. 13–36.

Hund, F. (1929). 'Chemical Binding', *Transactions of the Faraday Society* **25**, pp. 645–647.

Kohler, R. E. (1971). 'The Origin of the G. N. Lewis Theory of the Shared Pair Bond', *Historical Studies of the Physical Sciences* **3**, pp. 343–376.

Kronig, R. (1935). *The Optical Basis of Valency*, Cambridge University Press.

Kursanov, D. N., Gonikberg, M. G., Dubinin, B., Kabachnik, M. I., Kaveraneva, E. D., Prilezhaeva, E. N., Sokolov N. D. and Freidlina, R. Kh. (1952). 'The Present State of the Chemical Structural Theory'. Translation by I. S. Bengelsdorf in *Journal of Chemical Education* **78**, pp. 2–13.

Lennard-Jones. J. E. (1929). 'The Electronic Structure of Some Diatomic Molecules', *Transactions of the Faraday Society* **25**, pp. 665–686.

Lewis, G. N. (1923). *Valence and the Structure of Atoms and Molecules*, The Chemical Catalog Company.

McCrea, R. (1985). 'How Quantum Physics Came to England', *New Scientist* 17 October, pp. 58–60.

Moyer Hunsberger, I. (1954). 'Theoretical Chemistry in Russia', *Journal of Chemical Education* **80**, pp. 504–514.

Mulliken, R. S. (1928). 'The Assignment of Quantum Numbers for Electrons in Molecules. Part I', *Physical Review* **32**, pp. 186–222.

Mulliken, R. S. (1928). 'The Assignment of Quantum Numbers for Electrons in Molecules. II', *Physical Review* **32**, pp. 761–772.

Mulliken, R. S. (1929). 'The Assignment of Quantum Numbers for Electrons in Molecules. III', *Physical Review* **33**, pp. 731–747.
Mulliken, R. S. (1931). 'Bonding Power of Electrons and Theory of Valence', *Chemical Reviews* **9**, pp. 347–388.
Mulliken, R. S. (1932). 'The Interpretation of Band Spectra', *Reviews of Modern Physics* **4**, pp. 1–86.
Mulliken, R. S. (1932). 'Electronic Structures of Polyatomic Molecules and Valence. II. General Considerations', *Physical Review* **41**. pp. 49–71.
Mulliken, R. S. (1933). 'Electronic Structures of Polyatomic Molecules and Valence. V', *Journal of Chemical Physics* **1**, pp. 492–503.
Mulliken, R. S. (1935). 'Electronic Structures of Polyatomic Molecules and Valence. VI. On the Method of Molecular Orbitals', *Journal of Chemical Physics* **3**. pp. 375–378.
Mulliken, R. S. (1965). 'Molecular Scientists and Molecular Science: Some Reminiscences', *Journal of Chemical Physics* **43**, S7.
Mulliken, R. S. (1967). 'Spectroscopy Molecular Orbitals and Chemical Bonding', *Science* **157**, 7 July, 17.
Mulliken, R. S. (1989). *Life of a Scientist*, edited by B. J. Ransil, Springer-Verlag.
Pauling, L. (1928). 'The Application of the Quantum Mechanics to the Structure of the Hydrogen Molecule', *Chemical Reviews* **5**, pp. 173–213.
Pauling, L. (1928). 'The Shared-electron Chemical Bond', *Proceedings of the National Academy of Sciences* **14**, p. 359.
Pauling, L. (1931a). 'The Application of the Theory of Homopolar Valency to Polyatomic Molecules', B.A.A.S. meeting of 1931, pp. 249–254.
Pauling, L. (1931b). 'The Nature of the Chemical Bond I', *Journal of the American Chemical Society* **53**, p. 1367; '. . . II', *Ibid.*, pp. 3225–3237; '. . . III', *Ibid.* **54** (1932), pp. 988–1003.
Pauling, L. (1931c). 'The Nature of the Chemical Bond. Application Obtained from the Quantum Mechanics and from a Theory of Paramagnetic Suceptibility to the Structure of Molecules', *Journal of the American Chemical Society* **53**, pp. 1367–1400.
Pauling, L. (1939). *The Nature of the Chemical Bond*, Cornell University Press.
Pauling, L. (1940). 'A Theory of the Structure and Process of Formation of Antibodies', *Science* **62**, pp. 2643–2660.
Pauling, L. (1955). 'Modern Structural Chemistry' in *Les Prix Nobel*, Stockholm.
Pauling, L. (1960). *The Nature of the Chemical Bond*, Third Edition, Cornell University Press.
Rodebush, W. H. (1928). 'The Electron Theory of Valence', *Chemical Reviews* **5**, pp. 509–531.
Schweber, S. S. (1990). 'The Young John Slater and the Development of Quantum Chemistry', *Historical Studies in the Physical and Biological Sciences* **20**, pp. 339–406.
Sidgwick, N. V. (1927). *The Electronic Theory of Valency*, Clarendon Press.
Simoes, A. (1993). 'Converging Trajectories Diverging Traditions: The Chemical Bond Valence Quantum Mechanics and Chemistry 1927–1937', Doctoral Thesis University of Maryland.
Sopka, K. (1988). *Quantum Physics in America*, American Institute of Physics.

Tatevskii, V. M. and Shakhparanov, M. I. (1952). 'About a Machistic Theory in Chemistry and its Propagandists', *Journal of Chemical Education* **78**, pp. 13–14.

Thomson, J. J. (1924). 'Introduction to the Session on the Electronic Theory of Valency', *Transactions of Faraday Society* **19**, p. 450.

Van Vleck, J. (1928). 'The New Quantum Mechanics', *Chemical Reviews* **5**, pp. 467–507.

Van Vleck, J. (1970). 'Spin the Great Indicator of Valence Behaviour', *Pure and Applied Chemistry* **24**, pp. 235–255.

Van Vleck, J. and Sherman A. (1935). 'The Quantum Theory of Valence', *Reviews of Modern Physics* **7**, pp. 167–228.

Wheland, G. (1944). *The Theory of Resonance and its Applications to Organic Chemistry*, John Wiley.

MATTHIAS KAISER

EMPIRICAL VERSUS THEORETICAL PROGRESS IN SCIENCE

I. INTRODUCTION

Progress is a value-term. It denotes some form of improvement between at least two parameters. Thus it is also a two-place relation. The progress of knowledge that we believe the sciences achieve over time involves the relation between two intersubjectively shared belief-descriptions. Philosophers traditionally assumed that the only beliefs worthy of being compared in terms of a 'progress relation' were those characteristic of the presumably highest intellectual achievement: **theories** that explain the underlying mechanisms of the world. Intellectual achievement and progress were thus, albeit implicitly, denied with regard to the empirical basis of science.

I disagree with this assumption. The central thesis of this paper is the contention that **scientific progress can only be adequately understood if we distinguish systematically between theoretical progress on the one hand and empirical progress on the other**. It is maintained that empirical progress is no less of an intellectual achievement than the progress of high level theories. Yet, its characteristics are very different from what philosophers say about the progress of theories. It is also different from what the author thinks should be said about the latter.

The position that is sketched in the remainder of this paper has certainly a realist flavour to it. It is also a position that seems to fit in well with the recent interest that many philosophers of science take in the making of science. The laboratory, the experiment, the instrument, the data in science are now focussed upon by several philosophers. A fundamental view presupposed in this paper is the replacing of the dichotomy between theory and data by a **trichotomy** between **theory, phenomena and data**. This position is most clearly articulated in the work of Jim Woodward and Jim Bogen (Bogen and Woodward, 1988; Woodward, 1989; but also Kaiser, 1991a). On this view, a scientific phenomenon is a conceptual abstraction, sprung out of the available data, and far removed from sensible experiences. However, this tripartite

R. S. Cohen, R. Hilpinen and Qiu Renzong (eds.), Realism and Anti-Realism in the Philosophy of Science, 171–193.

view seems also to go well with the views of Nancy Cartwright and several other philosophers.

The evidence for the thesis to be defended here is basically a case study from earth science. Parts of this study have been described elsewhere, and shall not be repeated in detail. A brief sketch of the relevant features of this case study shall however follow. The full characteristics are developed in a book which appeared in German under the title **Aspekte des wissenschaftlichen Fortschritts** (Kaiser, 1993b).

II. CONTINENTAL DRIFT AND PLATE-TECTONICS AS A CASE-STUDY

Many philosophers agree in the assumption that the breakthrough of plate-tectonics in earth science during the late 1960's constitutes an example of progress in science. Interpretations of the history of this development vary with the preferred view of scientific change. Where some see a Kuhnian development (e.g., Hallam, 1973), and some a realist unfolding of new facts (e.g., Ruse, 1981), others see a Laudan-type rationalist story (e.g., Le Grand, 1988), or, yet others, a variant of cognitivistically interpreted naturalism (e.g., Giere, 1988). My reading of the history of this case is realist since I hold that the major developments that account for this theory change are a number of new phenomena that were newly discovered. However, I shall not defend this view here in any detail. Rather I shall outline some of the developments that are important with regard to the topic of the paper.

The early history of continental drift theory (as started by Alfred Wegener) was based on evidence from field work and interpretations of these findings. It was beset with many errors and difficulties. As Oreskes points out in an important paper (Oreskes, 1988), the quality of data pointing in the direction of drift changes dramatically from the early 1950's on. There is the new evidence from paleomagnetism (which I described in detail in Kaiser, 1991), and there is the new evidence from the sea-floor. Important developments took place in both areas in the 1950's and 1960's, developments that first re-opened the discussion about drift after approximately 30 years of virtual silence, and that later formed the basis for the acceptance of plate-tectonics. I shall now summarize these developments with a view to the thesis of this paper. Developments that contributed to the evolution of the **theory** of plate-tectonics will thus only play a marginal role here.

1. *New Instruments and New Methods*

In the newly emerging sub-field of earth science called paleomagnetism a new instrument makes possible the measurement of a very weak magnetization of rocks: remanent magnetization. This magnetization is typically acquired from the earth's magnetic field through cooling from high temperatures (as, e.g., in lava flows) and kept constant in magnetic declination and inclination through large changes of the surrounding magnetic field, and even maintained in processes of sedimentation. The significance of this lies in the fact that remanent magnetization can be regarded as a truthful record of the earth's magnetic field at the forming of strata or sediments. Thus, readings of remanent magnetization provided the earth scientist with hitherto unparalleled data from geological history. The importance of this new class of data was early recognized by a small group of leading scientists, many of them from England. It was made possible by the development of a highly sensitive instrument, the *astatic magnetometer* (cf. figure 1), which the Nobel laureate P. M. S. Blackett originally built for quite different purposes. The instrument promised precise and quantitative access to many geological phenomena which until then could only be inferred by very indirect qualitative data from field observations, as, e.g., fossil evidence in paleoclimatology. The researchers could thus utilize the ever so slight magnetic traces of minerals contained in rocks as direct evidence of past conditions in earth history. It is part of the story that this instrument was later replaced by others which were even more precise.

But remanent magnetization is not always easy to detect. Many rock samples have acquired an additional magnetization from their surroundings over long time spans, the so-called viscous magnetization. It is less permanent than remanent magnetization, growing logarithmically with time, superimposed on the prior remanent magnetization, and deflecting the measurements. The Icelandic researcher Ari Brynjolfsson developed a method of *magnetic cleaning* by alternating magnetic fields that managed to target the measurements to true remanent magnetization.

Other methods, instruments, and measurement techniques were developed during this time which in a similar manner gave scientists access to totally new data-sets. This is particularly true for studies of the seafloor. Backed by large research grants,[1] heavy machinery could now be employed in order to chart the seafloor and conduct seismological, topographical, gravimetric, and magnetic measurements. A huge amount of

Fig. 1. Astatic magnetometer, showing Helmholtz coils, for removing the horizontal and vertical components of the geomagnetic field, the magnetometer, and specimen holder (reproduced from: S. K. Runcorn, 'The permanent magnetization of rocks' *Endeavour 14* (1955), 152–159).

new data was produced that to start with had no obvious link to theoretical disputes in earth-science.

One last development deserves mention in this context. Data from the earth's past would not be very helpful if we could not associate the appropriate time period with them. This has always been a problem in earth science. For a long time one was aware of several methods to determine the relative age of strata, or other qualitative determinations. But in the late 1950's a new method emerged that allowed relatively precise quantitative measurement of geological age: the so-called '*radioactive clocks*'. The key principle of this method is the longevity of some radioactive substances. Slight traces of such substances are contained in all rocks.

Given that we know the rate of decay of these substances, we can determine the ratio between parent element and daughter elements, and thus calculate backwards to the beginning of the decay which coincides with the time of the forming of the rock. The potassium 40 – argon 40 method (as well as several similar methods) proved to be highly important for the understanding of the geological history of our planet.

2. *New Phenomena Emerged*

A small group of scientists, among them Blackett and Runcorn, started in the early 1950's to collect evidence of remanent magnetization in different parts of the world. At this time the theory of continental drift was considered but a mere aberration of the past, certainly not a current alternative for theory choice, studies of remanent magnetization in rocks (and sedimentations) set out unrelated to the hypothesis for which it was to become decisive later. However, the data revealed evidence of a clearly anomalous kind in the light of accepted theory at the time. First of all, the data indicated magnetic reversals, meaning that the polarity of the earth's magnetic field had changed several times from North to South and back. Secondly, if the readings were trustworthy, then the data indicated records of dramatic changes either in the earth's magnetic field (e.g., pole wandering), or in the relative location of continents and islands with regard to the magnetic north pole, or even in the nature of magnetism itself (bipolarity) (cf. figure 2). Due to the complexities of measurement involved and the anomalies with regard to textbook knowledge, many of the earth scientists who were aware of this new research (and those were but a few), typically without knowing much about its theoretical underpinnings, felt legitimized in doubting the trustworthiness of the data. Yet, the few who were actively engaged in this research soon found the means to support new phenomena that were at variance with established dogma in earth science. It took some time before these findings were incorporated into the discipline. Historians of science seem to agree on the two main reasons for this. First, paleomagnetism was still an esoteric field of research in earth science, unknown to most practitioners and ignored by the leaders of the profession. Second, the absence of a good theory that could provide a coherent account of the causal mechanisms explaining these phenomena made them difficult to swallow for other fields in earth science, as, e.g., geophysics.

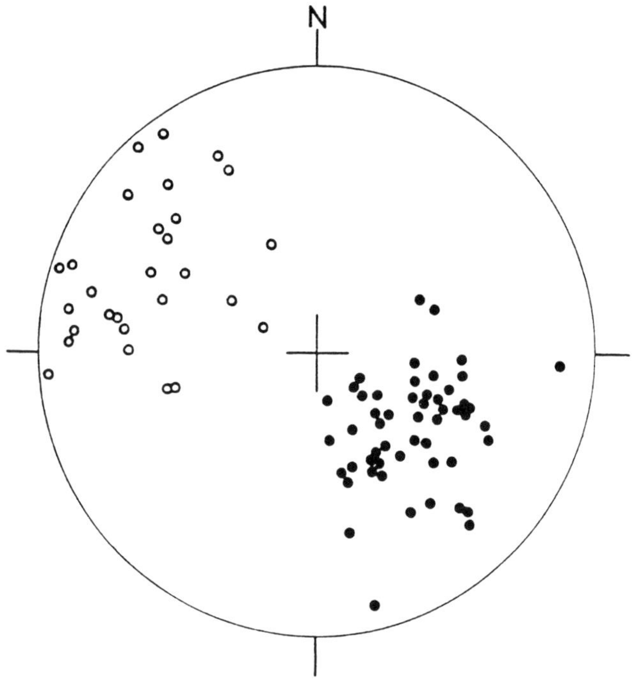

Fig. 2. Directions of magnetization of Torridonian sandstones. Open circles – upward directions. Closed circles – downward directions (reproduced from: S. K. Runcorn, 'The permanent magnetization of rocks' *Endeavour* 14 (1955), 152–159).

But the discovery of new phenomena was not restricted to paleomagnetism. Basically it was the result of the mutual support that different fields of study lent to each other.

What seemed a mere possibility in the light of one line of study became a near certainty in the light of the combination of several very different studies. And phenomena that, when seen in isolation, seemed incoherent fell into a coherent pattern when combined with other new phenomena. It is absolutely essential to my (realist) understanding of the theory change that occurred in earth science during the late 1960's that the changes in the **empirical basis** of earth science are duly recognized.

The most important new phenomena were the following:

(i) **drift**: the relative position of large land masses on the earth's crust has changed during geological history;
(ii) **magnetic reversals**: the magnetic field of the earth has changed its polarity from north to south and back several times over the geological past;
(iii) **pole-wandering**: the location of the magnetic (north) pole had moved along a certain path during geological history;
(iv) **transform-faults**: in addition to the known 'ridges'and 'trenches' (subduction zones) of the sea floor there exists a third class of faults in which the displacement suddenly stops or changes form and direction, Several types of transform-faults could be shown to exist (cf. figure 3);
(v) **succession and duration of magnetic reversals**: one could establish a timetable where 'events'of magnetic reversals were placed in larger periods (epochs) of geological history.

This is the list of empirical phenomena that together provided the basis for the acceptance of continental drift theory. It should, however, be mentioned that the theory which was supported by these phenomena also changed considerably in relation to the very first versions of it. The new theory of plate-tectonics had little in common with Alfred Wegener's

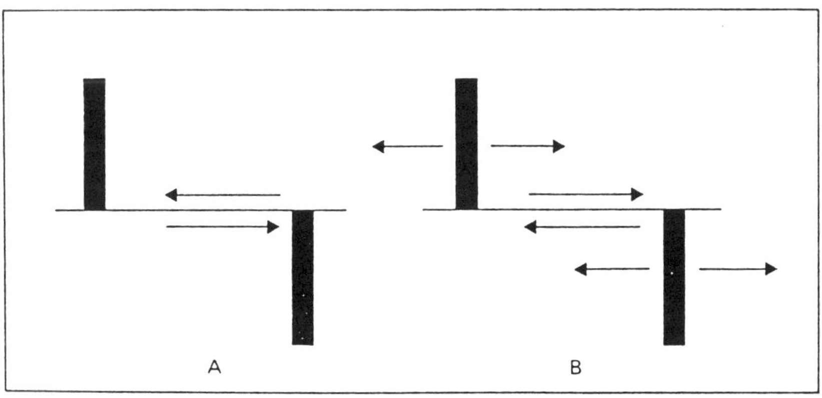

Fig. 3. Diagrammatic illustration of: A: The directions of movement along a transcurrent fault responsible for the off-setting of the black band; B: The directions of movement along a transform fault connecting off-set portions of a mid-oceanic ridge (black) (reproduced from: A. Holmes, *Principles of Physical Geology*, 3rd edition, Van Nostrand Reinhold (UK) Co. Ltd., Workingham, 1978).

theory of continental drift other than postulating some kind of drift. It was also different from Arthur Holmes's early and merely hypothetical speculations about drift. Basically the theory was shaped as a merger between Harry Hess's 'geo-poetry' of 1960 (in which a causal process of displacement was described), F. Vine and D. Matthew's theory of sea floor magnetization, and Tuzo Wilson's integration of transform-faults into the class of contact-zones between tectonic plates.

3. *How Phenomena were Established*

The claim that most if not all of these new phenomena were established in relative independence from explanatory causal theory, like plate-tectonics, is essential to my thesis. In addition, none of the phenomena mentioned were in any sense the result of some direct observation. They were constructed out of existing data by utilizing a highly developed theory, though not the very theory for which they constitute the empirical basis. The phenomena are abstractions in the sense that they constitute the facts of the world to which the data point. Where data are incomplete, local, and beset with complexities of their own (like, e.g., the super-imposition of several types of magnetizations in a given rock sample), phenomena are relatively stable, simple, and universal. Though why-questions may arise at different points in research, it is in relation to phenomena such as those mentioned that why-questions are posed that call for causal explanation by means of high-level theory. It is thus important that phenomena are established by means of reliable data, preferably by means of very heterogeneous data-sets, and utilizing reliable theory.

In this paper I have no intention of showing how the phenomena mentioned above actually were derived from different sets of data. This would by far exceed the given space limitations. However, I shall try to give a general outline of some typical developments which may shed some light on the central thesis defended here.

The phenomenon of drift was in some sense first indicated by paleomagnetism. Rock samples from different continents showed remanent magnetizations that were clearly at variance with the present magnetic conditions on earth. They also showed some relative uniformity when restricted to samples of the same age and same continent. There also emerged a dynamic uniformity when samples of different ages from the same continent were compared. This suggested the hypothesis that the continents may have undergone a change of relative location. Yet, other explanations were still reasonable, for instance the assumption of pole-wandering. Blakett and Runcorn – the latter was at that time still opposed to drift – realized that if data from Europe were compared to data from America one would be able to settle this question. When this finally was done, the data revealed a surprise to everyone: the data-set yielded a sensible uniformity only if one combined drift with pole wandering (cf. figure 4). In some sense, this was a first test of a prediction made on the basis of the drift hypothesis. Note that drift

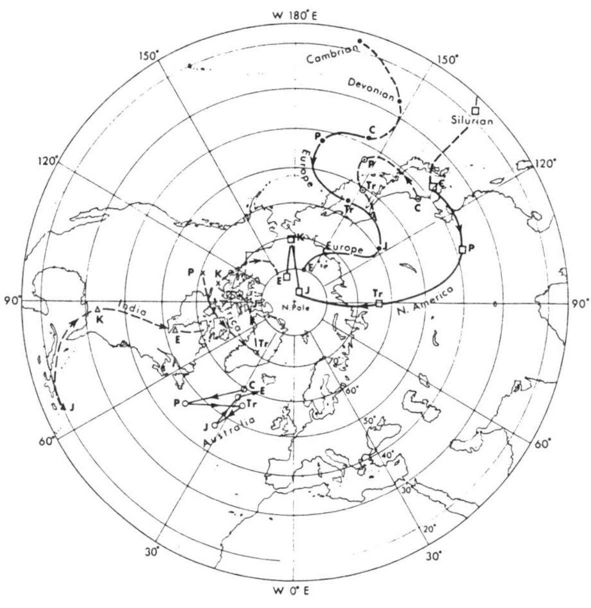

Fig. 4. Stereographic projection of the northern hemisphere from the N Pole to Latitude 16°, showing palaeo-magnetic polar-wandering curves for Europe (solid black circles) with an offshoot for Siberia (open circles with a central dot); North America (open squares); Africa (crosses); India (open triangles); and Australia (open circles). Geological ages of samples from which the data were obtained are indicated by letters: E, Eocene K, Cretaceous; J, Jurassic; Tr, Triassic; P, Permian; and C, Carboniferous. It should be noticed that to avoid congestion these curves are drawn through the mean positions of clusters covering considerable areas. To represent the data more accurately the curves should be broad bands instead of lines (reproduced from: A. Holmes, *Principles of Physical Geology*, 3rd edition, Van Nostrand Reinhold (UK) Co. Ltd., Workingham, 1978).

in this context does not denote some (causal) theory, like Wegener's or Holmes's theoretical outline of continental drift. It only refers to a supposed fact of nature, the relative displacement of large land masses.

If such a phenomenon is to be credible, it is reasonable to expect that data from other sub-disciplines from earth science will point in the same direction. E. Irving was the first to suggest that the paleomagnetic data should be linked to paleoclimatic data. If the continents occupied widely different positions on the earth's surface, then the distribution of climatic zones should also be expected to have changed considerably. This suggestion implied an important methodological point. Whereas on the one hand the data from paleomagnetism were comparatively uncertain due to the complexities of measurement and interpretation, data from paleoclimatology were relatively straightforward. And whereas, on the other hand, the data from paleoclimatology were relatively imprecise and rough, the data from paleomagnetism were relatively precise and quantitatively accessible. Field data from paleoclimatology did indeed indicate that drift could be a sensible hypothesis (e.g., Permo-Carboniferous glaciations, the distribution of salt deposits, etc.). This was in part true already at the times of Wegener, but in the 1950's it was even more pressing, particularly for young researchers who tried to be open-minded about the interpretations of their findings.

But not only paleoclimatology contributed to the phenomenon of drift. Studies of the sea floor had an even more important impact. The magnetic charts of the sea floor showed an impressive pattern of varying magnetization (cf. figure 5). It showed stripes of magnetization that seemed to follow along certain regular lines, until they suddenly were disrupted by fracture lines. These fractures could displace the pattern by regular several hundred miles. In 1963, in relation to the Carlsberg Ridge, F. J. Vine and D. H. Matthews suggested interpreting these stripes as result of magnetic reversals, in combination with sea floor spreading.

The phenomenon of magnetic reversals was also first shown in paleomagnetism. Vine and Matthews's paper permitted the use of the sea floor as a relatively continuous magnetic tape on which these reversals were recorded. The basic feature of it is that stripes of identical magnetization could be found along both sides of oceanic ridges in a strictly symmetrical manner. One could study the (magnetic) history of the earth,

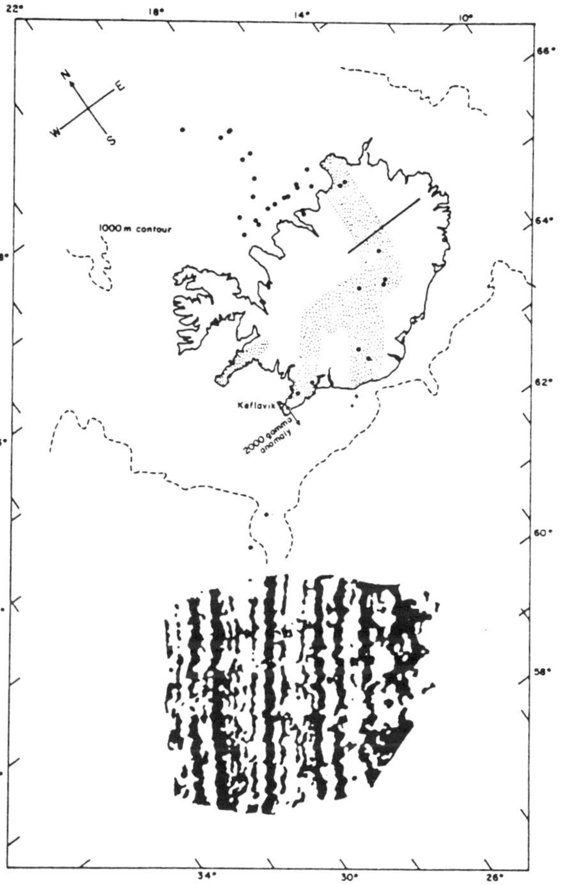

Fig. 5. Magnetic anomalies over the Reykjanes Ridge, south-west of Iceland, showing 'magnetic stripes' (positive anomalies black) and their bilateral symmetry (*After J. R. Heirtzler* et al., *1965. Reproduced from Heirtzler, 969, with permission of John Wiley and Sons*) (Reproduced from: A. Holmes, *Principles of Physical Geology*, 3rd edition, Van Nostrand Reinhold (UK) Co. Ltd., Workingham, 1978).

given that one had a good classification of fractures of the seafloor. Tuzo Wilson's introduction of transform faults provided just that. At this stage the sea floor studies mentioned were doubly relevant for the phenomenon of drift. First, they could confirm paleomagnetists' assumption of magnetic reversals, and, second, if the continents were carried on top of a spreading sea floor, as Hess's 'geo-poetry' suggested, then the postulated convection currents in the mantle could provide drift with a mechanism of movement.

The really compelling breakthrough in this development happened then in connection with the establishment of a time-scale for magnetic reversals. Cox, Doell, and Dalrymple published a first rough timetable of magnetic reversals in 1963. They distinguished between major magnetic 'epochs' (Bruhnes, Matuyama, Gauss, Gilbert), and minor magnetic 'events' (Olduvai, Mammoth) (cf. figure 6). This was made possible by utilizing the radioactive clocks mentioned earlier. The time-table was soon perfected by identifying other new magnetic events. One such event proved particularly decisive, the Jaramillo event. Given this new event, existing data from the research vessel Eltanin, specifically the Eltanin 19 profile, could now give rise to an interpretation of a uniform speed of sea floor spreading.

At the same time other, and quite different, studies fell into the same pattern based upon the phenomena mentioned above. Opdyke had completed a study of sea floor sediment magnetization, based not on the horizontal magnetization stripes, but on vertical probes. The same mag-

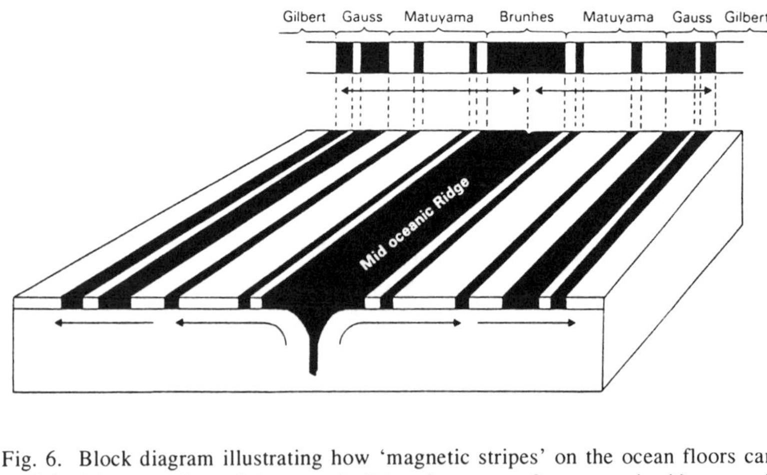

Fig. 6. Block diagram illustrating how 'magnetic stripes' on the ocean floors can be explained by spreading of the oceanic lithosphere away from oceanic ridges on either side, according to the Vine-Matthews hypothesis (*Adapted from A. Cox* et al., *1967*) (reproduced from: A. Holmes, *Principles of Physical Geology*, 3rd edition, Van Nostrand Reinhold (UK) Co. Ltd., Workingham, 1978).

netization pattern could be employed here (cf. figure 7). Later Sykes published her seismological studies which also supported the uniform pattern of dynamic change of the earth's crust that emerged from these other lines of research.

The phenomena mentioned were thus firmly established in earth science by the end of the 1960's. Together they formed the empirical basis for plate-tectonics. Many of them could be derived not only from very different sets of data, but also from very different lines of research. All of these phenomena constitute a complex intellectual achievement, and they are very much the result of advanced instruments and complex theory. Yet each of these phenomena enjoyed a relative independence with regard to the theory of plate-tectonics itself. Even if we were to revise our conception of *how* the continents are moved, their relative dis-

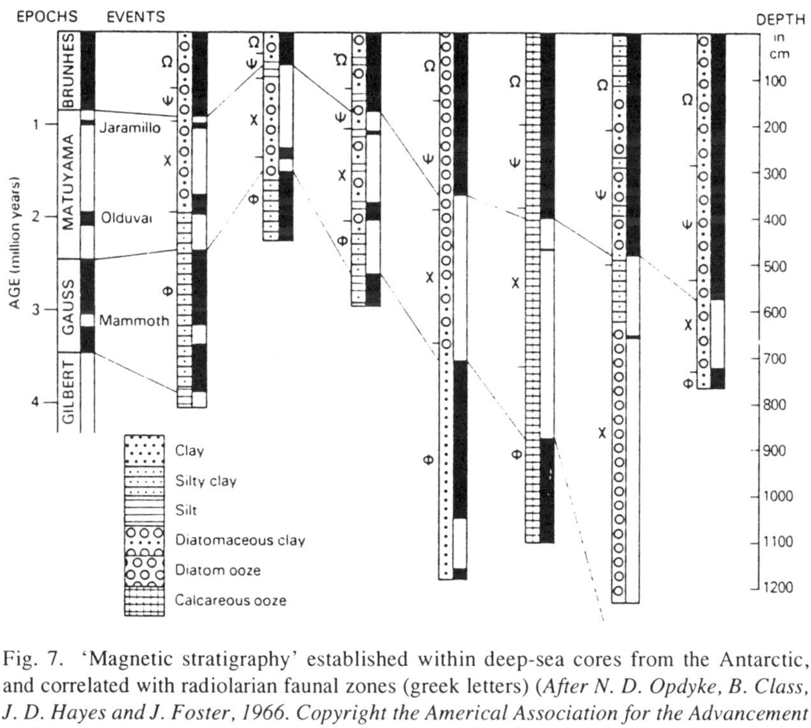

Fig. 7. 'Magnetic stratigraphy' established within deep-sea cores from the Antarctic, and correlated with radiolarian faunal zones (greek letters) (*After N. D. Opdyke, B. Class, J. D. Hayes and J. Foster, 1966. Copyright the Americal Association for the Advancement of Science*) (reproduced from: A. Holmes, *Principles of Physical Geology*, 3rd edition, Van Nostrand Reinhold (UK) Co. Ltd., Workingham, 1978).

location would still be an empirical phenomenon to take into account, as well as the wandering of the poles or the magnetic reversals. The discovery of each of these phenomena constitutes an instance of what I call 'empirical progress' in science.

III. FROM DATA TO PHENOMENA

Elsewhere (Kaiser, 1991a, 1993b) I have described in some detail how I conceive of the path from data to phenomena. I shall therefore restrict myself here to the provision of a rough outline of the basic view, and refer to my other works for further elaboration. I would also like to mention that I conceive of Woodward and Bogen's work as contributions to this topic along similar lines. Indeed, I owe my terminology to them.

The fundamental claim is that phenomena are highly theoretical constructs based upon data. In my view data must always be intersubjectively accessible. This means that in principle the raw data of a study could be handed over to another researcher for inspection. In their most rudimentary form, data are objects of our experience, they are things or recorded singular events. They may be the rocks of the paleomagnetist, the cell-cultures of the micro-biologist, the filled-in questionnaires of the social scientist, or the satellite pictures of the meteorologist. Data, I say, are **experientally anchored**. It is important to differentiate between the experiental anchor of a data set and the descriptions of it. Descriptions are interpretations in the light of theory and allow for error. Experiental anchoring allows at most for ostention, like 'this rock here'. Philosophers have rightly pointed out that we may be shown to be wrong in our descriptions ('this is not really a rock'), given that we got the reference right.

However, a scientist seldom publishes her raw data. Typically we look for descriptions that are both non-trivial and provide a maximum of information. We redescribe the raw material we deal with in research. I suggest that we conceive of this process as the provision of a **stepwise structuring**. A structure consists basically of a domain and a number of operations performed on the objects of the domain. What we do when dealing with data is both redescribe the kind of objects we deal with, and perform new operations (measurement, etc.) on them. Correct classification is a prerequisite of good science. It is also in many areas, such as in geology, a task that involves test-procedures or measurements of a complex nature. In some areas the intended classifications presuppose some successful prior classification that subsequently becomes irrelevant and is therefore thrown out again. For instance, after having classified a rock sample as 'red-sandstone sedimentation', the paleomagnetist may then go on to classify the data as a sample of 'remanent magnetization'. Later a specification of time may be added, etc. In this stepwise process, both the description of the domain and the operations performed on the sample change. They reveal higher levels of abstraction.

It is an important claim of my earlier paper on this topic that the process of stepwise abstraction is conducted under the constraint of **theoretical inference tickets**. These inference tickets come in two kinds: inference tickets for redescription and operational inference tickets. The term denotes the scientific rationale for describing the data set in a new manner, or for applying new kinds of operations, or measurements, to it respectively. Typically, some theory is referred to in these inference tickets, e.g., statistical theory or the physical theory of magnetism. At the most basic levels, the inference tickets employed in the dealings with particular data often need not be mentioned in the final publication. They are simply regarded as part of the standard procedure, and thus part of the training in the discipline. The higher in the level of abstraction, the more explicit the use of the inference tickets becomes. In the 'method part' of a scientific paper, most such inference tickets only receive brief mention, but some may be described in greater detail. Philosophically speaking, inference tickets provide the grounds for the contribution made to the phenomenal level. Thus, criticism in science is often levelled at the legitimacy of the inference tickets that are used.

One ends up with structures that acquire the status of **empirical phenomena**. These phenomena differ greatly in description and generality from the raw material from which they are inferred. Epistemologically one may think of this inferential process as similar in kind to the Aristotelian notion of induction. If one prefers, the term idealization seems equally appropriate here. Phenomena refer to those facts of nature which form the basis of higher level theories. The task of the theories is then to provide good causal explanations that link different phenomena together.

IV. EMPIRICAL PROGRESS

The basic progress relation can be conceived of as being of the following kind:

(i) **progress =: {⟨T′, T″⟩, . . .}**

Depicting progress in this manner amounts to saying that progress is a two-place relation linking two (scientific) theories. It should be an asymmetric and transitive relation, if one sticks to some very basic intuitions about the term. Since progress is a value term, the yardstick for comparing theories with regard to improvement is supposed to be constant, and is thus suppressed in the formulation of the relation. For a realist this yardstick is truth or truthlikeness. Ilkka Niiniluoto's work (Niiniluoto, 1984, 1991) provides a very good analysis of progress in terms of truthlikeness. If one holds other views about science, other aims, like problem-solving or empirical adequacy, may serve equally well under a purely formal point of view. The problem becomes more tricky if one holds that the aim is itself a matter of historical development in science (but cf. Kaiser, 1991b). I shall, however, not go into this matter here.

Progress conceived in this manner gives us a good idea of what it is we mean by the term 'progress'. It states what genuine progress amounts to, what we actually mean by the term. In this sense, it is a semantical explication of an idea, very much in line with the semantic conception of truth. We disregard what indications we may have for ascertaining whether or not such progress actually obtains between two theories. Any such estimation will be subject to human fallibility. We may believe that a certain theory change was progressive, but it may turn out that it was not. The ideal-semantic version of progress on the other hand elucidates what true progress is.

If, however, we try to capture those ascriptions of progress that deal with our estimations of progress, the progress relation changes from a two-place to a three-place relation. Basically it now looks like this:

(ii) **progress =: {⟨T′, T″, E⟩, . . .}**

Here 'E' denotes the evidence that is cited as a ground for believing that T′ is better than T″. If the available evidence changes, then our estimations of progressiveness may change with it. Whereas (i) represents an **ideal-semantical** notion of progress, or a metaphysical notion if you like, (ii) is an **epistemological** version of progress.[2] It is typical for the realist attitude to distinguish these two versions. This idea we find already employed in Popper's work. Truth is one thing, indications of truth are quite another. Similarly, true progress is one thing, our estimations of progress another.

For the historian of science, and indeed for scientists, the epistemological version of progress is of utmost importance. In it we try to capture those arguments that may be regarded sufficient for matters of dispute over theory preference. In philosophy of science I regard it as still an unsolved problem what kind of evidence one should expect for an adequate notion of epistemological progress in science. I shall not enter this discussion here, since it is not the main objective of this paper.

The lesson to be learned from the case study mentioned earlier is, I claim, that we should widen the scope of the above explications of progress. I think there is sufficient historical evidence from science to claim that we should integrate the following notion of progress into our picture:

(iii) **empirical progress =: $\{\langle \mathbf{P}', \mathbf{P}'', \mathbf{E}\rangle, \ldots\}$**

Here I replaced the basic entities that are to be compared in terms of improvement. Instead of dealing with theories, we now deal with two (sets of) phenomena. The rationale for this move lies in the fact that the conditions under which we would like to replace one set of phenomena with another are quite different from those that apply in the case of high-level theories. Basically, phenomena are judged on the basis of data and the inference tickets used in their reformulation. Theories, however, are judged on the basis of phenomena. Fundamental to this way of viewing the matter is the conviction that there is no deductive relationship between high-level theories and low-level data. Our general theories simply never reach the level of things in our world or their immediate description. Since I have argued that phenomena constitute an impressive intellectual achievement of science, it is natural to conclude that progress in phenomenal change should complete our conception of scientific progress.

The task then remains to explicate under what conditions such empirical progress obtains. I have done this in some detail in the work mentioned, and I shall now present the basic tenets of my view.

It seems to me that empirical progress occurs in at least two, possibly in three ways. First there are those cases where we somehow manage to establish new phenomena, i.e., phenomena which did not earlier figure within the empirical basis of our theories. The discovery of new types of galaxies or stars may serve as an intuitive example from astronomy. Secondly, there are those cases where we manage to 'fine-tune' some known phenomena, i.e., specify them in new and better ways. This includes cases of merging or dividing known phenomena. Thirdly, there are cases where new methods or instruments provide us with access to hitherto inaccessible areas of reality. The introduction of radio telescopes or electron-microscopes may serve as prominent examples here. I am a little uncertain about this third area of empirical progress, since it seems hard to specify in more precise terms. From an intuitive point of view, however, several aspects indicate that it is reasonable to include this as empirical progress, if for no other reason than that many scientists themselves commonly regard such developments as progressive breakthroughs in their discipline.

In sum, I want to suggest the following general definition of empirical progress:

> **In a given discipline a set of phenomena P′ is to be judged as empirical progress of the general form {⟨P′, P″, E⟩, . . .} in light of the available evidential situation E if:**
> **(i) P′ contains the discovery of new phenomena in relation to P″, or**
> **(ii) P′ corrects known phenomena in relation to P″, or**
> **(iii) P′ represents new methods or instruments in relation to P″ by means of which new types of phenomena become empirically detectable which hitherto were beyond the reach of scientific investigation.**

It will perhaps be noted that I relativize empirical progress to disciplines. This has to do with my general view of epistemological progress in science. I have argued elsewhere (cf. Kaiser, 1991b) that at least some methodological standards of theory acceptance that we find in scientific practice are designed for the particular purposes of a given discipline, and may be at variance with standards from other disciplines. The degree to which they are in conformity with those very general standards that the methodologist is normally interested in, is a matter of empirical investigation. In methodology I consider these disciplinary methodological standards as epistemologically prior. In any case, taking a close look at the accepted standards of disciplines allows us to take seriously the justifications that the scientists themselves inside a discipline provide for a given scientific change. We need not artificially look for reasons that may not be there, or may not have played a role at all. Thus, judgements of progress do not result in a proof or disproof of scientific rationality. Instead we start with an initial assumption of scientific rationality. I see scientific rationality and scientific progress as intertwined, as two sides of the same coin, if you wish. One should note, however, that this view gives one room to deviate from an insider judgement. We may still disagree with the scientist with regard to a particular scientific change. For instance, we may find that the methodological standards accepted in the discipline do not validate the change. Or we may find that the standards do indeed validate this particular change, but the scientists were unaware of it, and accepted the change for irrelevant reasons. Of course, one may also have reason to criticize the accepted standards of a discipline, but this is a different matter that I shall leave aside here (but again cf. Kaiser, 1991b). It is my claim then that in judging cases of alleged empirical progress in science (as well as indeed cases of theoretical progress), we have to utilize those methodological standards that characterize a discipline at a given time.

All three cases mentioned in the above definition should be clarified further. To some degree this can be done by utilizing the model of the relationship between data and phenomena that I briefly described above. I shall try to elucidate some of the main characteristics of it.

What precisely could it mean to say that we have discovered a new phenomenon? In terms of my model, I think it can mean one of three things. First, the new phenomenon P may emerge as the abstraction of a new set of data, utilizing legitimate inference tickets. This is, for instance, the case with the magnetic reversals of the earth's magnetic field. Reversals were shown to have occurred by having available data (and methods) of a totally new kind, i.e., data of remanent magnetizations. Secondly, a new phenomenon may be established by higher abstraction from given phenomena when new inference tickets back up such a move. In our example, the emergence of a precise timetable of reversals was a development that could be utilized as inference ticket to infer a uniform spreading of the sea floor. Thirdly, a new phenomenon, P, may be the result of merging several lines of abstractions from data to phenomenon, where each individual line employs some inference tickets that yield uncertainty with regard to P. Yet when 'merged', these different lines of inference may compensate for these uncertainties. To some degree this was already the case by relating the paleomagnetist's with the paleoclimatologist's findings. Later, when the findings from the sea floor were also included, the phenomenon of drift became a virtual certainty.

Now, what could it mean to say that P′ corrects some phenomena in relation to P″? Again I think three cases can be distinguished. First, there are cases where a discipline realizes that certain inference tickets are much more problematic than assumed earlier. The evidential situation E may change in such a way that it can be shown that certain methods just cannot be employed as intended. For instance, Wegener believed that certain measurements, made by himself and others, showed a westward drift of Greenland. Not only could one soon show defects in the carrying out of those particular measurements, but one could also show that the theoretical limit of such measurements was such that one either had to assume a tremendous rate of movement, or one had to compare them with nonexisting measurements from some time before human history. Thus, the data could not yield the phenomenon in question.[3] Secondly, there are cases where known phenomena have been described differently by utilizing new inference tickets or new sets of data or both. Most typically this happens in those cases where phenomena which originally were described in qualitative terms are subsequently described in quantitative terms. The time-scale for magnetic reversals was such an example representing an improvement of the general phenomena of reversals as such. A previously known phenomenon was not replaced, but represented in much more precise terms. A third case of correction may occur through the combination of these processes. One may replace a known phenomenon by utilizing a new classification (structure) which is better supported by the known data and inference tickets. The introduction of transform faults is, I suggest, an example. The known fracture zones of the sea floor were originally described as either ridges or trenches. Wilson showed that transform faults are not really a class of the known transcurrent faults, and thus improved the set of phenonomena that together described sea floor fracture zones.

It is much more difficult to specify the third alternative of empirical progress, the introduction of new methods or instruments that reveal new 'layers' of reality. I shall therefore abstain from attempting this here, and rather point out that leaving out this alternative would make the definition less in accordance with the intuitive judgements of scientists.[4] In the case study, it seems that the astatic magnetometer, the radioactive clocks, etc., were recognized as important breakthroughs inside the speciality even before they actually produced data that revolutionized the discipline.

V. A CONCLUSION

I believe that there are many good reasons to differentiate between empirical and theoretical progress in science. The reasons are both of an empirical and a philosophical kind. As to the first, it seems to me that the case of the recent revolution in earth science, indicates that such a differentiation is in line with many important intuitions of scientists. It also structures the whole history of this revolution in a manner which makes it more easily amenable to philosophical descriptions. The theoretical change that occured after 1967 is incomprehensible if we ask what particular set of data brought about this sudden conversion. There was, for example, nothing in the *Eltanin 19* profile that would warrant such a dramatic conversion, if the profile is not seen in relation to the number of new phenomena that already were established or largely supported (cf. also Kaiser, 1993a). It would be wrongheaded to assume that, e.g., the phenomena of magnetic reversals, transform faults, and drift were accepted then because of the acceptance of the theory of plate-tectonics. The theory was accepted because of these phenomena. Phenomena enjoy a considerable independence with regard to the theory they support. Still, the theory itself was more than just the sum of these phenomena. The theory consists fundamentally in the provision of a causal mechanism, the sea floor as conveyor belt of the moving continents, and physical models of how plates collide. If one agrees with my idea that theoretical progress in science (in its epistemological version) consists basically in the provision of causal mechanisms (processes) with explanatory relevance, then this change of theory constitutes a good example of theoretical progress. It was made possible by the previous empirical progress that consisted in the discovery of important new phenomena.

As to the philosophical reasons for differentiating between empirical and theoretical progress in science, it seems to me that the most important reason has to do with the recognition of the essential distinction between data and theory. We have to appreciate how phenomena are made out of data, and how theories connect not to data but to phenomena. In terms of epistemology, the 'upward' movement from data to phenomena is met with the 'downward' movement from theory to phenomena. Furthermore, realizing the important intellectual achievement that goes into the construction of phenomena, the theoretical inference tickets for abstractions, we realize that if we have reasons to assign the value-term 'progress' to theory change, then we also have reasons to assign this very same term, though with different specifications, to phenomenal change.

With regard to the realist position it seems worthwhile to point out that the view sketched in this paper fits in well in several respects. Truth still plays the dominant role. Our division between true progress and estimated progress is in line with a realist attitude. The realist rejection of the traditional division between observational terms and theoretical terms in scientific theories is backed up by an explanation of why this distinction is inadequate. Furthermore, the so-called 'entity realism' may be considered a sub-case of some phenomenal realism. Most entities, like electrons or quasars, are phenomena of nature that are inferred along similar lines as described here. However, some phenomena may not have entity character. Magnetic reversals are not entities or things, but relatively stable, universal (in time and space) re-occurring events. Like electrons that one can spray and then detect their effect, the effects of the reversals can be detected. I thus conclude that a realist has many good reasons to adopt the differentiation between empirical and theoretical progress in science which has been presented in this paper.[5]

The National Committee for Research in Ethics in Science and Technology (NENT), Oslo, Norway.

NOTES

[1] The money came from the US Office of Naval Research (ONR) and the National Science Foundation (NSF), and must be seen in connection with the military's recognition of the importance of submarine activity for future warfare between the super-powers.

[2] The same distinction between true and estimated progress is utilized in the work of Ilkka Niiniluoto. I owe the terminology to him. Cf. Niiniluoto, 1984, 1991.

[3] This is not really an example of empirical progress since few if any practioners of the field lend any credibility to the phenomenon in question to begin with.

[4] Among other things this seems to deviate from the intuitions that guide the Nobel prize committee.

[5] I gratefully acknowledge helpful criticisms and comments by Craig Dilworth and David Olridge.

BIBLIOGRAPHY

Bogen, J. and Woodward, J. (1988). 'Saving the Phenomena', *The Philosophical Review* **97**, pp. 303–352.

Giere.R. (1988). *Explaining Science – A Cognitive Approach*, Chicago/London.

Hallam, A. (1973). *A Revolution in the Earth Sciences*, Oxford.

Kaiser, M. (1991a). 'From Rocks to Graphs – The Shaping of Phenomena', *Synthese* **89**, pp. 111–133.

Kaiser, M. (1991b). 'Progress and Rationality: Laudan's Attempt to Divorce a Happy Couple', *Inquiry* **34**, pp. 433–455.

Kaiser, M. (1993a). 'Philosophers Adrift? – Comments on the Alleged Disunity of Method', *Philosophy of Science* **60**, pp. 500–512.

Kaiser, M. (1993b). *Aspekte des wissenschaftlichen Fortschritts*, European University Studies XX/398, Frankfurt a. M./Berlin/Bern/New York/Paris/Wien.

LeGrand, H. E. (1988). *Drifting Continents and Shifting Theories*, Cambridge.

Niiniluoto, I. (1984). *Is Science Progressive?*, Dordrecht.

Niiniluoto, I. (1987). 'How to Define Verisimilitude', in T. A. F. Kuipers (ed.), *What is Closer-to-the-Truth*?, Amsterdam.

Niiniluoto, I. (1991). 'Realism, Relativism, and Constructivism', *Synthese* **89**, pp. 135–162.

Oreskes, N. (1988). 'The Rejection of Continental Drift', *Historical Studies in the Physical and Biological Sciences* **18**, pp. 311–348.

Ruse, M. (1981). 'What Kind of Revolution Occurred in Geology?', in P. D. Asquith and I. Hacking (eds.), *PSA 1978, Proceedings of the 1978 Biennial Meeting of the Philospohy of Science Association*, Volume Two, East Lansing, Michigan.

Woodward, J. (1989). 'Data and Phenomena', *Synthese* **79**, pp. 393–472.

CHIN-TAI KIM

SCIENCE AS IDEOLOGY

I

The idea of science as *theoria* has since Aristotle dominated the philosophical consciousness of the West. Theoretical science, according to Aristotle, is the pursuit of knowledge of universal and necessary truths about the world for its own sake; if there is any interest that should motivate theoretical inquiry, it is the interest in knowing, nothing else. Aristotelian defenders of science will therefore eschew and debunk as ideology[1] any cognitive enterprise that either has an explicit non-theoretical end or evinces an implicit non-theoretical interest. We should ask the following question to assess the idea of science as *theoria*: Can theoretical interest be as sharply distinguished from the non-theoretical as Aristotle thought? And, given a distinction, can such interest alone adequately justify theoretical inquiry? The answer to each question should be negative.

The Aristotelian thesis that knowledge of universal and necessary truths about the would should be sought for its own sake implies that such knowledge is intrinsically valuable or worth seeking for its own sake.[2] The thesis cannot mean that theoretical knowledge is inapplicable, hence cannot be sought for anything else. Such knowledge, by Aristotle's own admission, has wide relevance and applicability. If the thesis, on the other hand, implies that such knowledge is actually sought by all humans for its own sake and for the sake of nothing else, it is likely to be false. We shall not here address the question as to how being worth seeking is to be differentiated from being actually sought by all humans. We only note that, whatever the proper analysis of being worth seeking for own sake may be, such knowledge must be thought worth seeking for its own sake by defenders of science as *theoria*.

Is knowledge of universal truths about the world the only thing that is intrinsically valuable? The affirmative answer to this question is difficult to defend. Are life, beauty and happiness, for instance, not also things that are intrinsically valuable? In point of fact Aristotle himself

R. S. Cohen, R. Hilpinen and Qiu Renzong (eds.), Realism and Anti-Realism in the Philosophy of Science, 195–205.

subscribes to a pluralism of intrinsic values. If a listing of plausible candidates for being intrinsically valuable other than such knowledge is not an effective argument against the claim that it is the only thing having intrinsic value, then by the same token the claim to that effect cannot be grounded on intuition either as in Aristotle.

Sensible defenders of the idea of science as *theoria* are not likely to say that their commitment to this idea commits them to the view that theoretical knowledge is the only thing that is intrinsically valuable. Let us assume that it is not. Suppose that something which is not theoretical knowledge is intrinsically valuable also. Its pursuit will have as much justification as the pursuit of theoretical knowledge unless some of its properties other than its intrinsic value are relevant to the justification of its preferential pursuit.

Defenders of the idea of science as *theoria* must claim that intrinsic values come ranked, that the value of theoretical knowledge is to be ranked as the highest or as one of the highest at least. There indeed is no inconsistency in saying that two things are both intrinsically valuable but one of them is contextually preferable to the other. A promising argument for the preference of theoretical knowledge will have to rely on the consideration that more than anything else it contributes to the realization of some greater good such as personal happiness or the welfare of society or humankind. It can be said that the judgment as to what proportions of various ends should go into the constitution of the greater good must be an application of relevant knowledge; that the realization of those proportions of those ends requires technology; that the most effective technology is again an application of knowledge. Making these points however, is trying to justify the preferential pursuit of theoretical knowledge by appealing to its utility, abandoning the Aristotelian dogma that such knowledge has intrinsic value only, but no extrinsic value.

The situation that unfolds itself is as follows: Theoretical knowledge is likely to be only one of many things that have intrinsic value if there are such values, though it may be an indispensable element of and a necessary instrument for realizing some greater good that transcends whatever intrinsic value it may have. So even if the initial pursuit of knowledge can be explained by its having intrinsic value, a full justification of its pursuit should be ultimately grounded in its instrumental relationship to some greater value. If ideology is to be thought of as pursuit of knowledge for use in realizing a non-theoretical end, the above consideration shows that theoretical science cannot be clearly differentiated from ideology.

II

The conclusion of the foregoing discussion still leaves intact the idea of *theoria* as a privileged sort of human endeavor and achievement – search for and discovery of truths. Apart from the question of whether *theoria* is free of values other than its intrinsic value, there is the important question as to what it is and how it is possible.

There is an interpretation of *theoria* according to which it is a mental representation of things independent of human consciousness in existence and characteristics, and the cognitive apparatus of humans has a natural affinity with potential objects of knowledge so that the apparatus can be made to serve the end of representing things faithfully if only due methodological care is taken.

Much of the history of Western epistemology has been a battle against these assumptions. The critique of such realism typically took the form of denying its metaphysical thesis by denying its epistemological thesis: no knowledge of objects is possible if they are metaphysically external; but there is possible knowledge of physical objects; therefore, physical objects are not metaphysically external. The core of the reasons supporting the first premise is that there is no cognitive access to metaphysically external reality except through subjective vehicles of representation, ideas or concepts, so that there is no possible knowledge of the agreement of subjective representations with external reality. The denial of the metaphysical externality of the physical opened up a range of possible views, e.g., forms of idealism and phenomenalism. What is common to the older forms of anti-realism is the thesis that the physical world at least in part is a human construction though they differ on the precise nature of the elements of the construction and the principles by which they are organized.

Kant's transcendental idealism deserves special consideration because it was the most thorough response to the problem of representation raised by metaphysical realism, of which Cartesian dualism gave a clear symptom, and because a perception of its difficulties gave rise to trends of thinking some of which are alive today. While denying the human possibility of representing a world of things-in-themselves, thus siding with the subjective idealists and the phenomenalists, Kant held that there are *a priori* forms of sense intuition and a system of *a priori* concepts and principles that must pre-structure, 'make possible,' all human representations. 'Objectivity' can have one of a least two different senses: structural correspondence between things-in-themselves and their representations, or the universal and necessary validity of subjective forms, concepts and rules of representation that make the representation possible. Kant replaced the first sense with the second, his only concession to metaphysical realism, "transcendent realism," being that there must be some unknowable things-in-themselves that somehow are responsible for sensations, the matter of representations.

But the notion of the synthetic *a priori*, which can be interpreted as the invariable species-universal presuppositions of knowledge – has gradually eroded. Kant's idea of a transcendental subject constructing nature without being in it leaves the subject outside the domain of possible scientific knowledge. And his transcendental theory of experience was not preceded by a reflective investigation of its own possibility. The idea of science as a comprehensive inquiry about the world demands that the human cognitive apparatus and its use should themselves be scientifically accessible. The human subject of knowledge needs to be viewed as a being in nature to make scientific knowledge of it possible. Since Kant's time there has been a steady movement toward detranscendentalization and naturalization of the human subject and its cognitive projects and acts. With the naturalization of the subject, its cognitive competence comes to be thought of as a property whose realization throughout the species and persistence through time is to be taken as matter of contingency to be discovered, not to be postulated *a priori*.

The placing of the human subject in nature provides a basis for placing it in history as well. Being in nature makes history possible. Humans create history and culture while existing in nature. Once the human subject is placed in nature and history, however, there arises a dialectic of the two. Humans are situated in nature yet nature in turn becomes an object of their interpretation and construction. That with which humans interpret nature, theory, is a product of nature in a broad sense but it, more proximately, is a product of culture: its genesis, its acceptance, and its use cannot be fully explained apart from the history of human inquiry into nature. The history of natural science, and indeed that of all cognitive pursuit, becomes a proper reference in its interpretation, critique and justification.

A revised Kantian view that the knowable nature of the world must depend on a variable theory constructed by humans existing in nature and history underlies modern-day anti-realism. The current realism/anti-realism controversy differs from the older, metaphysical controversy. Today's realists and anti-realists accept a broadly naturalistic metaphysics that places humans within physical nature, so that the old question as to whether the physical is metaphysically external and, if so, how human consciousness can ever have cognitive access to it is no longer their issue. Also the realists and anti-realists alike consider the history of scientific inquiry relevant to an epistemological critique of science. Scientific theories are recognized by both parties, willingly or reluctantly, to be cultural artifacts of historical beings – beings subject to limitations of the human mind and the senses and influences of language, tradition, interests, values, prejudices and societal forces.

Despite such agreements, the realists and the anti-realists still differ in the theory of science. Today's realism, although it has shorn the physical of its metaphysical *chorismos* from human consciousness, adheres to the idea of truth as correspondence and knowledge as representation: a physical theory is true if and only it represents the physical world. If we consider the fact that the metaphysical externality of the physical world was rejected primarily to ensure the possibility of its representational knowledge, we can understand the continuing attachment to the idea of knowledge as representation that seemed to have been legitimated.

The realists, however, recognize that the representational adequacy of theory can only be conceived as an ideal and that still developing theories can only be conceived as approximations to true representation. In order to prevent this ideal from becoming empty, realists have to devise an epistemology and a methodology of science that promise to ensure a continuous progress of science to ever closer approximations to adequacy. But a gap will always remain between a semantic ideal of true representation on the one hand and an epistemological and methodological account of its possible achievement on the other. And the question arises as to how the epistemology and methodology themselves are to be grounded. It will be a reversion to transcendentalism to try to construct them *a priori* from an a-historical idea of scientific knowledge. It will be only half-heatedly historical to try to construct them on the basis of an analytical and critical reflection upon the current state of science with an uncritical assumption that current science is a culmination of a continuous process of evolution. A fully self-conscious epistemology and methodology of science should be grounded in a critique of the history of scientific practice with different epistemologies and methodologies that have legitimated and regulated its varieties – a critique that shows that the past norms of scientific practice have indeed been superseded. Part of the burden of such a critique will be to show that any significant change in the history of scientific practice is a case of genuine progress that preserves the continuity of science.

Confronted with the historical fact of changes in science, realists may say that there is a constant model of science with a body of methodological canons and try to explain theory changes, rejections and replacements as being due to new observations or corrected observational errors, that it is the content, not the structure, of theory that changes. The realists can hope to place on this basis their claim that the history of scientific inquiry is a continuous progress toward a true theory. Today's realism thus centers around the key theses of a possible correct representation of the real and the adequacy and diachronic constancy of the criterion of adequacy of theory.

Current anti-realism expresses a skepticism about each of the realist theses on various grounds. Going beyond rejecting the old metaphysical view of the physical world as a reality independent of human consciousness yet capable of being accurately represented, they impugn the concept of representation itself. One of the grounds of such impugning is the relativity of the criterion of correspondence to a conceptual scheme, which will imply the untenability of the view that the world has one intrinsic structure to be uniquely represented. Each conceptual scheme may still avail itself of the idea of truth as correspondence but that will not mean that realism with its original intent can be preserved, for a realism that deserves the label must claim that there is only one true theory. The realist idea of truth as correspondence should be either relativized to theories or replaced with an epistemic concept that can countenance a plurality of incompatible theories, such as coherence of beliefs.

Anti-realism also draws support from reflections on the history of scientific practice. There is no norm of science constant through the history of science, nor can one be instituted *a priori*. A rigid distinction between the structure or form and method of science on the one hand and its content, or between theory-free observation and theory on the other, is untenable. The Aristotelian theory of science and Aristotelian methodology consisting of syllogistic and intuitive induction, for instance, cannot be separated from his essentialist realist metaphysics. And although an ultimate convergence of theories can be an object of hope, even faith, the history of scientific inquiry should not be assumed to be a continuous progress such that a later phase of science supersedes all of its earlier phases. Any discontinuities or revolutionary changes, 'paradigm shifts,' in the history of science should be acknowledged to be such after the exhaustion of an honest effort to find hidden continuities. The future states of science exemplifying new norms cannot be predicted.

A key issue between current realism and anti-realism thus turns out to be how the historical career of science should be brought to bear upon the explanation and justification of scientific practice. The verdict on them is still pending. But it would not be unfair to say that the realist burden of defense against the anti-realist attacks is heavy.

III

Current realism and anti-realism have different implications on the issue of the value-neutrality or otherwise of science. Even the realists cannot honestly deny that non-theoretical interests have partly motivated the pursuit of knowledge and the conduct of inquiry, or that science has served non-theoretical interests, but they would say that an objective critical reflection is always possible that can identify intrusive non-theoretical interests and correct their negative influences. Reason is prone to influence by interest but is ever able to reclaim and exercise its autonomy with the strength of its self-interest.

Norms of science appear to differ from ethical norms. The existence of a plurality of conflicting ethical norms and people's choice of different ones tends to be taken for granted as a human condition, while a plurality of conflicting norms for doing science discomfit scientific sensitivity. The explanation is that the goal of science is thought to be a true representation of the world, and that a plurality of conflicting truths of the same world is thought impossible, while it is thought that even with a universal acceptance of the same factual account of the world people can differ on the norm for action. That is, science is still thought 'objective' while ethics is thought to have an uneliminable 'subjective' dimension. This differential intuition about science and morals may appear to support realism. However, the anti-realists can counter: There is not one world with many representations or 'versions' but there are many world-representations or word-versions; *the* world can be no more than a world-representation that will ideally be accepted by a global scientific community at the limit of possible inquiry. But that world is something that cannot now be predicted. It is arguable that a convergence of human morals also is a desideratum and that such a convergence will be a function of a convergence of the human sciences. The divergence of human morals in significant part is due to the conflicts among the theories of human nature and human society. The point is that science and morals do not radically differ.

It was argued, in Section I, that even science conceived as *theoria* is value-oriented. Two points were made: first, science as *theoria* is conceived by its advocates to have intrinsic value, so that its pursuit expresses a commitment to a value – the epistemic value if you will; second, the pursuit of science must be ultimately justified by the instrumental connection that the product of theoretical activity has some good that transcends its epistemic value. If value-neutrality is to be understood as the total absence of commitment to a value, even science as *theoria* is not value-neutral. And if value-neutrality is to be understood as the justifiability of the given practice apart form a consideration of its instrumental connection to a value that transcends whatever intrinsic value the practice may have, even science as *theoria* cannot be considered value-neutral.

With the discarding of the idea of science as representation of a metaphysically independent world by a subject wearing a protective transcendental shield, the issue of the value-neutrality or otherwise of science becomes more complex. The human subject can no longer be viewed as a pure knower, with no interest but the theoretical; it must be viewed as a being subject to forces and influences of nature and history that it tries to understand and cope with. The question no longer is "Is science value-oriented or value-neutral?" but "What value commitments explain or justify scientific practice?" The avowal, even an honest one, by the practioners of science, that they have nothing but a general commitment to the epistemic value and a specific commitment to the regulative norms of their practice, should not be taken as sufficient evidence that their practice is neutral as to all other values. There are reasons for caution other than a psychological one.

A practice is not a totally isolable segment of a web of practices. The total web of practices within a culture is permeated with value-commitments. The influence of one practice upon another, that of religion upon science in the Middle Ages, for a notable example, attests to an osmosis of values and beliefs from one practice to another. The fact that natural science has steadily gained autonomy from religion need not be denied, but this fact implies that the epistemic values thought internal to it have gained precedence over values of other sorts, and/or that the interests of the political and social forces opposing religious authority and institution have coincided with the interests of science, not that science has become value-neutral. Even if for argument's sake science is supposed to have nothing but an internal epistemic value, the practice of science must still be justified in reference to a system of values that transcend it.

Science must be ideological at least in a broad sense. Indeed all human practices are. The idea of a value-neutral practice expresses an ignorance of the organic interconnections among the domains of the cultural and historical existence of humans. It is one thing to be immersed in a practice; it is another to have a reflective understanding of it in its existential and historical context. Fully self-conscious science is one that acknowledges and takes seriously its ideological nature. Science distinguishes itself from other human practices by its commitment to the achievement of rational consensus on its goal and method. Anti-realism gives support to a conception of science as a practice having value commitments and subject to values by noting that it is a practice in a system of practices that values permeate. Just as anti-realism must impugn the distinctions between the given world and its variable representation, between the constant form of science and its variable content, between a diachronically valid norm of science and its errant application, it must also impugn the distinction between the value-free knowledge of the world and value-oriented distortions of its true picture. The issue between realism and anti-realism here is not so much whether non-epistemic values have actually influenced scientic practice as whether science should be so conceived as to require its freedom from such values. The issue is a humanization of science.

Current anti-realism helps uncover the ethnicity or, better, cultural specificity, of science and raises an issue about its universality. Even the realists would find little difficulty in accepting the fact that customs, morals, laws, institutions, religions, and philosophies (taken broadly as world views) are relative to cultures. But they retain their sanguinity about the universality of science. But the science whose universality they claim is their science, a practice that has been shaped in Western culture and civilization. All of the problems and disputes about Western science are fully significant only in the conceptual space of that culture and that civilization. Anti-realism seems to offer an antidote to cultural chauvinism and ethnocentricism. If even within the same Western civilization the idea of science is, and has been, an issue, then the alleged universal validity of that idea across civilizations should all the more be an issue.

Upon admission of the historical and cultural identity and boundaries of Western science, two extreme attitudes should be avoided. One is that of assuming the total impermeability of non-Western civilizations to Western science and resolving to do Western science in an isolationist way; the other is assuming the superiority of Western science, indeed of Western civilization, and trying to convert the non-West to Western ways of thinking with missionary zeal. Neither attitude seems justifiable. A third option, though apparently constructive, is equally untenable, namely assuming the rationality of all civilizations and the possibility of a universal meta-discourse in which Western science and non-Western analogues can be comparatively interpreted and assessed. There is no such meta-discourse even within a civilization. What is taken to be such a discourse may only be a rhetorical modification of a favored object discourse devised for its justification.

The correct attitude, I propose, is that of promoting studies of civilizations by one another and assimilation of the foreign by each, and hoping for expansions of inter-civilizational spaces for conversation. Such promotion requires as much mutual openness as self-reflection, as much pride in one's own tradition as respect for another, as much valuation of unity as recognition of the historical facticity of diversity, and as much appreciation of agreements after conversation as tolerance of differences that persist.

Case Western Reserve University,
U.S.A.

NOTES

[1] The label 'pseudo-science' tends to be applied to cognitive practices that in the judgment of the critics who use the label employ a wrong method or have false substantive assumptions or presuppositions. The pejorative label, however, has no specific implication that

what is so labelled is motivated by a non-theoretical interest, though it may well be. The label 'ideology' on the other hand explicitly impugns the motive of a cognitive practice as impure.

2 See Aristotle, *Met.*, Bk. I, 982^{a}15. The word 'theoretical' in Aristotle primarily concerns the motive of inquiry. But a classification of knowledge by motive for its pursuit fails to inform us of the objects for different classes of knowledge. First Philosophy, mathematics, and philosophy of nature ('physics') are the major branches of theoretical science as Aristotle conceives it. Aristotle fails to make clear whether these sciences are theoretical because of the nature of their subjective matters or the concept of theoretical science implies that it should be about those objects that sciences concern.

MARX W. WARTOFSKY

THREE STAGES OF CONSTITUTION: HISTORICAL CHANGES IN THE ONTOLOGICAL STATUS OF THE SCIENTIFIC OBJECT

In this paper, I muse very briefly on realism and constructivism not simply as conceptual or philosophical alternatives in the interpretation of the relation between theoretical knowledge and its object, but rather *as philosophical articulations of historically different modes of scientific practice, in which the relations between knower and known are in fact different.*

What I would like to pose for discussion is a thesis on the historical changes in the relation between the subject and the object in science. That is to say, that there are in fact situations in scientific inquiry which are, so to speak, "realist", in which the object of knowledge is entirely independent of the subject, as were the stars and celestial phenomena in classical astronomy. There are on the other hand objects of science which exist only in being constituted in the course of experimentation itself – i.e., either as artifacts of research activity, or as transformations of such objects as these are changed by scientific-technical practice.

It seems to me that there is a certain dialectic that develops in this regard; that, in the natural sciences, where we began by eliminating all supernatural and mythic anthropomorphisms, we now return more and more to a humanized naturalism. That is to say, classic realism, which proposed an absolute breach between the subject and the object [in asserting the ontological independence of the object], has to be modified, but not because of any philosophical decision. It is more a matter of a change in the mode of scientific production, i.e., in the relations between scientific theory, on the one hand, and experimental practice and observation, on the other. We are now in need of a theory of scientific production as theoretical production and that means not only the production of ideas, as some hypostatized, objectified "third world" entities of the sort Popper proposes, in *Objective Knowledge*, but rather the reproduction of the world in theory, as an actual transformation of *matter* into *object*, of the *an sich* into the *für uns*.

This production is a process of scientific construction that proceeds

R. S. Cohen, R. Hilpinen and Qiu Renzong (eds.), Realism and Anti-Realism in the Philosophy of Science, 207–218.

historically from the level of pure speculation, to the level of passive reflection, to the level of active transformation of the object of knowledge by intervention. Nowadays, one does science not merely by interpreting the world, but by changing it.

1. THE PROBLEM AND ITS GENESIS: THE BOHR-EINSTEIN DEBATE

One of the crucial debates in contemporary physical theory that has been as explicitly 'philosophical' as it has been 'scientific' – I use scare-quotes here because, as you may know, I don't believe there is any sharp demarcation between the two – is the Bohr-Einstein debate over the completeness or incompleteness of Quantum-mechanics. The disagreement centered, originally, over the physical interpretations of the wave-function, ψ, specifically, as to whether its statistical form was to be taken as a description of the average behavior of a sub-atomic system, and therefore as an incomplete account of the individual phenomena or states making up the ensemble; or whether it was instead to be taken as a complete description of an objectively probabilistic phenomenon, in which an exhaustive measurement of the simultaneous values of the classical parameters – the canonical conjugates – of individual systems or of particles was in principle excluded, by the Heisenberg uncertainty relation, and by the formalism of Quantum-mechanics, e.g., the non-commuting operators. In terms of physical interpretation, the question was one concerning the locality or non-locality of the individual subsystems or particles constituting the ensemble which the wave-function described, and hence, also the time-evolution, in terms of continuous functions, of such systems. Locality and causality (defined for a certain ontology of the physical world) were apparently put in question by the Bohr interpretation. In their place, what was proposed was the local non-separability of quantum-systems. In the classical case, if two parts of a system – two particles, e.g., electrons or photons which have interacted – separate and go in different directions, it is possible to correlate the states of the two particles at a distance from each other, given the initial energy or momentum of the system, in the sense that a measurement on one of the particles will not have any effect upon a measurement on the other. Given the conservation law. the energy or momentum of the unmeasured particle can be inferred from measurement on the other. In effect, each is locally separable from the other. If, on the other hand, the two systems are locally non-separable, any measurement on one will affect a measurement on the other, so that the two subsystems are then not independent of one another. Metaphorically, we may say that one subsystem 'knows' instantaneously what is happening to the other (what Einstein dismissed as "telepathy"); or alternatively, what would require action at a distance, in contravention of the constraints of special relativity, where signal velocity cannot exceed the velocity of light. This quantum-correlation of the measurements taken upon two such subsystems at a relatively large distance from one another violates the notion of locality, which Einstein held to be crucial to a complete physical description.

Bohr's response to the charge of incompleteness was to take Einstein to task for what he called "the essential ambiguity involved in reference to physical attributes of objects when dealing with phenomena where no sharp distinction can be made between the behavior of the objects themselves and their interaction with the measuring instrument".

2. STRIFE ABOUT REALISM

From this and from the way in which the discussion of the so-called 'Copenhagen Interpretation' of Bohr and Rosenfeld progressed, in later years, it would seem that there are ambiguities about 'physical reality' on both sides, however. Bohr, and especially Rosenfeld, eschewed the charge of 'positivism' and of Berkeleyan phenomenalism that Einstein leveled at what *he* called the 'orthodox' interpretation of Quantum-mechanics – namely, that it reduced reference to nothing but the phenomena of measurement, 'observables', and thus denigrated any talk of 'physical reality' beyond such observables, as empty or unwarranted metaphysics. But, Bohr objected, what we cast in doubt is not our access to 'physical reality' but rather a certain characterization of that reality, or of our relation to it in a given mode of scientific practice. In effect, Bohr rejected the accusation that he was not a realist in physical theory.

In short, we may speak here of two alternative proposals for constituting this 'physical reality' as an object for science; and we will also have to speak of an ambiguity, in the distinction between the reference of the term 'scientific object' and the term 'physical reality' or 'physical object' as it is used by Einstein and by Bohr. It is true that Bohr sometimes sounds like a reductive or eliminative positivist, asserting that reference ends at the measurement itself, that what is, is the value of a variable, determined by measurement, so that the entity, in physics – or the 'physical' object, so-called – is no more than what the *phenomenon* of measurement yields. It is this point of view that Einstein reacts against in his argument concerning the time of radioactive decay, which he offers as a crucial argument for the continuing incompleteness of Quantum-mechanics. The issue, in that case, concerns whether it makes sense to talk about the time-instant of the disintegration of a radioactive atom, or whether, on the contrary, it makes any sense to *deny* that there is such a time instant at which disintegration taken place. Einstein argues that the quantum theorist is forced to take a theoretically implausible position, in this case – another argument of the original EPR sort – and to deny that the notion of *time-instant of radioactive disintegration* has any physical significance, or any physical correlate in reality. Einstein writes, "If . . . the individual atom has a definite disintegration time, then as regards the individual atom, its description by means of the ψ-function must be interpreted as an incomplete description" (Schilpp, p. 668). In the mouth of a hypothetical quantum-theoretical critic, Einstein puts the following positivist language: "This consideration stands or falls with the assertion that there actually is such a thing as a definite time of disintegration of the individual atom (an instance of time existing independently of any observation). But this assertion, from [the Quantum-Mechanical] point of view, is nor merely arbitrary but actually meaningless. The assertion of the existence of a definite time-instant for the disintegration makes sense only if I can in principle determine this time-instant empirically. . . . The entire alleged difficulty proceeds from the fact that one postulates something not observed as real' (*Ibid.*, pp. 668–669).

Einstein replies: (to his own version of the 'positivist'): "What I dislike in this kind of argumentation is the basic positivist attitude, which from my point of view is untenable, and which seems to me to come to the same thing as Berkeley's principle, *esse est percipi*" (*Ibid.*, p. 669).

But is Bohr 'positivistic' in this sense? And, on the other hand, is Einstein 'realist' in the sort of metaphysical realist way his statements sometimes suggest? In the Bohr quotation earlier, he distinguishes between "the objects themselves" and "the measuring instruments", where, presumably, reference is distinct, between one and the other, in straightforwardly classical terms. And yet, of course, Bohr's reference to "the objects themselves", is incomplete, since the "object" can be described only within the limits of quantum measurement. But the measurement apparatus is also a physical object, which in the activity of making measurements interacts with the first object in some way so that the two together form a system. Yet, we do not 'observe' *this* system by means of still another measurement apparatus, which measures *it*, but rather interpret the measurements of the apparatus in terms of the framework of the (necessary) interaction between object and apparatus.

Thus, the 'physical object' or 'physical reality' as the object of measurement must 'exist' in some sense, to *be* measured, but cannot be known apart from the conditions of measurement and the theory of that measurement, which interprets it physically. But surely, in talking about 'physical reality', Bohr is not simply indulging in pre-quantum-theoretical and hence anomalous usage, resorting, so to speak, to folk-language. Or is he? Bohr, of course, insisted that the facts of measurement – the marks on strips of tape, the clicks of counters, the tracks or patterns on photographic plates, etc. – had to be dealt with in terms of classical physics. If, on the other hand, Quantum-mechanics is a fundamental theory *and* complete, then the classical talk is relativized to these quantum-foundations, and the dynamic variables of the classical macroworld are simply the 'phenomena' yielded (as we approach the classical limit), by the essentially statistical or objectively probabilistic 'reality' at the level of the quantum microworld.

Thus, the very notion that the quantum-description is incomplete depends on smuggling classical criteria of completeness into a world – or into an ontological level of the physical world – where they are anomalous. Quantum phenomena, in this sense, are completely described *the way they really are in themselves*, for their *being-in-themselves is* the *being-for-one-another* of their constituent subsystems, that is, in the interaction. There's nothing positivistic about that. Rather, it is just ontologically queer, and it may seem queer because we've only just opened up, in the last sixty or seventy years, the *real* domain of the quantum, which would have been there all along, had we gotten to the territory earlier. Ontologically speaking, we may say – and of course it had been said – that we're dealing with a world fundamentally characterizable, physically, as a set of (graded) potentialities – the probabilities assigned to the eigenstates by the wave-function – waiting to be actualized by interactions. Thus, Bohr may claim to be a kind of realist, given the latitude of classically queer reformulation of the ontological status of the scientific object.

The confrontation of the two views thus centers on a sharp difference concerning not only the characterization of the nature of the object of inquiry – in terms of criteria of locality and separability – but on alternative construals of the ontological status of the scientific object, i.e., whether it exists objectively, as independent of the interaction of measurement, or whether it is constituted in this interaction itself. The first position is realist, with whatever qualifications or ambiguities there may be in Einstein's formulation of it. The second is constructivist, in the sense that the object is taken to be constituted as an artifact of the practice of inquiry itself. Now it may be said that artifacts are no less 'real' than non-artifacts, and that 'nature' or 'physical reality' contains both. But the same thing cannot *be* both. The only escape between the horns, here, is to distinguish between the object as construed in a physical theory, as a 'theoretical' artifact, and its reference to or correspondence with some physical object, established by some criterion of 'fit' between predicted values and measured values, which justifies us in taking the theoretical object as a true description of the 'real' object; or, more holistically, which justifies us in accepting the theory which yields such successful predictions as a true theory, or as a putatively true theory. But it is precisely this claim which is thrown in question in the context of quantum measurement.

Let me turn then to the way in which this problem, posed in the Einstein-Bohr debate, bears on my thesis about historical changes in the constitution of the scientific object.

3. PRELIMINARY STATEMENT OF THE THESIS AND CLARIFICATION OF TERMS

The thesis I propose and examine in this paper is that the ontological status of the scientific object changes historically, in the sense that it is constituted differently by changing modes of scientific praxis; and further, that we may mark, in a general way, three distinctive stages of scientific praxis, the first of which I will characterize as an *anthropic* or *projective* mode, the second, as a *deanthropic* or *objective* mode, and the third as an *interactive* or *constructive* mode, I argue that in each of these stages, there are both practical and theoretical changes in the relation between scientific inquiry and its object.

Let me offer a rough sketch of the three stages, and a quick characterization of the modes of scientific praxis that are associated with them. The first stage, I called anthropic or projective. I take this to be the 'classical' stage of Greek science, in which the earlier modes of the mythopoetic understanding and explanation of nature are replaced by explanation in terms of ostensibly 'natural' categories, i.e., in terms of physical elements, and depersonified natural processes. This mode of scientific praxis I will characterize as a speculative construction, in which modes of technological and social action are depersonified, negated, abstracted, mediated – *aufgehoben* is the word – and reconstrued as accounts of physical process. In this context, I will distinguish the anthropomorphic from the anthropic, and suggest that this anthropic mode is also mediated by an objectivist mode of explanation, which already prepares the ground for the second stage. The second stage is the stage of the classical realism of the scientific revolution – the objective mode in which the physical world, as the object of knowledge, is construed as an independent reality which can be known by passive contemplation or observation, and in which the knower stands to the known as a spectator. There is no interaction with the object, except as it stands in causal relation to the knower, in perception, and as it has a rational or mathematical structure which is correlative with human reason. Here, the development of experimental science, as it involves the manipulation or transformation of nature in the practice of empirical tests, creates a mode of interaction, which poses difficulties for such a contemplative realism, and points to the third stage. The third stage, which I have characterized as interactive or constructive introduces the notion of the scientific object as an artifact, produced by or transformed by the mode of inquiry and hence, not only theory-dependent, but praxis-dependent, in the preparation of the contexts of measurement. The conceptual alternatives of realism and constructivism are seen here as results of different modes of scientific practice, in the second and third stages, respectively.

Pre-scientific mythopoetic understanding of nature is distinctively anthropomorphic, in the explicit sense that natural forces, events, processes are taken to be the activities of persons, or person-like entities – gods, or spirits, or even, more abstractly, principles; and the actions of these personified forces are understood in terms of the motives, purposes, or effective responses of such beings. Anthropomorphic explanations, in general, project human modes of production and social interaction upon nature, reading nature in the image of human activity, and empathetically exercising 'control' over the physical world, or bending to its constraints and necessities in terms of models of human making and doing.

The first stage of *scientific* practice is, in effect, a negation of this anthropomorphism, but one which retains essential features of it in a transformed way. I am thinking here of Greek science, from the pre-Socratics through the classical modes of Parmenideanism, Democriteanism, Platonism, Aristotelianism. The break with pre-scientific modes includes the rejection of the personified nature-gods in favor of more abstract principles: the natural 'qualities' or fundamental elements, in motion or change: the wet, the hot, the dry, the cold, in combination, or in just proportions; matter in motion, in its uncuttable form as atoms; potencies realizing themselves; natural kinds doing what comes naturally, and behind all, form, structure, mathematical essences bodied forth in the *show* of themselves that the senses perceive, but which reason sees through, to the objective, hidden reality behind the appearances.

The Greek mode is complex, as we know. Along with the retention of anthropisms, in their abstract, mediated form, there is also the deanthropic, objectivizing side of Greek thought.

This alternative, deanthropic aspect in Greek science, lies in the notions of blind necessity, implacable and immune to human entreaty or action, to be recognized as an otherness which stands beyond the power of human agency. It also lies in the notions of truth and being as constituting a reality which reason can attain to, and may contemplate, but to which it is related in only a spectative, passive way, though by virtue of a correspondence between the structures of reason and those of being.

Thus, in Greek science and philosophy, there is a tension between anthropic and deanthropic modes, between projective, sympathetic models of a physical nature in the image of human action, and objective models of a transcendent reality which can only be contemplated, but not acted on, or transformed.

The mode of speculation which reflects on human practices, and retains their normativeness in its construction of the scientific object, thus combines with a mode of 'pure' speculation, divorced from all activity or intervention in nature, active instead as 'pure' thought, *theoria, contemplating* the universal forms or principles, the "*archei*" of things; or *observing* nature, for the sake of classifying the *many* in some unified, hierarchical order which will reveal the essential rational structures of being.

The second stage constitutes the scientific object as an ontologically independent, self-subsisting physical world, apprehended by the senses, but known by thought. It is the deanthropic moment of the first stage in its full realization and dominance. In it, the last obvious element of anthropism is eliminated – i.e., the notion of teleology – when Aristotelian "final causes", and any notions of providential causality are assigned either to supernatural miraculous intervention, or to the asylum of ignorance. Determinism is, typically, taken to be an objective feature of the relations among things and things remain self-identical through transformation. Space-time is the universal matrix of the physical world: Space is homogeneous and isotropic, each part like each other, and the whole the sum of its parts, in a uniformly metrical extension. Time flows equably. Measurement is therefore dependable, and locally indifferent: Objectivity is not the view from nowhere. It is the view from anywhere and everywhere; The object of science is, classically, impassive. It bears contemplation with no ill effect. Observation is transparent: nothing muddies the view or distorts it, which cannot be removed by correction or cancellation. Everything is clear and distinct; or can be made so. The order and connection of things is the order and connection of thoughts. The truths of nature are simply the explication of these correspondences of thought to thing. We are in good old metaphysical realism country, and the primary properties – number, shape, size, position and time, tell all and hide nothing.

Now this is clearly an idyllic, untroubled version of Cartesian or Galilean science, a movie version of the novel. Yet, it reveals some crucial features. The scientific object is a space-time object and as such fully accessible to measurement. Measurement, moreover, exhausts the account of all of the properties of this scientific object, so it is wholly knowable, in principle, though not wholly known. The theory describing it in these terms is essentially complete, therefore. The formalism – ultimately, differential equations – is congruent with the continuous magnitudes of space and time, and leaves no pockets of discontinuity in the manifold, due to gaps in the language, and what it can name.

Neither observation nor the activity of measurement disturbs the scientific object in any way: it is, ontologically, independent of the knower. Therefore, realism, in its classical form, is an accurate account of this relation.

But isn't this just a realist version of physics? How then could realism *fail* to be accurate, since its appropriateness is circular. Not quite. Take the paradigmatic object of science of this scientific realism: it is, in the first place, the heavens. Astronomy has, as its object, that with which we, in practice, *cannot* interact. All we can do is look at the heavens from a distance, *contemplate* them. They are the ideal object for a realist epistemology, since measurement cannot disturb the stars, though disturbances *of* measurement can take place, and need to be corrected. This seems to be common-sensically incontrovertible. (At least in the 17th century, and before Goodman.)

In short, in the hopeful picture of stage two, characterized as deanthropic and objective, a robust realism would seem to be the epistemology of choice. The paradigm case of astronomical observation and measurement clearly leaves the object of scientific knowledge undisturbed. This ideal case, is, on the one hand, the contemplative model of Greek science, mathematized and experimentalized, and brought up to date. In that sense, 17th century realism recapitulates the intellectualist model of Greek thought, in which detached spectation without intervention is the ideal of a gentlemanly science. Astronomy, mechanics, mathematics are a clean game: "No shadows. Everything on the surface, everything clear" as Neurath liked to say.

Steam, coal and urea are, by contrast, dirty games, as are fermentation, fertilization of the soil, syphilis, and surgery without anesthesia. Thermodynamics, industrial chemistry, pharmacology are messy sciences, in which the scientific object is not so clearly detached from the conditions of its observation, as are the heavens. At the very least, conditions of experimentation are such that the experimental situation is not there for the contemplation; it has to be prepared. The experimenter rearranges nature, tortures it, subjects it to transformations, in which its patterns, its truths, its secrets are revealed. Yet, here too, it seems we may ideally translate the messier phenomena into our clockwork heavenly model. But there are shadows, no larger that a man's hand. And thus we turn (or turn back) to stage three.

The issue between Einstein and Bohr, in their confrontation over the completeness or incompleteness of quantum-mechanics, is the way in which the difference between stage two and stage three develops, in actual fact. Einstein is no simple realist, though he preserves the ideological core of stage two thinking. Bohr holds out against the adequacy of stage two, and for stage three as the fundamental model of the ontological status of the object of physical science. But Bohr, in principle, accepts Einstein's classical realist requirements as holding for the macroworld. For *that* level of physical reality, our classical picture is not only adequate, but essential. As Arthur Fine points out, Bohr is the conservative and Einstein the radical, in terms of willingness to change or reformulate the classical physical world picture to be adequate to our experience. But for Einstein, there is a deep intuition that the changes, if any, should retain locality and separability, hence also causal determinism, and that there should be continuity in these ontological criteria across relativistic, classical and quantum mechanical frameworks. Bohr is ready to opt for discontinuity, and that's what complementarity is all about.

Now, all of this focuses on questions of measurement, the conditions of measurement, and the ineliminability of the effects of measurement upon the object of knowledge. The scientific object, therefore, is essentially an interactive one, or one in which what is taken *to be* is projected by the conditions of measurement. 'Taken to be' here is not to be read as simply an act of mental representation, or as the *way* we happen to understand something, in our reflection upon it. Rather, 'taken to be' connotes an activity of construction, in which "taking the object to be x" is an actual transformation of something, or a projection of something – a projective transformation then – into x. A measurement, at the quantum level, requires a preparation: not simply the existence or presence of a measurement apparatus, e.g., a diaphragm, an emulsion, a screen, a cloud chamber, a geiger counter which registers the quantum-events, but also the generation of quantum-events under controlled conditions, or under conditions of selection and screening, on the basis of physical theory, e.g., the use of polarizers, of optical switches, of slits; of linear accelerators, under the tutelage and direction of particular experimental hypotheses: Nature theorized, so to speak, but not inside the head; rather in the objectified, externalized, constructed world of experiment: the arena in which the scientific object comes into being, emerging on the stage of scientific observation, to try out the role written for it in the scripts of the theorists, the mathematicians, the imaginers of 'thought experiments'.

In this sense of construction, the scientific object is not a 'theoretical construct' as some conceptual entity, 'existing', so to speak, as the 'meaning' of the theory, or as some sort of dispositional mental entity, e.g., what would be the intentional object of thought thinking the theory. Instead, the scientific object is an actual experimental transformation of what, for want of more neutral terms of reference, we may call the natural world. In this sense, we may speak of changing the world by 'theorizing' it.

Department of Philosophy,
Baruch College & the Graduate Center,
The City University of New York,
New York, U.S.A.

JOHN WATKINS

SCIENTIFIC REALISM *VERSUS* COMMON-SENSE REALISM?

Adopting or rejecting scientific realism may make a big difference to one's world-view, and so may adopting or rejecting common-sense realism; it therefore becomes an important question whether these two kinds of realism are mutually exclusive, as has often been supposed, so that one cannot adopt them both.

What do they respectively assert? Scientific realism is primarily concerned with the invisible infrastructure of things. Let a generally accepted scientific theory say that the sun is largely composed of hydrogen atoms, and exerts a gravitational pull on the earth and the other planets that keeps them in orbit round it. Scientific realism says that if that theory is true, then the sun *is* largely composed of such atoms and *does* exert a gravitational pull. Now common-sense realism, as I understand it, does not deny that the sun has an internal structure, though it leaves it to science to discover what this is, rather as it leaves it to science to estimate the sun's size and distance from the earth. But common-sense realism does insist that the macroscopic sun which we perceive exists independently of our perceptions of it: all perceivers of it might perish without it perishing. It is a general presumption of common-sense realism that where there are intersubjectively shared perceptual experiences of a macro-object, these experiences constitute perceptions of something "out there", something which exists independently of these perceptions of it. In the case of rainbows and mirages it has been obliged to withdraw that presumption; but it stands by it with respect to mountains, buildings, elephants, oranges, and beetles.

That common-sense realism should be repudiated in favour of scientific realism, is asserted by the doctrine of the Omnicompetence of Physics. This doctrine, which has sometimes sailed under the banner of 'Unified Science',[1] claims that all meaningful questions about matters of fact are in principle answerable by a completed physics; if there are questions which appear to be unanswerable by physics this is only because physics is as yet far from being completed. It is a doctrine which draws a large cheque on the future. When philosophers support their metaphysical predilections with forecasts about future scientific devel-

R. S. Cohen, R. Hilpinen and Qiu Renzong (eds.), Realism and Anti-Realism in the Philosophy of Science, 219–226.

opments, they engage in a sort of circularity, since their forecasts are obviously inspired by their metaphysical predilections. But this appeal to future developments renders the present doctrine irrefutable; its defenders can always say of a question apparently unanswerable by physics that it is so only because physics is not yet complete. Nevertheless I will now mention some meaningful, factual questions, which it seems very doubtful that physics could ever answer.

Question 1: What divides the past from the future? The obvious answer is the *present*, or the *Now* as Einstein called it. Is this something that physics will ever comprehend? In his 'Intellectual Autobiography' Carnap reported some conversations with Einstein, in one of which

> Einstein said that the problem of the Now worried him seriously. He explained that the experience of the Now means something special for man, something essentially different from the past and the future, but that this important difference does not and cannot occur within physics. That this experience cannot be grasped by science seemed to him a matter of painful but inevitable resignation. . . . [T]here is something essential about the Now which is just outside the realm of science' (1963, p. 38).

Carnap disagreed; although he had not pressed the point against Einstein at the time, he now commented: "Since science in principle can say all that can be said, there is no unanswerable question left." But that is merely to summon up the Omnicompetence doctrine to sweep away this counter-example to it. I think that Einstein was right. For each of us, at any moment in our waking lives, nothing could be more real than the *now*, or specious present. But how could science ever capture this elusive and shifting datum? Science can locate events, to employ McTaggart's distinction, only on the B-series, with its *earlier-later* relation. If it ties t_0 to a momentary present, t_0 is then fixed, and a moment later has ceased to be *now*.

Question 2: How did consciousness arise? On this issue I like the following exchange between two distinguished evolutionists. Sewall Wright: "Emergence of mind from no mind at all is sheer magic" (1964, p. 278); Dobzhansky: "If this is 'sheer magic', it is a kind of magic the world is full of" (*1967*, p. 31). I think that both men were right, if by 'magic' is meant 'unexplainable in principle by physics'; there is something magical about the trick so effortlessly performed each time a newborn infant grows into a normal child with a mind of its own; we certainly seem to have emergence of mind from no mind here. The conviction that that is impossible has driven some tough-minded thinkers to very strange conclusions. Ernst Haeckel, the nineteenth century evolutionist who did for Darwin in Germany very much what T. H. Huxley did for him in England, had a mind that was usually as tough as old boots; but it turned pretty soft in a chapter in his *1899* on the 'embryology of the soul'; he there announced that when a sperm and an ovum come together at conception, they each bring with them their own "cell-soul"; these two mini-souls then coalesce into one. I regard this as doubly absurd, first because gametes surely have no mentality, and second because I hold it to be a fundamental feature of the mental that separate centres of mentality cannot coalesce. But Haeckel had not yet gone far enough: these gametes would have needed to derive their soulfulness from an earlier source if soulfulness was not to have emerged ex nihilo with them. So Haeckel added that unicellular protozoa, the earliest and most rudimentary form of life, have a simple cell-soul (p. 53). But these protozoa would in their turn have needed to have acquired their soulfulness from a still earlier source if soulfulness was not to have emerged ex nihilo with them. Once again Haeckel obliged: 'even the atom is not without a rudimentary form of sensation and will' (p. 80). William James had gone straight to the conclusion to which one is driven if one admits consciousness and denies emergence: "If evolution is to work smoothly, consciousness in some shape must have been present at the very origin of things" (*1890*, p. 152, italicized in the original). I am glad to say that James repudiated this panpsychist conclusion (p. 164).

Question 3: How did humour arise? So far as I know, only human beings (and by no means all of them) possess humour. Some animals possess rough equivalents of those of our smiles and laughter that go with high spirits and playfulness (see Darwin, *1872*, pp. 196f); but it seems that they all lack even a rudimentary or incipient form of anything like our chuckling over something funny. Will physics ever explain why there are jokes in our lives but not in theirs? It occurred to me that when Darwin was writing *The Descent of Man* he would have been glad to find evidence of humour in other animals, since it was a main thesis of that work that 'the various emotions and faculties, such as love, memory, attention, curiosity, imitation, reason, & c., of which man boasts, may be found in an incipient, or even sometimes in a well-developed condition, in the lower animals' (p. 193), and he would surely have liked to add humour to the list. The second (1874) edition has a very thorough index, running to 77 pages, and I have combed it carefully, There are no entries for 'jokes', 'laughter', or 'wit'; the only entry I could find that seemed relevant is the following: 'Humour, sense of, in dogs, 743'. I turned eagerly to page 743, but Darwin is there dealing with the brilliant plumage of certain male birds, and there is nothing about either dogs or humour. So I turned to the long entry under 'Dogs'. Among its many sub-entries are ones on dog's reasoning faculties, moral qualities, and possession of conscience, but none on their sense of humour. I began to wonder whether he had inserted that page 743 entry as a joke.

Question 4: How did physics arise? I will here take it for granted that its growth has involved intellectual creativity and was not just an inductive accumulation; so we could frame question 4 more generally: will science ever explain how some particular product of human creativity, whether in science or art, came to exist? Kant had a view about creativity which seems right and which implies that science cannot explain such comings-to-be. Let x be the occurrence of some unprecedented event. Thus x could be Michelangelo painting the creation of Adam; or it could be the first explosion of an atom bomb. A scientific explanation of x, if there is one, consists of laws of nature in conjunction with a set (perhaps a very large set) of initial conditions. If true, such an explanation shows both how x was in fact produced and, at least in principle, how it could have been produced at an earlier (or later) time; for it implies that x occurs whenever such a set of initial conditions is assembled. In short, a scientific explanation of x would provide a recipe for the production of x.[2] It might have been impracticable to assemble such a set of initial conditions at any earlier time; but a recipe for a cake doesn't stop being a recipe just because some of the ingredients it calls for are presently unobtainable.

So a scientific explanation of how Michelangelo's picture of the creation of Adam came into existence would be a recipe (perhaps a very cumbersome one), not for duplicating the picture subsequently which would be philosophically uninteresting, but for its first creation. Now Kant insisted that there can be no such recipe for a product of genius; that is something 'for which no definite rule can be given' (*1790*, §46). I think he was right.

We cannot refute the Omnicompetence of Physics doctrine, but we can and should put a large question-mark over it. But that still leaves open the question of the reconcilability of science and common sense.

Let us hark back for a moment to the seventeenth century, when a serious split developed in the realist camp. It seemed to many leading thinkers on both sides of the divide that there is an irreconcilable conflict between science and common sense: adopting the new science meant repudiating one's common-sense convictions even where these are not under specific challenge from science (in the way in which, for instance, the geocentric hypothesis was). Nearly all those in the vanguard of the scientific revolution, such as Galileo and Boyle, together with their philosophical allies, such as Descartes and Locke, believed that the new science had overthrown the common-sense world-view: must not the world be essentially different from what it appears to be if it consists, in reality, of hard, massy, odourless, and colourless corpuscles, obeying only the laws of mechanics? Its apparent colours, sounds and smells must be subjective qualities projected by us onto external objects. These men, Berkeley declared, "lead us to think all the visible beauty of the creation a false imaginary glare" (*1713*, p. 211). He sought to turn the tables by arguing that it is not the world but *science* that is not what it appears to be; it appears to give us information about realities behind the phenomena; but since there is nothing behind the phenomena for it to get hold of, the most science can do is to give us rules for correlating and predicting phenomena.[3]

I say that what generated this conflict was a wrongful assumption shared by both sides, namely that what is real is *all at one level*; at the bottommost level, according to the scientific realists, and at the surface, according to Berkeley (for whom the only other realities were souls and God, which are not science's business). Remove this assumption, allow that reality is multi-leveled, and the conflict dissolves: quarks, atoms and oranges may coexist.

The thesis that only entities at the bottommost level are real had a certain plausibility so long as classical atomism survived intact; it declared any complex thing to be, at bottom, just a swarm of particles. But classical atomism has of course been swept aside by modern physics; as Patrick Suppes put it, "it is not swarms of particles that things are made of, but particles that are made of swarms" (*1984*, p. 122). Popper had previously made a similar point with clouds and clocks (*1972*, chap. 6): in the heyday of classical mechanics, with its atomism and determinism, seemingly "cloudy" things could be assumed to be made up of bits that behave like clockwork; but now, in the aftermath of quantum theory, it seems to be the other way round, with seemingly clocklike things really being statistical aggregates of "cloudy" bits. Suppes continued: 'we cannot have a reduction of subject matter to the ultimate physical entities because we do not know what those entities are' (p. 123).

It is not just that we do not at the present time know what they are; the assumption that science will ever reach a bottommost level, or even that there is such a level for it to fail to reach, is in doubt. As John A. Wheeler put it: "One therefore suspects that it is wrong to think that as one penetrates deeper and deeper into the structure of physics he will find it terminating at some *n*th level" (*1977*, pp. 4–5).

So a scientific realist who restricts the title 'real' to the ultimate components of matter would have to admit that nothing so far postulated by science is real. If we shy away from that, and allow that atoms are real despite having graduated to a non-ultimate level, then we must surely allow that molecules are real; and so we can go on working up towards the macro-level. And why should we not eventually allow that diamonds are real? They are, after all, a lot more stable than, say, radon atoms. I say that just as common-sense realism is under no internal compulsion to deny that things have an invisible infrastructure, so scientific realism is under no internal compulsion to deny the reality of macro-objects.

It is a mistake to suppose that a scientific explanation of properties of things at one level, in terms of the properties and relations of things at a deeper level, *explains away* the former properties, or *reduces* them to the deeper ones. Bohr's model of the hydrogen atom, with its electron circling a positively charged nucleus, explained something not explained by Rutherford's model, namely the stability of the atom. There was not the slightest tendency for its stability to be thereby explained *away* (though a deep explanation may have revisionary implications for its explanandum; thus Bohr's model did suggest that an atom is not, after all, absolutely stable, but might conceivably be split). The account that nuclear physics gives about what happens at the micro-level when an "atom bomb" explodes has no tendency at all to deprive the macro-event of its reality. Likewise, a social science explanation of a social phenomenon, such as inflation, as the unintended outcome of the decisions and activities of individual people has no tendency at all to deprive the social phenomenon of its reality. Reality is multi-leveled.

A plausible but invalid argument from scientific realism to the unreality of macro-objects was put forward by the late Grover Maxwell, in his 1968: – You are looking at an orange, say; then please make an inventory, in the light of the best available scientific knowledge, of the kinds of entity involved in the causal processes that issue in your orange-like perceptions. Such an inventory will include all sorts of things like photons, neurons, atoms, and so on, but no *orange*: 'the stimulation of the retina is accomplished solely by photons emitted from the atoms comprising the surface of the material object in question. The material object just *is* this collection of submicroscopic particles and the relations that subsist among them' (p. 151); in seeing an "orange". then, you are seeing something that isn't really there, an illusion generated by the interplay of submicroscopic particles which you don't see and are really there.

This argument fails to take account of the fact that animals' brains *decode* the input from their sense-organs. My television set has often presented me with images of Mrs Thatcher. An inventory of the impulses reaching the TV aerial, and of the processes within the set, will include photons and electrons, but no Mrs Thatcher, Does that mean that nothing real corresponded to these Thatcherite images? No; a television set decodes signals encoded at the transmitting end, and these images corresponds pretty well (apart of course from their two-dimensionality) with certain realities at the other end. Now consider a dog sniffing a piece of meat. The meat is giving off signals; not deliberately encoded signals, to be sure, but signals that the dog's brain is able, as a result of natural selection, to decode. The dog turns away, because the meat smells bad. And this corresponds pretty well with certain realities at the other end; the meat *is* bad.

Even so, it will be said, I must admit that *colours* are subjective qualities with which we daub physical objects. In my youth I would have agreed with this, but now I am not so sure. Suppose that you are told that there is something, call it x, which usually appears to be homogeneous; however, when x is subjected to a certain treatment, call it P_1, it is decomposed into components $a, b, c, \ldots$; and when $a, b, c, \ldots$ are subjected to a certain treatment, call it P_2, x is reconstituted, this whole process occurring independently of perceivers. Would you not conclude that $a, b, c, \ldots$ are not mind-dependent but objective? But this account fits an experiment of Newton's where x is sunlight, P_1 and P_2 are suitably positioned prisms, and $a, b, c, \ldots$ are the colours of the spectrum.

The ontology of common-sense is revised and enriched, but not abolished, by science.

Centre for the Philosophy of the Natural and Social Sciences,
The London School of Economics,
London, U.K.

NOTES

[1] Thus Otto Neurath included physicalism and unified science in the so-called 'scientific conception of the world' (*1973*, p. 417).
[2] See Briskman 1980.
[3] See Popper, *1963*, chaps. 3 and 6.

BIBLIOGRAPHY

Berkeley, George (*1713*). 'Three Dialogues Between Hylas and Philonous', in A. A. Luce and T. E. Jessop (eds.), *The Works of George Berkeley*, 9 vols, 1948–57, Nelson, London.

Briskman, Larry (1980). 'Creative Product and Creative Process in Science and Art', *Inquiry* **23**, pp. 83–106.

Carnap, Rudolf (1963). 'Intellectual Autobiography', in Schilpp (ed.), *The Philosophy of Rudolf Carnap* (The Library of Living Philosophers), Open Court, La Salle, pp. 3–84.

Darwin, Charles (*1871*). *The Descent of Man and Selection to Sex*, Murray, London, first ed. 1871, second ed. 1874.

Darwin, Charles (1872). *The Expression of the Emotions in Man and Animals*, 1872; reprinted University Press, Chicago, 1965.

Dobzhansky, Theodosius (*1967*). *The Biology of Ultimate Concern*, New American Library, New York.

Haeckel, Ernst (*1899*). *The Riddle of the Universe*, trans. by Joseph McCabe, Watts, London, 1904.

Kant, Immanuel (1790). *Critique of Judgement*, first ed.; trans. by J. C. Meredith, Clarendon Press, Oxford, 1928.

Maxwell, Grover (1968). 'Scientific Methodology and the Causal Theory of Perception', in Lakatos, Imre and Musgrave, Alan (eds.), *Problems in the Philosophy of Science*, North-Holland, Amsterdam, pp. 148–160.

Neurath, Otto (*1973*). *Otto Neurath: Empiricism and Sociology*, Robert S. Cohen and Marie Neurath (eds.), Vienna Circle Collection, vol. 1, Reidel, Dordrecht.

Popper, Karl R. (*1963*). *Conjectures and Refutations*, Routledge & Kegan Paul, London.

Popper, Karl R. (*1972*). *Objective Knowledge: An Evolutionary Approach*, Clarendon Press, Oxford.

Suppes, Patrick (*1984*). *Probabilistic Metaphysics*, Blackwell, Oxford.

Wheeler, John Archibald (1977). 'Genesis and Observership', in Butts, Robert E. and Hintikka, Jaakko (eds.), *Foundational Problems in the Special Sciences*, Reidel, Dordrecht, pp. 3–33.

Wright, Sewall (1964). 'Biology and the Philosophy of Science', *The Monist* **48**, pp. 265–290.

JOHN WRIGHT

METAPHYSICAL REALISM AND THE EXPLANATION OF THE SUCCESS OF SCIENCE

One argument for Scientific Realism is from the success of science. But it is not clear whether this argument supports Metaphysical Realism with respect to the entities of science, or merely some weaker form of Realism such as Hilary Putnam's 'Internal Realism'.[1] In this paper it will be argued that *if* the argument supports any form of Realism, it supports Metaphysical Realism. More specifically, it will be argued that *if some* notion of truth is needed to explain the success of science, then that notion of truth must be a non-epistemic, Realist notion.

We can, on the face of it, draw a distinction between the *strong* predictive success of a theory and *weak* predictive success. A theory is strongly predictively successful if it successfully predicts *novel* regularities – that is, if it successfully predicts regularities that were not used in actually formulating the theory. If a theory only succeeds in predicting more instances of regularities which *were* used in its formulation, then it is only weakly predictively successful.[2] The main claim argued for in this paper is that strong predictive success cannot be explained by an epistemic or verificationist conception of truth. This form of success can only, I think, be explained by a Realist conception.

1. SOME QUALIFICATIONS TO THE ARGUMENT

In this paper it will be *assumed* that in order to explain the strong predictive success of a theory, we need to say that at least a part of a theory is true, or close to the truth in *some* sense of truth.[3] However, three qualifications to this claim need to be strongly emphasised.

(a) It is not claimed, and neither is it for the argument presented here necessary to assume, that if a theory is strongly predictively successful then *all* parts of it are true. It only needs to be assumed that certain parts, very directly involved in the derivation of the prediction, are true.[4]

(b) Strictly speaking, it is not even necessary to assume that the parts of a theory directly involved in the derivation of the prediction are

R. S. Cohen, R. Hilpinen and Qiu Renzong (eds.), Realism and Anti-Realism in the Philosophy of Science, 227–243.

true tout court. It can be sufficient to say that they are, in some sense, 'close to the truth'.[5]

(c) Even the claim that some part of a strongly successful theory is close to the truth is not claimed to be certain. It is merely claimed it can be something that is *reasonable to believe*, but which may be refuted by future experimental results.

Although I will not argue for the claim that *some* notion of truth is needed to explain the success of science, one possible objection to this claim will be considered. This is what may be termed 'the redundancy objection'. It is that in order to explain the success of a theory it is not necessary to state that the theory is *true*, it is enough merely to state the theory. For example, in order to explain the success of the theory that electrons have negative charge it is not necessary to state that the theory is *true*; it is sufficient to simply say: 'Electrons have negative charge'. Call this the *redundancy explanation* of the theory's success.

Even if it is granted that some examples of success can be given a redundancy explanation it does not follow that this type of explanation is adequate for *all* phenomena connected with the success of science. Some phenomena cannot, it will be argued, be given a redundancy explanation. Suppose, for example, that scientists are observed to manipulate some ink marks on paper in a particular way, and at the end of this process they produce the observational prediction, 'There will be a yellow spark in region R'. Let us assume that this is a 'novel' observational prediction. Subsequently, the scientists do indeed observe a yellow spark in region R. We will call this phenomenon – that is, the phenomenon of the manipulation of ink-marks yielding a novel prediction that is subsequently verified – phenomenon S. If the novel observational prediction is a surprising one, then phenomenon S will surely require explanation. But it needs to be emphasised that to say that phenomenon S requires explanation is *not* to say that it is *tokens* of sentences that are the primary bearers of truth. To admit that a link between the manipulation of tokens and success needs explanation is compatible with saying that the bearers of truth are, for example, propositions. One *possible* explanation of phenomenon S is that the ink symbols manipulated by the scientists expressed true propositions. (Perhaps some of them expressed the true proposition that electrons have negative charge.) This enabled the scientists to derive the true observational prediction that there will be a yellow spark in region R. Here I will not be concerned to argue either that this is the only, or even the best, explanation of the scientists' success. But let us ask, 'Can it be given a redundancy explanation?' Clearly, it is no explanation of phenomenon S to merely say 'Electrons have negative charge'. That does not explain why *the scientists who manipulated those ink marks* made a successful prediction.

It might be suggested that a redundancy explanation would be adequate if that which required explanation were re-expressed: 'Why were the scientists who manipulated symbols that *expressed the proposition 'Electrons have negative charge'* successful? It may appear that a sufficient explanation of *this* re-expression of the explanandum is that electrons have negative charge. However, this line of thought does not show 'true' to be redundant in the explanation. Suppose that drugs A and B are both necessary, and jointly sufficient to cure a particular disease. One explanation of why patient P was cured of the disease is 'Because he was given drugs A and B'. Now, suppose we change the explanandum to 'Why was P, who was given drug A, cured?' It might be thought it would be sufficient to say 'Because he was given drug B'. But this is *not* sufficient: we still need to add, as a part of the *explanation*, that P was given drug A. After all, the fact that P had been given drug A might have been included in the explanandum as an irrelevant bit of extra information. So it is not otiose to include it in the explanans, since doing so tells that it is an *explanatorily relevant* factor. Its mere inclusion in the explanandum does not tell us that.

Now, suppose the explanandum is: 'The use by scientists of tokens that expressed the proposition that electrons have negative charge led to success'. Is a sufficient explanation of this: 'Because electrons have negative charge'? The considerations of the above paragraph indicate that it would not be sufficient if the fact that the tokens *expressed the proposition that electrons have negative charge* is *relevant* in leading to the success. And it surely is relevant. If the tokens used by the scientists had expressed something different – if, for example, they had referred to protons instead of electrons or they had made claims about electrons that were false – then they would have yielded observational predictions which we have no reason to suppose would have been confirmed by tests. The semantic properties of the tokens *are* relevant in producing the empirical success that follows the manipulation of these tokens. Consequently, the redundancy objection – that the semantic properties of a theory need not be mentioned to explain the success of the theory – fails if what is to be explained is the success that follows from the manipulation of tokens.

Of course, to say that the redundancy objection fails is not to say anything about which semantic notion is needed to explain the success of science. In particular, it is not to say whether it is a Realist, or a verificationist conception of truth (or truthlikeness) that is needed.

There is another objection to the general enterprise of arguing for a realist conception of truth on the grounds that only it can explain the success of science. This is the objection that the strong predictive success of science does not *require* explanation. The claim is that since so many scientific theories are advanced, it would rather be surprising if some of them did not, simply through lucky chance, turn out to be strongly predictively successful. But there are, I think, two reasons why this objection is unconvincing.

(a) Some types of strong predictive success simply have too many positive instances to be merely due to lucky chance. For example, scientists are able to correctly predict the ways in which artificial, trans-uranium elements will chemically combine with thousands of other substances. It is beyond the bounds of plausibility that such a large number of successes could simply be a fluke.

(b) If on a particular occasion, a novel prediction made by a theory is confirmed, scientists will re-test the theory to ensure that the apparent success is not an anomalous result but is a *repeatable effect*. So what requires explanation is that the ability of a theory to correct predict a novel observation is a repeatable effect. Establishing that a result is a repeatable effect makes it less likely that it is merely due to chance.

B. van Fraassen has offered an evolutionary explanation of the success of scientific theories.[6] Van Fraassen says that a theory will be retained in the body of accepted science only if it successfully predicts observations. If it is not successful it is removed. As a result of this process science has come to contain only successful theories. However, it is clear that van Fraassen's suggestion at most only provides an explanation of weak predictive success. It does not explain how a theory correctly predicted observations that were only made after the theory was formulated, that is, it provides no explanation of strong predictive success.

2. IS AN EPISTEMIC NOTION OF TRUTH ABLE TO EXPLAIN THE SUCCESS OF SCIENCE?

Broadly speaking, there are, I think, three epistemic notions of truth. I will call these Strict Verificationism, Liberal Verificationism and the Epistemic Ideality Conception of Truth. Roughly, *Strict Verificationism* asserts that a theory is true iff it can be verified by the techniques we currently possess. *Liberal Verificationism* says that a theory is true iff we have, *or could acquire*, the ability to verify it. The Epistemic Ideality Conception of Truth says that a theory is true iff it is a part of an epistemically ideal theory. I do not claim that these three notions exhaust all the conceptions of epistemic truth that have ever been advanced by anyone. But I do think they represent currently popular conceptions.

2(a). *Strict Verificationism and the Explanation of Strong Predictive Success*

A more careful statement of Strict Verificationism is:

> A sentence S, as uttered by a speaker U at time t, expresses a true proposition at time t iff it is possible for U to establish its truth using only techniques available at t.[7]

Note that a Strict Verificationist need not say that if we acquire new techniques of verification then some *propositions* may change their truth-value. He/she can say that, with the acquisition of such new techniques, a previously unverifiable sentence may come to express a new proposition. So, although the sentence 'Diamonds and coal are made of the same substance' was, perhaps, once unverifiable, the Strict Verificationist can say that with new techniques of verification that sentence came to express a different proposition that was both verifiable and true. No *proposition* changed its truth-value.

There are at least some instances of the strong predictive success of science that cannot be explained by Strict Verificationism. Let us say that M is a theoretical property. At some time, say 1950, M was only detectable by a particular range of tests $t_1 \ldots t_n$. Now, suppose T is a theory of the nature of M that makes the *novel* prediction that M is reliably indicated by phenomenon e_n. Subsequently, in 1951 (say), this novel prediction is verified.

In order for the argument against Strict Verificationism to go through we only need to say that the part of T which asserted

(1) M is reliably indicated by e_n

expressed a true proposition prior to its verification. What will be argued is that the sentence (1) needed to be true in a sense not available to the Strict Verificationist.

Note that (1) does not refer to unobservables, or to anything else not very directly involved in producing T's success.

Now, let us consider what an explanation of T's strong predictive success would look like. We are assuming that strong predictive success is to be explained by saying that *some part* of the successful theory T is true. Let us refer to this part of T as T_D. Then the explanation of T's success will consist in the derivation of (1) from T_D.

Now, if we are to *explain* the strong predictive success of T by saying that the special part T_D is *true*, then it must be asserted that T_D was already expressing a true proposition *prior to* the verification of (1) in 1951. If T_D only came to express a true proposition as a result of the observations confirming (1), then its truth could not be used to *explain* those observations. But if T_D was expressing a true proposition prior to 1951, then so must (1) have been expressing a true proposition, since it is entailed by T_D. As can be seen, these considerations do not require us to make any specific assumptions about what part of T is true; they do not require us to say anything about T_D. The only specific sentence that needs to be assumed to be true is (1).

Now, suppose that an object X was found in 1950 which was known to cause e_n, but which could not be verified by any means *available in 1950* to have M. From 'X has e_n' and the claim that (1) expressed a true proposition in 1950 it logically follows that

(2) X has M

expressed a true proposition in 1950. But (2) was not verifiable at that time. So, the Strict Verificationist cannot say it expressed a true proposition at that time.

The two claims required to derive the claim that (2) expressed a true proposition in 1950 are that (1) expressed a true proposition in 1950 and that 'X has e_n' expressed a true proposition in 1950. Since the advocate of Strict Verificationism is unable to say (2) expressed a true proposition, he must reject one of the two statements that together imply it. But, we are assuming, 'X has e_n' was *known by observation to be true.* Therefore, what the advocate of Strict Verificationism must *deny* is that in 1950 (1) expressed a true proposition. But this claim must be asserted if the strong predictive success of T is to be explained. So, there are at least some cases of strong predictive success that cannot be explained by the Strict Verificationist.

Since the above argument against Strict Verificationism is rather intricate, it is worth giving a briefer, more intuitive account of it. If the success of a theory T in correctly predicting some novel phenomena is to be explained, it needs to be asserted that at least some part of T was true prior to the observation of the novel phenomena. In the above example, it was argued that the sentence 'Property M is reliably indicated by e_n' is such a sentence. But if this was *true* prior to its verification, it is easy to imagine situations in which there is some object X that has M (since it has e_n) but cannot be verified to have it by available techniques. Since the Strict Verificationist cannot allow this, he/she cannot say that 'Property M is reliably indicated by e_n' was true. But since the truth of this sentence is required to explain the strong success of T, in this case the advocate of Strict Verificationism is rendered unable to explain T's success.

2(b). *Liberal Verificationism and the Explanation of Strong Predictive Success*

The argument of the previous section attacks only the strict form of verificationism which asserts that a statement made at a particular time is true iff its truth can be established by the techniques of verification available at that time. But, it may be objected, an anti-Realist need not accept this. Although the techniques for verifying 'X has M' did not, by hypothesis, exist in 1950, it would be easy enough for a scientist working at that time to acquire the ability to verify it. All he has to do is to confirm inductively that whenever an object has been established by other methods to have M, it has the capacity to produce e_n. Once the scientist has inductively confirmed this, he will be justified in making the inference from 'X causes e_n' to the conclusion 'X has M'. And perhaps the anti-Realist will say that in order for a sentence to be true at time t, it is sufficient that a scientist at that time *be able to acquire* the ability to verify it. We will call such a doctrine 'Liberal Verificationism'.

The difficulty with Strict Verificationism was that it implied that certain sentences, the truth of which was required to explain the strong predictive success of science, were *not* true. But Liberal Verificationism is confronted with a different difficulty. The Liberal Verificationist *is* able to say that the required sentences are true, but (it will be argued) this notion of truth does not give us an *explanation* of strong predictive success. Roughly, the difficulty is that the Liberal Verificationist's account of what it is for a strongly predictively successful theory to be true simply gives back to us the very thing that it is meant to *explain*.

A theory is verifiable, on the Liberal Verificationist's account, in virtue of the fact that a user of the theory has, or could acquire, techniques that would enable him to verify it. Let T be a theory that is not verifiable using techniques *currently available* but which is verifiable in the Liberal Verificationist's sense. More specifically, assume that we *could acquire* some experimental technique E, which would yield observations O, and the obtaining of these results would verify T. Now, since E is a technique we do not currently possess, E $\rightarrow$ O is a *novel* observational regularity. And if E $\rightarrow$ O verifies T, then it is a novel regularity which is predicted (perhaps in conjunction with other theories) by T. Since T is (we are assuming) verifiable in the Liberal Verificationist's sense the novel observational regularity E $\rightarrow$ O would in fact be obtained. And I am assuming that the best explanation of this is that T – or at least the part T_D of T directly responsible for the derivation of 'E $\rightarrow$ O' – is true. But now, on the Liberal Verificationist's account of truth, T is true *in virtue* of the fact that it is possible for us to acquire the techniques that would verify it. In particular, T is true in virtue of the fact that we could acquire technique E, *and* that if we applied E, we would obtain O. That is, the Liberal Verificationist's account of that in virtue of which T is true mentions as one of its conjuncts the very thing that is to be explained by saying that T, or some part of it, is true. But if the truth of T, or a part of it, is to *explain* certain results, then it cannot be the case that it is true *in virtue* of the fact that those results would be obtained. This is merely an instance of the general principle that if some state of affairs S holds in virtue of condition C, then the condition C cannot be explained by state of affairs S. This can be illustrated by a few simple examples. There is a disease of horses called *Glanders*. A horse with Glanders exhibits certain symptoms such as swelling, the production of mucus, etc. But we cannot say that the horse has Glanders in virtue of exhibiting these symptoms, since there are a number of diseases which could be associated with those symptoms. Rather, the horse has Glanders in virtue of being infected with the Actinobacillus *Pfeifferella mallei*. But now, suppose we want an explanation of why the horse is infected with *Pfeifferella mallei*. It is clearly no explanation to say 'because the horse has Glanders'. Rather, what is wanted is an account of how the horse came to be infected. This account may be because the horse drank infected water, or because the horse came into contact with another infected horse. Or consider another example. We can say that a utensil is a spatula rather than a knife in virtue of having a broad flat blade with neither edge sharper than the other. But it clearly does not *explain* why neither edge is sharper than the other to say it is a spatula.

There is a simple conceptual reason why the condition C in virtue of which some state of affairs S holds cannot be explained by S. To cite some condition C in virtue of which S holds is to say what it is for S to hold. We can perhaps say that C *constitutes* S. But to explain C is to mention some other event or state of affairs to which C is related. Usually, this relation is a causal relation: If C* explains C, then C* is – at least very often – a cause of C. But in general the explanation of why C holds must refer to some other state of affairs or event that features in an account of how C came to be. If we are simply given the state of affairs that C constitutes, then C remains unexplained. The difficulty with Liberal Verificationism is that it says our theories are true in virtue of the fact that they would be confirmed. But then the Liberal Verificationist cannot explain this confirmation by saying our theories are true.

In the previous section it was argued that Strict Verificationism cannot explain strong predictive success because it is incompatible with the truth of certain sentences required for that explanation. But we can now observe that there is another difficulty that Strict Verificationism shares with Liberal Verificationism. If Strict Verificationism says that a theory is true in virtue of being verifiable by currently existing techniques, then it cannot explain the empirical success of the theory by saying it is true.

It may be objected that the argument given above against Liberal Verificationism relies on the assumption that the strong predictive success of a theory is to be explained by saying that the theory is *true*, whereas all that may be needed is the assumption that the theory is in some sense close to the truth. However, it is hard to see how this could aid an advocate of an epistemic conception of truth. If truth is defined in epistemic terms, then so must the notion of closeness-to-the-truth.[8] Hence, a verificationist would explicate the claim that a theory is close to the truth in terms of our ability to (perhaps partially) verify the theory. But then the claim that the theory is close to the truth would fail to explain its subsequent confirmation for exactly the same reason as would the claim that the theory is *true* (in the verificationist sense of 'true').

2(c). *Truth as the "Epistemically Ideal" and the Explanation of Strong Predicative Success*

In a number of places, Hilary Putnam has argued that if a theory is *epistemically ideal*, it could not fail to be true. According to Putnam, a theory is epistemically ideal if it is 'complete, consistent . . . predict[s] correctly all observation sentences (as far as we can tell), . . . meet[s] whatever "operational constraints" there are, [and is] "beautiful", "simple", "plausible", etc.'[9] In this section it will be argued that we do not explain the strong predictive success of a theory by saying that it is epistemically ideal.

To explain why a theory T is strongly predictively successful is to explain why it was able to predict successfully some regularity N, instances of which were not used in its formulation. On the face of it, this is not *explained* by saying that T is epistemically ideal. The only part of what it is for a theory to be epistemically ideal that *might* have a role in explaining this is that an epistemically ideal theory correctly predicts *all observation sentences*. So the proposed 'explanation' would amount to: 'T successfully predicted the novel regularity N because T correctly predicts all observation sentences'. Intuitively, that does not seem to be an *explanation* of T's predictive success: It seems merely to give back to us that which requires explanation.

Against this, it may be objected that the proposed explanation is legitimate since it can be represented in deductive-nomological form.

(i): For any theory, *t*, if *t* is true, then a novel prediction made by *t* will be successful.
(ii): T is a true theory that makes novel prediction N.
So: T's prediction of N will be successful.

If truth is interpreted as epistemic ideality, then the generalisation (i) will certainly be true. And (i) and (ii) together deductively entail that which is to be explained. So the proposed explanation has the *logical form* of a deductive-nomological explanation. But if it is actually to be such, the generalisation (i) must be a law. In this section it will be argued that if truth is given a Realist interpretation, then (i) will be lawlike and explanatory, but if truth is interpreted as epistemic ideality, it does not explain the strong predictive success of science.

One test of the lawlikeness of a statement is that S is lawlike if we can infer from the fact that S holds true in actual situations that it would continue to hold true in (at least some) non-actual situations. If this test is applied to (i), and truth is interpreted Realistically, then (i) is lawlike and intuitively seems to be explanatory. To say that a theory is true in the Realist sense is at least to say that it is true *independently* of our forms of perception or of our means of testing theories. A theory that is true in the Realist sense would remain true even if we had different sensory organs, or developed ways of testing theories very different from those we currently possess or will develop in the future. So, such a theory would still be empirically adequate even if we developed sensory organs, or modes of testing, different from those we have or will have. That is, if a theory is true in the realist sense, not only will it actually be empirically adequate, but we may assert that it would still be empirically adequate in possible *counterfactual situations*, such as those in which we developed modes of testing very different from those we actually will develop. Consequently, we may assert that if a theory is true in the Realist sense it has a lawlike tendency to be empirically adequate, and hence that if 'true' is interpreted realistically, (i) will be lawlike. Therefore saying that a theory is true in the Realist sense provides us with an *explanation* of its empirical success. On this view, it is the realist idea of truth being *independent* of our modes of perception or testing that confers upon the Realist notion of truth its explanatory power.

To say that the truth of a theory is independent of all types of test not only provides an explanation of individual events of a theory passing tests, it also provides an explanation of why a theory regularly passes *types* of tests, since if a theory is true in the Realist sense it would continue to be empirically adequate even if we subjected it to *types* of test very different from those we will ever develop.

Let us now consider the question of whether the notion of epistemic ideality provides us with an explanation of strong predictive success. First we should note that, quite plausibly, *some* counterfactual statements are supported by the claim that T is epistemically ideal. Putnam says that a theory is 'epistemically ideal' only if it 'correctly predicts all observation sentences (as far as we can tell)'. Now, the observations predicted by a theory will depend, at least in part, on what means of testing the theory there are. The development of new apparatus and experimental techniques can lead to new means of testing a theory and to new observational predictions made by a theory. Similarly, if we had sensory organs different from those we actually possess, then what would count as an 'observation-sentence' would, quite plausibly, be different from what it actually is, and so the observational predictions made by a theory would be different from what they actually are. Let us initially assume that a theory 'correctly predicts all observation sentences' if it successfully passes all the tests to which we actually have, or actually will subject it. Later we will examine the consequences of rejecting this assumption. To say that a theory passes all the tests to which we actually have or will subject it certainly gives us good reason to suppose that *if we had* performed on the theory tests of the same type as those, but at locations of space and time different from the ones we actually used, the theory would still have passed those tests. So there are, very plausibly, *some* counterfactual statements supported by the claim that a theory is epistemically ideal.[10] But it will be argued that this does not enable the notion of epistemic ideality to explain strong predictive success.

At this point it is worth considering again exactly what it is that needs to be explained. A theory is accepted as being strongly predictively successful, not merely if it correctly predicts one instance of some novel regularity N, but if scientists have become satisfied that the production of instances of N is a *repeatable effect*. So, what needs to be explained is not just the fact that the theory has passed one, or even a number of *instances* of test N, but that the theory has a lawlike tendency to pass tests of this *type*. That is, what needs to be explained is not just a number of individual events, but a regularity. We refer to this *regularity* when we say that a theory's ability to pass a type of test is a repeatable effect.

If T is epistemically ideal then we have good reason to suppose that if it had been subjected to instances of test N at *any* location of space and time, it would have passed those tests. This supports the idea that T has a lawlike tendency to pass instances of N, and this in turn makes it possible to explain *instances* of T passing tests of type N by saying it is epistemically ideal. But this is not sufficient to explain everything that needs to be explained about the strong predictive success of T. On the contrary, it is T's lawlike tendency to pass tests of type N that needs to be explained. The lawlike component implicit in the claim that a theory is epistemically ideal does not *explain* the strong predictive success of the theory; it is rather an expression of what it is that needs to be explained.

We can now see that our original intuitive assessment of the epistemic ideality of truth is correct. Like Liberal Verificationism, it does not *explain* strong predictive success, but merely re-states the very thing that requires explanation.

To say that a theory is true in the realist sense, that is, true independently of all perception or means of verification, supports the claim that it would have a lawlike tendency to pass whatever tests to which it was subjected. In particular, it supports the claim that it would have a lawlike tendency to pass whatever *new* type of test to which it was subjected. So, saying that a theory is true in the realist sense does provide an explanation of strong predictive success.

Earlier on we tentatively defined an epistemically ideal theory as one that passes all tests to which it is actually subjected. We also noted that if a theory passed those tests then it would be reasonable to assert that it would also pass tests of the same type performed at different locations of space and time. But it may be objected that this conception of an epistemically ideal theory is unnecessarily restrictive. In the remainder of this section it will be considered whether a less restrictive definition of 'epistemically ideal' might enable us to explain strong predictive success.

Might an epistemically ideal theory be defined as one that would pass all possible types of test? There are, I think, at least three things wrong with this suggestion.

(1) The notion of all possible types of test is extremely unclear. In the previous section we noted that the observational predictions made by a theory depend at least upon experimental apparatus and the sensory organs of users of the theory. But what would count as a possible type of experimental apparatus, or a possible type of sensory organ? It is doubtful that these ideas are clear enough to aid our understanding of the concept of truth.

(2) On this suggestion a theory would be epistemically ideal iff it would pass all *possible* tests. But there almost certainly are tests which, although possible, could never be performed by us. So defining epistemic ideality in this way has the effect of making truth inaccessible to us. And one of the motivations for adopting an epistemic conception of truth is that it ensures that truth *is* accessible to us. So the proposed definition of epistemic ideality deprives the epistemic conception of truth of one of the features which has been thought to make it attractive.

(3) It is not clear that, as defined, the epistemic definition of truth would be any different from the Realist conception. It was noted in (2)

above that if truth was defined in this way it may transcend verification by us.[11] On this suggestion, a theory is true iff it would pass all possible tests. But might not tests capable of identifying the (Realistically interpreted) truth-value of a theory be *possible*? If such tests are possible this version of the epistemic conception of truth is no different from the realist conception.

Other ways of defining the notion of 'epistemically ideal' might be suggested, For example, it might be suggested that a theory is epistemically ideal if it would pass all physically possible tests, or all practically possible tests. But such definitions have difficulties that are familiar. The notion of a 'physically possible' test, and especially the notion of a *practically* possible' tests, are not clear. The notion of a practically possible test is defined with reference, not only to the laws of physics, but also with reference to facts about human physiology and the availability of resources. But it is not very plausible that our concept of truth should be defined with reference to facts about the availability of resources or the laws of physics.

In any case, there is a more fundamental difficulty with the proposal to define a theory as epistemically ideal iff it would pass all possible types of test – whether those tests are construed as logically, physically or practically possible. *Perhaps* saying that a theory is true in one of these senses would explain the strong predictive success of the theory, but a new problem would arise: *why* would the theory pass all *possible* types of test? That this is something that would very much require explanation can be brought out by considering a simple example. Suppose T predicts that the value of a particular physical constant **k** is 1.06758. Suppose it is also claimed that T is epistemically ideal in the sense of passing all possible types of test. This means that all possible types of test would give the value of **k** as 1.06758. This remarkable agreement across types of test would surely require an explanation. As far as I am aware, no advocate of an epistemic conception of truth has even attempted to provide an explanation of why a theory should pass all types of test.

So, the advocate of an epistemic conception of truth would only acquire the ability to explain the strong predictive success of science by embracing a claim which is at least as much in need of explanation as the phenomena it explains. The Realist, of course, has a simple and natural explanation of the agreement between all possible tests: **k** has its value *independently* of human perception or techniques of verification and this is why all the techniques of verification are led to assign the same value. Of course, one very important problem remains for the Realist: given the underdetermination of theory by actually obtained data, how have we managed to find theories that are true in the Realist sense? However, that is an epistemological problem.[12] It is very different from the ontological or semantic problem: How is the claim that our theories are true to be *interpreted* if it is to enable us to explain their strong predictive success?

CONCLUSIONS

In this paper it has been argued that neither Strict Verificationism nor Liberal Verificationism nor the 'Epistemic Ideality' conception of truth can explain the strong predictive success of science. It has also been argued that a Realist conception of truth, according to which truth is independent of our perception, modes of testing or techniques of verification, is able to explain strong success. Whether this will be regarded as a good argument for a Realist conception of truth may depend on the attitude adopted to abduction as a form of inference. It is beyond the scope of this paper to assess the strength of abductive arguments. However it is possible to construe the argument for Realism from the success of science, not as abductive, but as an instance of *modus tollens* in which the consequent is probabilistically falsified. It should also be noted that a recent study has defended the legitimacy of inference to the best explanation.[13]

Department of Philosophy,
The University of Newcastle,
Australia.

NOTES

[1] Putnam's notion of Metaphysical Realism is developed in much of his writing since 1976. See especially his *Reason, Truth and History* and *The Many Faces of Realism.*

[2] I will not attempt to clarify the notion of strong predictive success, or of a theory making *novel* predictions, in this paper. But a good discussion of the notion, and a defence of its viability, can be found in J. Worrall 'Scientific Discovery and Theory-Confirmation' in J. C. Pitt (ed.) *Change and Progress in Modern Science*, especially pp. 301–331. One author who has emphasised the importance of novel predictions in lending credibility to a theory is Alan Musgrave in his 'The Ultimate Argument for Scientific Realism' in Nola (ed.) *Relativism and Realism in Science*, esp. pp. 229–252.

[3] This claim is argued for in my *Science and the Theory of Rationality* (Aldershot, Avebury, 1991), especially pp. 10–25.

[4] A way of singling out the parts of a theory directly involved in the derivation of a prediction is given in my *Science and the Theory of Rationality* (*loc. cit.*).

[5] I recognise, of course, that the notion of 'closeness to the truth' is an unclear one. But (a) it is not *necessary*, for the argument presented here, to say that theories are close to the truth. Rather, it is claimed that *even if* it is said that theories are in some sense merely 'close to the truth', still some form of the argument presented here would remain sound. (b) The notion of closeness to the truth is one used by many authors including K. Popper, W. H. Newton-Smith and R. Boyd, and may be necessary to give an adequate account of scientific progress. (c) Recent work on the notion of closeness to the truth gives cause for optimism regarding our ability to understand this notion (See G. Oddie, *Likeness to Truth* (Dordrecht: Reidel, 1986) and Ilkka Niiniluoto, *Truthlikeness* (Dordrecht: Reidel, 1987). See also Theo A. F. Kuipers (ed.) *What is Closer-to-the-Truth?* (Amsterdam: Rodopi, 1987) in which a number of approaches to the concept of truthlikeness are discussed.)

[6] See B. S. van Fraassen *The Scientific Image* especially pp. 39–40.

[7] There is reason to suppose that, at least at one stage, Dummett saw anti-Realism as involving Strict Verificationism. For example, in his 'The Social Character of Meaning' he says that the adoption of a technical means of identifying gold would change the sense of 'gold' and it is clear from the context that this might also involve changing its reference. But this means that some sentences of the form 'This is gold' would change their sense, and maybe even their truth value, with the acquisition of a new technique. See Dummett's *Truth and Other Enigmas*, p. 429. Also, Dummett's suggestion that according to anti-Realism, a statement has a truth-value only if *we possess* an effective method for deciding it, suggests Strict Verificationism. However, Dummett also sometimes says that we can display knowledge of what it is for a sentence to be true if we can 'recognise it wherever it obtains, or [get] *ourselves in a position to do so*' (my italics). (See, for example, *Truth and Other Enigmas*, p. 225.) The italicised phrase suggests Dummett would allow that we could know what it is for a statement to be true if we *could acquire* a technique for verifying it. This would make Dummett what I have called a 'Liberal Verificationist' (see below). Dummett's frequent use of the word 'capacity', in his claim that, according to anti-Realism, a statement is true or false if we have a *capacity* to determine whether it is true or false, can perhaps be interpreted as supporting either Strict or Liberal Verificationism.

[8] This is obviously so if closeness to the truth is identified with Popper's *verisimilitude*. Popper's definition of verisimilitude explicitly uses the concept of truth. (See Popper's *Conjectures and Refutations*, p. 392) According to the 'similarity' approach to truthlikeness advocated by Niiniluoto, a claim S is 'truthlike' iff the state of affairs allowed by S is similar to that allowed by a *true* sentence. Similarly, Oddie says that a claim is like the truth if the state of affairs it allows is close to that allowed by a true sentence. So, if true were to be defined in epistemic terms, so would closeness to the truth.

[9] See Putnam's *Meaning and the Moral Sciences*, p. 125.

[10] I owe this observation to Professor Gregory Currie. But we should also note that while the claim that epistemically ideal theories are projectible in this way seems very plausible for theories of physics, it is not so plausible for theories of biology. For example,

that all organisms of a particular type that we have so far observed have immunity to bacterium Y does not lend great support to 'Organisms of that type in some unobserved location also have immunity to bacterium Y'.

[11] In at least some of his writings, Dummett has seen the possibility of verification-transcendent truth as one of the hallmarks of Realism. See, for example, his paper 'Realism' in *Truth and Other Enigmas*. Crispin Wright, in his discussion of Dummett's work, has even seen the possibility of verification-transcendent truth, rather than bivalence, as constituting the core of Dummett's conception of Realism. (See Crispin Wright's paper 'Dummett and Revisionism' in Barry Taylor (ed.) *Michael Dummett: Contributions to Philosophy*, pp. 1–31, especially p. 4.)

[12] I have attempted to give one possible answer to this problem in *Science and the Theory of Rationality*.

[13] For a defence of abductive inference, see P. Lipton *Inference to the Best Explanation*. For the possibility of construing the argument for Realism from the strong predictive success of science as an instance of *modus tollens*, see my *Science and the Theory of Rationality*, pp. 14–15.

BIBLIOGRAPHY

Dummett, M. A. E. (1978). *Truth and Other Enigmas*, Duckworth, London.

Dummett, M. A. E. (1991). *The Logical Basis of Metaphysics*, Duckworth, London.

Fraassen van B. (1980). *The Scientific Image*, Clarendon Press, Oxford.

Kuipers, T. A. F. (ed.) (1987). *What is Closer-to-the-Truth?*, Rodopi, Amsterdam.

Lipton, P. (1991). *Inference to the Best Explanation*, Routledge, London.

Niiniluoto, I. (1986). *Truthlikeness*, Reidel, Dordrecht.

Nola, R. (ed.) (1988). *Relativism and Realism in Science*, Kluwer Academic Publishers, Dordrecht.

Oddie, G. (1980). *Likeness to Truth*, Reidel, Dordrecht.

Pitt, J. C. (ed.) (1985). *Change and Progress in Modern Science*, Reidel, Dordrecht.

Putnam, H. (1978). *Meaning and the Moral Sciences*, Routledge and Kegan Paul, London.

Putnam, H. (1981). *Reason, Truth and History*, Cambridge University Press.

Putnam, H. (1987). *The Many Faces of Realism: The Paul Carus Lectures*, Open Court, La Salle, Illinois.

Taylor, B. (1987). *Michael Dummett: Contributions to Philosophy*, Martinus Nijhoff Publishers, Dordrecht.

Wright J. (1991). *Science and the Theory of Rationality*, Avebury, Aldershot.

WU QIONGBING

IS NATURAL SCIENCE FREE FROM MORALITY?

I. INTRODUCTION

It has been frequently asserted that science is distinguished by its objectivity. Science, real science, deals only with what is, and has no concern with what ought to be. Science is antiseptically devoid of any involvement with human values. Science, in its way of looking at the matter, is so purely objectively and narrowly factual in its concerns that it can, and indeed should, be wholly insensitive to the emotional, artistic, and ethical values of human life. The activities of science are morally and socially value-free and so are the laws and facts of science. As regard the subject of scientific activities, the scientist goes about his work in a rigidly impersonal and unfeeling way, unmoved by any emotion other than the love of knowledge and the delights of discovering the secrets of nature. A scientist makes no value judgement in his whole scientific research, he is someone painstakingly obtaining objective data, testing every side of a question and disregarding personal interests. Although the applications of science may be good or evil, these have nothing to do with science itself. The two-edged sword of science is fashioned for whomsoever will pick it up and wield it, the scientist carries no special responsibility for those applications, save as a normal citizen.

This image of science has gained for itself a distinctive label as the thesis of the value neutrality of science. It has been widely accepted not only by the scientifically uninformed and scientific outsiders, but also by many authorities within the scientific community itself.

The view of the value neutrality of science also has been subjected to quite a lot of attacks. For instance, some people point out that in the whole course of the intellectual history of human beings, from Aristotle and his predecessors to Descartes, Newton, Kant, and Einstein, science has been a part of the cultural tradition in its larger sense. Throughout the whole course of the development of our civilization, science has always merited the historic epithet of "natural philosophy". No matter how much our way of describing the facts may change, there is little doubt that this basic circumstance of the formative role of science in

R. S. Cohen, R. Hilpinen and Qiu Renzong (eds.), Realism and Anti-Realism in the Philosophy of Science, 245–264.

molding all of our areas of thought will remain invariant. Natural science and human or social science, as parts of human culture, are unavoidably inseparable and interdependent, they permeate each other. Some people propose that scientific activity is a kind of human activities, and since human activities are purposeful and goal-oriented, scientific activity inevitably becomes morally and ethically coloured. No matter how the purpose is understood, whether by appealing to the logic of scientific development, or to the needs of society, or to the desires of the employer, the act of cognition is immersed in a value-rich rather than value-neutral atmosphere. Some people sustain the great role of values within science by indicating the intellectual values, such as simplicity, regularity, uniformity, comprehensiveness, systematicity, etc., in scientific knowledge, or by examining the ethical values, such as independence, originality, dissent, tolerance, in scientific activities. Some analysts, on the other hand, try to bridge between "is" sentences and "ought" sentences by language analyses. In our country, many authors criticize the view of value-neutrality of science on the basis of Marxist philosophy, by analyzing the course of the historical development of science, by showing the interaction between science and other dimensions of society, that is, the great role science plays in society and the effects of society imposed on science, etc.

There is something reasonable in all these ideas. However, it is regrettable to see that the view of value-neutrality of science still remains in fashion despite so many diverse criticisms. Why?

II. THE KEY TO THE PROBLEM

In my opinion, this is because most of the criticisms fail to reach the heart of the view. There are various types of values, such as cognitive, aesthetic, theological, political, ethical, economic, etc. In this issue, values mainly refer to social-ethical values concerning rightness and wrongness of conduct. I think that whether science is value free or not primarily relies on the way we understand the concept of science, on how we look at science. So in criticizing the view of the value-neutrality of science, we should first find out the way it understands science. In order to get the nub of the issue, I had better trace back to the origin of this view.

The issue of value neutrality comes into existence with the advent of modern philosophy and science, especially the scientific revolution of the seventeenth century. It is the product of the so-called mechanical picture of the world (especially classical Newtonian science), and the scientific revolution, most especially the epistemological and methodological revolution in science and philosophy inspired by its main architect, Rene Descartes (1596–1650).

The mechanical world-picture regards nature as a vast machine governed by quantitative laws and relationships (nature is written in the language of mathematics); the objective features of the world turn out to be those features – matter, motion, and physical magnitude – which constitute the nuts and bolts of the machine, together with the laws governing it. Only such features of experience are truly objective. Thus a rational methodology for inquiring about the machine's working – i.e., for acquiring knowledge – must take into account only those features which can be quantified, i.e., written in nature's language. The very essence of the world is given by the objective properties just mentioned, together with the mechanical laws which govern them (These essential features of the world are dubbed 'primary qualities' by Galileo and Locke). So, in order to acquire reliable (i.e., 'objective') knowledge of the world by the use of reliable ('rational') methods of inquiry, we must make a sharp distinction between what is objective and what is subjective. Science reflecting the objective world certainly belongs to the category of objectivity, value is subjective, and thus does not belong in an objective account of the world. Science and morality cannot conflict (they are 'complementary') since they have nothing to do with each other: they govern different spheres of experience (those related to the differences between 'man as object' and as actor). Science, which is filled with certainty and characterised by quantitative qualities, is, and must be, value neutral. Value judgements are merely subjective and irrational acts of the will, and scientists should, and indeed must, take no account of anything 'subjective' in their way of arriving at objective truth. Scientists have declared the neutrality of the knowledge of science since the 17th century, and extolled this kind of neutrality as a credit relating to the objectivity of the knowledge of science and to the powerful authority of science itself.

The way we look at science, in my opinion, is the key problem in this issue. Thus before I express my ideas, I would like to examine the concept of science first.

III. OUR CONCEPT OF SCIENCE

What is science? The English word 'science' is derived from the Latin world '*scientia*', which simply means knowledge or intellect. The German word '*Wissenschaft*' means systematic, organized knowledge. Thus, science is generally considered to be theoretical knowledge concerning the natural world both in living and non-living forms. The course of the historical development of science and the new characteristics emerging in modern science have proved that this kind of understanding of science is far from enough to embrace the contents of science. According to the *Modern Science Dictionary* (Franklin Publishing Company, Palisade, New Jersey 1975), science is "A systematic arrangement of scientific data, secured by controlled observation, experience, or experiment, including the method of obtaining and testing these data". So science is not only a kind of orderly knowledge but also a kind of cognitive activities to achieve this product of cognition. In fact, modern science is notably different from that before the early period of the 20th century. Science has been speedily regularized, industrialized and performed on a large scale. Now our cognition in science is still changing, for science has played such a great role in society that many disciplines of metasciences, whose research objects are science itself, emerge: history of science, philosophy of science, sociology of science, psychology of science, etc. These disciplines give science different definitions respectively in terms of their own interests in science – This is determined by the multisidedness of science itself – such as 'an instrument to resolve problems' which focuses on its instrumental property, or 'organized knowledge' in its archival dimension, or 'a series of procedures for obtaining objective knowledge concerning the natural world', in its methodological character, or 'discoveries made by those who own special research abilities', in its professionalism, etc. These definitions only describe some particular characteristics of science from different angles.

In respect to the general definition of science, this is just as J. D. Bernal said, that such important changes in science have taken place in the whole of human history that it is very difficult to give it a suitable definition. The nature of science cannot be fixed by definition forever through our historical research into science. Indeed, while we cannot give science an absolute and perfect definition for science itself is always changing, but this does not mean that we cannot give it a relative one.

Because our understanding of science in any time must have some marks of the developmental characteristics of science in that era, then our science concept should be based not only on the historic analyses of the development of science but also, and it is more important, on the features of our modern science. Unprecedented features have taken place in the development of modern science: 'small science' turns into 'big science', science becomes a profession, scientists are engaged in organized scientific activities and receive remuneration, etc. Bernal pointed out in his work *Science in History* (C. A. Watts, London 1954, pp. 5–6):

> Science may be taken, (I, 1) as an institution; (I, 2) as a method; (I, 3) as a cumulative tradition of knowledge; (I, 4) as a major factor in the maintenance and development of production; and (I, 5) as one of the most powerful influences moulding beliefs and attitudes to the universe and man.

These descriptions synthesize primary respects of science which seem to be different. Here, by making reference to Bernal's descriptions, I try to give science a descriptive definition: Science is: (1) a growing body of knowledge – not knowledge in a sense implying certainty but rather knowledge fallible in principle, and knowledge consists of scientific statements, explanations and theories; (2) above all a special social activity, i.e., research, governed by specific rules or norms of intellectual conduct; (3) a social institution: it is interplay of tradition and rational criticism contained in the network of institutions such as universities, scientific societies, and journals, which constitute the scientific community. The existence of such a community is a precondition for the possibility of that special activity which we call 'research'.

From the previous section, we can see that the value-neutral view derives from the so-called mechanical picture of the world, based on the classical Newtonian science of the 17th century. It rests on the so-called 'conception of small science', which refers to traditional science aimed at the accumulation of knowledge and characteristic of individual free research. The argument that science is ethically neutral originated precisely from this understanding of science. The neutral view essentially identifies science with classical natural science which inclined to mechanism and tried to explain all phenomena and objects in terms of Newtonian physics. As a result, the guidelines of scientific cognition at a certain stage in the development of science (although even at that time their authority was by no means absolute) are misrepresented as specific features of scientific cognition in general, and a conclusion is reached that science inevitably brings a disinterested narrow view of reality and kills creative imagination: the scientist merely describes objectively what happens in his research work.

This comprehension of science is too narrow to be suitable for our present concept of science. It exaggerates some properties of the development of science in the Newtonian period and identifies them as the fixed nature of science, paying no attention to the development of modern science. In fact, today's science becomes more and more dependent on society and the public; science has not been merely the individual preference of the scientist, or a pure search for truth just for the sake of science itself. To start, scientific research must have some value or utilitarian demands, and not merely aim at the accumulation of knowledge. And the solitary scientist laboring in isolation in his study or laboratory has given way to the institutionalized laboratory. The scientist nowadays usually functions not as a detached individual unit, but as part of a group, as a 'member of the team'. Science's research objects are not confined to the physical world, for it is increasingly turning man into an object of research (it is in this sphere that the 'headaches' of today's science are concentrated which may not be understood without socio-ethical analysis). Now the intension and extension of the concept of science have expanded quite a lot, the image of science which the neutral view stems from is far from enough to embrace the present contents of science, for it is just the product of a certain historic period and has become out-dated.

We should not always keep our eyes on the past and pay no attention to the present. Here I would like to emphasize our concept of science, that is, a growing body of knowledge, a special social activity, a social institution, and propose the very problem which will be discussed in detail in the following sections: Is natural science free from morality?

IV. TWO DIMENSIONS OF THE PROBLEM

The relationship of science to value has two dimensions: the issues of the 'value of science' and of 'value in science'. The former takes place in the course of interaction of science and other dimensions of society, concerning the relations between the uses (and abuses) of science and our moral beliefs and principle. The latter happens within science itself, concerning whether the scientist makes value judgements in his scientific research and whether there exist value judgements in the knowledge of science.

It is well known that science as a whole is closely related to human morality, in its interaction with society. The starkest forms appear in the moral dilemmas in nuclear physics, in issues of science and the global problems of human genetics, in freedom of research and the scientist's socio-ethical responsibility, etc. These problems are still under debate. Some authors criticize the neutral view by indicating these widely known facts, but I think that this kind of criticism does little to help undermine the neutral view since the neutral view also acknowledges that science may interact with society: the uses of science may be good or evil, science may be consistent with or conflict with our moral beliefs and principles. The question is: does it insist that scientists carry no special responsibility for the application of science?

That scientists have to be responsible for their research findings or not, in my opinion, relies to a large degree on how to interpret the ethical values. If we understand values in terms of a subjective interpretation (from which the neutral view derives), that values originate from the interests and desires of the subject (human beings), then the scientists have nothing to do with the uses or abuses of their research findings. If, in terms of an objective interpretation that values are the properties of the object, then 'counterculture' seems to gain its rational base, scientists should be responsible for the opening of 'Pandora's box', Einstein should be responsible for the nuclear disasters in Hiroshima and Nagasaki in August 1945. Here I prefer the Marxist definition of value: value is a specific form or manifestation of the relation between subject and object in which the properties of the object are appraised in terms of their ability to satisfy the needs of the subject. The interaction of object and subject is the decisive element in the determination of value, and values are likewise indices of subject-object interactive states. Then the scientists have to undertake, together with society, the socio-ethical responsibility for the uses of science; here the social responsibility of scientists is in itself a condition which, though indispensable, is unable to exclude the possibility of the misuses of scientific findings. The problem can only be resolved by the entire course of progressive social development.

All of the various moral problems of what is done with scientific discoveries are not problems that arise within science, and are not ethical choices confronting the scientist himself. I will not be concerned with them.

In this article I will concentrate on the issue of 'value in science', namely, the ethical problems in science. According to our definitions of science, we may discuss these ethical problems respectively in three dimensions. But I think, the ethical problems in science as a social institution and as a special activity are bound up with society, for the members of scientific institutions and the subjects of scientific activity are scientists themselves; thus these ethical problems are just what they have to face to in their scientific research. The existence of such a scientific institution is a precondition for the possibility of that scientific activity – just like a 'form of life' – in our era. It is difficult (and even impossible) for me to distinguish these ethical problems that arise in scientific institutions from those that arise in scientific activity. Therefore, here I discuss only the ethical dimensions in science *qua* social institution and special social activity as one part, and that in science *qua* systems of knowledge as another part.

V. ETHICAL DIMENSIONS IN SCIENCE *QUA* AN ACTIVITY AND AN INSTITUTION

Science as a special social activity and a social institution is a human enterprise, it is carried out in society, and is performed for and by human beings, and scientific research therefore inevitably exhibits some ethical complexion. Scientists – like other members of distinct professional groups, such as lawyers, doctors, soldiers – do have specific behaviour patterns which constitute the 'rules' and 'norms' of the scientific community. A professional scientist must be familiar with these rules and must be ready to abide by them in practice. The fact is that these rules are not all formerly codified and precisely formulated. R. K. Merton first put forward a set of norms in 1942 which may be expressed as follows:

Communalism: Science is public knowledge, freely available to all. That is to say, the results of research do not belong to individual scientists, but to the world at large. Scientific discoveries should be communicated immediately to the community by publication in the open literature, which anyone may draw upon for their own further use.

Universalism: There are no privileged sources of scientific knowledge. Truth should be judged in terms of intellectual criteria, criteria that are considered valid in the particular branch of science, and not in terms of the attributes – nationality, race, religion, class, age – of the person who produces it.

Disinterestedness: Science is done for its own sake, scientists should undertake their research, and present their results with no other motive than the advancement of knowledge. They should have no personal stake in the acceptance or rejection of any particular scientific idea.

Originality: Science is the discovery of the unknown. That is to say, scientific research results should always be novel. An investigation that adds nothing new to what is already well known and understood makes no contribution to science.

Scepticism: Scientists take nothing on trust. That is to say, scientific knowledge should be tested both empirically and logically, and should not be accepted on the basis of authority.

The Mertonian norms are much debated and are often 'honoured as much in the breach as in the observance'. They describe a set of ideal patterns of behaviour. What the neutral view takes interest in is the norm of disinterestedness, it claims that scientists have by nature, or acquire by nurture, a 'scientific attitude' which makes them peculiarly honest, objective, independent of mind, sceptical, rational, etc. Just as laws exist because of the existence of behaviours which the laws explicitly forbid, there must be some behaviours violating the ideal scheme which the neutral view asserts.

In this section I want to question the 'objective attitude' of scientists which the neutral view describes, that is, that scientists' activities and efforts are directed towards the extension of scientific knowledge, and that regardless of their personal interest, scientists make no value judgement in their research work. On the contrary, scientists have to be confronted with many ethical problems and have to answer them (i.e., make value judgements) in their whole research, just from the initial inception of the research to the ultimate reporting of its completed findings. A scientist is a flesh and blood man instead of a detached and antiseptic one. By examining the ethical problems emerging in the whole cognitive course of science, and through some case analyses, I want to prove that scientific activities in scientific institutions cannot be free from morality; the scientist really has to make value judgements in his research. Ethical problems, that is, issues regarding the rightness and wrongness of conduct – arise out of people's dealing with each other, and pertain necessarily to the duties, rights, and obligations that exist in every kind of interpersonal relationship. The limits of this paper prevent me from discussing in detail all those ethical problems at every junction of scientific research, such as the motives of engaging in scientific work, the choice of research goals, the staffing of research activities, the selection of research methods, the specification of a standard of proof, the evaluation of research findings, the dissemination of scientific information, the allocation of credit for research accomplishments, etc. Here I primarily discuss some ethical problems in the scientists' motives, the evaluation of research findings, the allocation of credit for research accomplishments, which are most acute in science, and just mention roughly those of other junctions.

(1) *Ethical Problems Regarding the Motives of Scientists*

Ethical problems regarding the motives of scientists, in my opinion, are the most critical of all, most of the ethical problems arising in scientific activities are directly or indirectly associated with them. Does the scientist go about his work in a rigidly impersonal and unfeeling way, unmoved by any emotion other than the love of knowledge and the delights of discovering the secrets of nature? Perhaps he would if he were isolated from the external world. In his speech in the celebration meeting of the sixtieth birthday of Max Planck, Einstein proposed that there are mainly four kinds of motives of inquiry (Max Planck: *Where Is Science Going?*, with a preface by Albert Einstein, London, George Allen & Unwin 1933, p. 18):

> Many kinds of men devote themselves to Science, and not all for the sake of Science herself. There are some who come into her temple because it offers them the opportunity to display their particular talents. To this class of men science is a kind of sport in the practice of which they exult, just as an athlete exults in the exercise of his muscular prowess. There is another class of men who come into the temple to make an offering of their brain pulp in the hope of securing a profitable return. These men are scientists only by the chance of some circumstance which offered itself when making a choice of career. If the attending circumstance had been different they might have become politicians or captains of business. . . .
>
> What has led them to devote their lives to the pursuit of science? That question is difficult to answer and could never be answered in a simple categorical way. Personally I am inclined to agree with Schopenhauer in thinking that one of the strongest motives that lead people to give their lives to art and science is the urge to flee from everyday life, with its drab and deadly dullness, and thus to unshackle the chains of one's own transient desires, which supplant one another in an interminable succession so long as the mind is fixed on the horizon of daily environment.
>
> But to this negative motive a positive one must be added. Human nature always has

tried to form for itself a simple and synoptic image of the surrounding world. In doing this it tries to construct a picture which will give some sort of tangible expression to what the human mind sees in nature. That is what the poet does, and the painter, and the speculative philosopher and the natural philosopher, each in his own way. . . .

In Einstein's view, there are many kinds of motives for scientists to devote themselves to science. Some are for the purpose of gaining self-satisfaction from their activities, some are for obtaining profitable returns, some are for fleeing from the dullness of everyday life, some are for constructing a simple and synoptic image of the natural world, etc., which indicate the complication of scientists' motives. In ancient times, science was an individual preference and an activity of a cultured, moneyed and leisure class; the English term 'scientist' did not emerge in Britain until the mid-nineteenth century; the phrase actually used was "a cultivator of science". At that time, the motive of the scientists was more to enjoy themselves in the research than any others. Just as the neutral view declares, such scientists go about their work for the love of knowledge and the delights of discovering the secrets of nature. However, that is far from the only motive of scientists. In modern society, things are more complicated. Scientists are engaged in their work to enhance their social position, or for money, or for patriotism, or for individual interest, or for all of these, etc., which constitute the complex of scientists' motives. For many centuries, one of the strongest motives driving scientists to take up scientific research is, in my opinion, to get the acknowledgement of other people (maybe expressed in other words, such as glory, credit, priority, etc.), which is almost always neglected. For example, the dispute in the history of science between Newton and Leibniz and their followers regarding priority in the invention of the calculus is such a case. Modern scientific research and scientific institutions are sponsored by governments or by industrial or financial magnates, and there are fierce competitions within scientific institution, which impose quite a lot of pressures on scientists. They have to do their best to acquire the research results in order to gain acknowledgement inside and outside science. For that kind of recognition will make them more easy to realize other ideals (i.e. to satisfy other motives).

The complication of scientists' motives leads to many complicated ethical problems. For scientists are flesh and blood men, they have all the general properties of a human being, and it is impossible for all scientists to realize their purposes in a 'normal' manner. In America, the integrity of the scientific research process is seriously challenged by the revelations of a series of misconducts of scientists. Misconduct includes such deviant behaviour as fabrication, falsification, or plagiarism, in proposing, performing, or reporting research. Different types of misconduct can be illustrated roughly as the following (cf. *False Prophets* by Alexander Kohn, 1986):

Forgery: reports experiments that have never been carried out, but which the researcher may feel he needs for the support of his or her hypothesis. In some cases forgers have not only invented experimental data, but have related these non-existent data to non-existent manuscripts allegedly published or accepted for publication.

Plagiarism: consists of using the data or ideas of other investigators without reference to the source, or even verbatim copying of a text written by someone else. A number of plagiarizers obtained the necessary data either from a grant proposal or from a manuscript submitted for publication.

Trimming: which is based on amplification of an experiment: the investigator describes accurately the nature of experiment and its controls, but reports a greater number of trials than have actually been performed, 'adds' or 'removes' animals to or from experimental or control groups, or misrepresents the variance, although using genuine numbers and means.

'Cooking': the researcher omits aberrant values, misreports actual conditions of the experiment or alters ancillary data. A kindred misdemeanour is omission of whole experiments which yielded negative or contrary results to the hypothesis under test.

There are many examples of such 'deviant behavior' in scientific research in which 'objectivity' was lost. Misconduct, regardless of whether conscious or unconscious, was recognized centuries ago, is evident today, and will no doubt continue into the future. I think that such kinds of conduct directly or indirectly result from, or are related to, the complicated motives of scientists, especially their purpose to gain recognition.

A torrid controversy in America, begun in 1989, about the discovery of cold nuclear fusion may be a good case in point (cf. *American Scientist, 80*). The principle of fusion is simple. It is the joining or fusing of two light nuclei (usually with a mass number below eight) to make one larger nucleus, and energy is released in the process. But fusion needs a temperature of about 100 million degrees Celsius to overcome the electrostatic repulsion. Cold fusion, however, requires only room temperature and no other extraordinary condition. It offers the potential for an inexpensive, inexhaustible and clean source of energy and thus received a great deal of attention. The controversy took place between two research groups independently working toward cold nuclear fusion – Electrochemists B. Stanley Pons of the University of Utah and Martin Fleischmann of the University of Southampton in England directed one group (PF group), and Physicist Steven Jones of Brigham Young University directed the other group (J group). Although the basic experiments and the theory they were based on were similar, the two groups obtained radically different results. The PF group detected more heat and no neutron emissions, the J group detected no heat but neutron emissions. In order to win the prestige of being the first to publish results on cold fusion, they all announced their findings hastily to the press before getting strong support or evidence. Therefore, cold fusion was doomed from the start when a race to be first took precedence over the desire to be right.

The analysis of what followed will demonstrate that the scientist is a real person, he has to confront, whether willingly or not, many ethical problems in his whole research process. Even scientists may lose 'objectivity' in the pursuit of truth.

(2) *Ethical Problems Regarding Research Goals*

The problems arise in such a way as to how to allocate limited resources (both human and material) to the prosecution of research efforts. The scientific community has to offer a promising and practicable scheme for the governing of itself. The setting of research goals always has some practical or utilitarian considerations instead of those of pure search for the truths about the natural world.

(3) *Ethical Problems Regarding the Staffing of Research Activities*

These problems relate to the allocation or even appointment of different research duties of scientists in a collective research or institution: Which role best fits a scientist, in just accord with his desire?

(4) *Ethical Problems Regarding Research Methods*

These problems arise perhaps most acutely in biological, medical or psychological experiments involving the use of experimental animals. After the second world war, people were so shocked to find that scientific experiments may be performed in such a terrible and brutal way, when the Nazi war criminals were crossexamined in Nuremberg. Scientific experiments must be carried out subject to be a series of strict rules which apply independent of people's subjective will; however, once scientists turn humans or animals into their actual objects, the arising of ethical problems becomes unavoidable.

(5) *Ethical Problems Regarding Standards of Proof*

The problems arise in this way: At what juncture should scientific evidence be reasonably regarded as strong enough to give warrant for a conclusion, and how would the uncertainties of this conclusion be presented? This relates to the motives and attitudes of scientists. The cold fusion controversy is just such a case.

(6) *Ethical Problems Regarding the Dissemination of Research Finding*

This kind of ethical problems is linked with plagiarism before the publishing of findings in the scientific community: How to deal properly with the circulation of information from private communications.

(7) *Ethical Problems Regarding the Evaluation of Research Findings*

The problems are closely bound up with the boundary line between proper science and pseudo-science. There is no doubt that the scientist has a duty to protect both his own colleagues in other specialties and the lay public against erroneous research findings, and has an obligation to maintain the professional literature of his field at a high level of content and quality. Can the scientist evaluate research findings in a purely objective, wholly rational and entirely open-minded way? This is quite questionable. It is an important phenomenon in the historical development of science: a scientist or the scientific community itself has often opposed genuine scientific findings as being pseudo-scientific on the one hand, and has overrated some strictly fraudulent and pseudo-scientific findings on the other hand. One man's interesting possibility may be another man's pseudo-science, and good ends do not justify questionable means. We know, for instance, the nineteenth-century English chemist J. J. Waterson; his groundbreaking papers on physical chemistry anticipating the development of thermodynamics by more than a generation, were rejected by the referees of the Royal Society for publication in its *Proceedings*, with the comment (among others) that "the paper is nothing but nonsense". As a result, Waterson's work lay forgotten in the archives of the Royal Society until rescued from oblivion by Rayleigh some forty-five years later. For another example, the French physicist Rene Blondlot allegedly discovered 'N-rays', which were supposed to be something like X-rays. This curious finding won him a prize from the French government. But the so-called "N-rays" proved to be just a delusion. If the scientist evaluates a scientific finding with some individual purpose or interest, the cases will become more intricate, and I will not discuss such in detail here.

(8) *Ethical Problems Regarding the Allocation of Credit for Research Accomplishments*

The allocation of credit for research accomplishments is particularly sensitive. The distinguishing feature of modern science is the collectivization and collaboration of research. Collaboration can immensely broaden a person's scientific perspective and advance work far beyond what can be accomplished alone. But it also can generate tensions between individuals and groups. Collaborative situations are far more complex now than they were a generation ago. A notable feature of 'big science' is the development of large-sized experimental instruments whose establishment and use cost dearly. The normal use of the big science apparatus needs the cooperation of a large number of well-trained scientists who play narrow and special roles respectively in a scientific project. For example, every experiment in the field of high energy physics generally needs the collaboration of more than one hundred scientists who are respectively responsible for the design and establishment of the instrument, dealing with the information provided by computers, adjustment of rays of light, supervision of the operation of the experiment, the interpretation of data, etc. Finally, the experimental result is announced by publication of a primary thesis, and each collaborator will seek individual 'acknowledgement' of his contribution to science. Here a delicate problem arises: how to allocate credit for research accomplishments in a fair and reasonable way? The problem refers to the listing of so many authors when the paper is published. Decisions about how credit is to be allotted for these and many other contributions are far from easy and require serious thought and discussion. Some research groups and journals simply list authors alphabetically in order to avoid these knotty decisions.

Another problem in respect to this is the apportioning of credit between junior and senior researchers. The attribution of credit for a new idea or discovery made by a professor and student, or by the student under guidance of his supervisor, or by the student entirely on his own, can not be viewed as a black and white situation. It is common (or even conventional) that a mentor takes away the credit from a student who takes part in the research. The following is just such a case.

Robert A. Millikan, a famous American physicist of the University of Chicago, won the Nobel prize in 1923 for his experiments of 1910 in which he determined the electrical charge of the smallest component of the atom, the electron. The basic idea to measure the charge of the electron on droplets was proposed by Millikan. The original experiment used water droplets, but they evaporated so quickly that observation of their fall was possible for a few seconds only. At this stage of research, a graduate student, Harvey Fletcher, advanced a breakthrough idea to substitute oil droplets for water droplets. Fletcher also made the discovery in his experiment; Millikan, who was away from the laboratory that day only saw Fletcher's set-up the next day, and was very much excited. From then on he continued working closely with Fletcher. They made some improvements in the design of the apparatus and within six weeks the discovery was announced publicly. Millikan published the paper in 1910, and although he gave credit to Fletcher in the text of this paper, he did not let him be a joint author on this first paper, which effectively led to Millikan's being awarded the Nobel prize.

I have illustrated from the above that ethical problems crop up at numerous points within the framework of scientific research, from the initial inception of the work to ultimate reporting of its completed finding. The course of producing scientific knowledge is morality-rich rather than morality-free. But I do not mean that all and every point in scientific activity is morality-laden. In traditional physics or chemical experiments, when the scientist puts on his coat, he must act strictly by the experimental rules; the course, we may say, is value free. In the case when the experimental object is related to animals (including human) or the experiment itself is directly bound up with the health and welfare of human beings, its moral colour is striking.

Here I have just listed some main ethical problems scientists have to face in their actual scientific activities, I have not given some general or universal ethical rules for scientists to abide by, that is, offered answers to these problem, for my article aims to prove that science is morality-laden but not to propose a absolute ethical rule for science. I sincerely agree with the view that the scientist should go about his work in an objective and rational way, but 'what should be' is one thing, and 'what really is' is another one.

VI. ETHICAL DIMENSIONS IN SCIENCE AS KNOWLEDGE

As far as science as knowledge is concerned, we should deal with this problem with caution. According to the neutral view, science aims at the discovery of causes and regularities in the physical world independent of us, and the facts and laws of science are valid irrespective of the nation, race, politics, religion or class position of their discoverer. The velocity of light is the same whoever carries out the experiment which measures it. It is based on the dichotomy of fact and value; scientific knowledge is expressed by factual judgements, that is, sentences concerning 'what is', and on the contrary, morality is expressed by value (evaluative) judgements, namely, sentences concerning 'what ought to be'. There is no logical leap from 'is' to 'ought', nor of course from 'ought' to 'is'. That the world appears to us is irrelevant to the way it really is. Then scientific knowledge must be value neutral.

Just as whether the scientist make value judgements in his scientific activity is the key problem about whether science as an activity and institution is value free, I think, whether there are value judgements in scientific knowledge is the key problem about whether science as a system of knowledge is value neutral. The problem relates to how to interpret the term 'truth', and touches on the everlasting issue between realists and anti-realists. I would not now plunge into such an issue.

It seems that there is no logical problem for the neutralists to assert the neutrality of scientific knowledge. I consent to the view that scientific knowledge (perhaps truly and correctly) reflects nature and its course of movement, its task being to describe and explain what happens in the natural world. However, the defect of the neutral view is, in my opinion, that it fails to recognize the new development of modern science.

The distinguished feature of modern scientific research is its collectivization and collaboration, and correspondingly, with the wide and deep advancement of modern science, a notable feature of scientific knowledge is the emergence of new disciplines, i.e., interdisciplines and borderline disciplines, some of which bridge the natural sciences and social sciences. Scientific research objects are not confined to the physical world; science is increasingly turns man into a object of research. The word 'science' now does not merely refer to the traditional physics, chemistry, astronomy and biology. I think that there are three kinds of research objects of science: (1) the physical world (excluding human being); (2) man as physiological body; (3) the relationship of the physical world to man. I adopt the letters K_1, K_2, K_3 that respectively stand for knowledge about these three objects. There are two levels in scientific knowledge, one which has been historically and fully confirmed by experiments and universally accepted to be the 'correct' descriptions of nature. I represent it as A; and there is one which is still in issue, represented as B. My idea can be illustrated by the following figure:

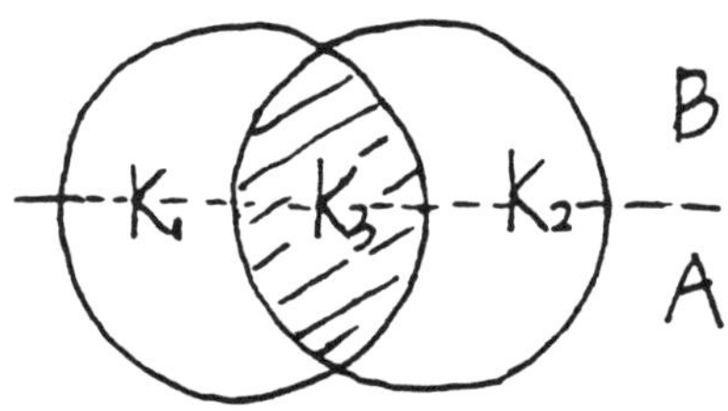

Fig. 1.

I think that K_1 is farthest from ethical values. $K_1 \wedge A$ is believed to truly describe and explain what the physical world really is, and may be value-free, that is, has no value judgement by the scientist. This part embodies the natural laws and theories of traditional physics, chemistry, astronomy, etc. $K_1 \wedge B$ needs to be approached further. There are many hypotheses or explanations or the same natural phenomenon, and which one can be entitled to be called 'true' is still unknown, so I had better keep silent as to it.

K_2 and K_3 are open to ethical questions, especially $K_2 \wedge B$ and

$K_3 \wedge B$. Scientific theory consists of concept, statement, and hypoth-esis. In the body of K_2, there are few terms and statements that are probably inextricably linked with values. In physiology, terms such as 'normal', 'abnormal', and 'deviant' carry value implications. In biology, the term 'adaptation' may also be value-laden; as it is often based on the assumption that survival and propagation are the good results which distinguish adequate from inadequate adaptation. In medicine, for example, the simple statement 'Smoking is harmful to human health' pre-supposes a general value context. Most of the knowledge statements in K_2, such as knowledge which 'truly' describe the internal properties of the human body, are value free. In respect to K_3, I think that the relationship of the human to the physical world can be distinguished by two types, one non-ethical, the other ethical. The former consists, in fact, of the relational properties of the physical world which manifest themselves in the interaction of the human and the physical world. They are beyond human control, but will not exist without human beings, such as colour, taste, voice, etc. Knowledge about this type of non-ethical relationship is value neutral. The latter is the extension of ethical relations of human beings. Traditional ethical relations exist only in the direct interaction of human beings; this type of ethical relation of human and physical world is actual an indirect interaction of human beings, through the medium of the physical world. These ethical relations can be expressed in the following figures.

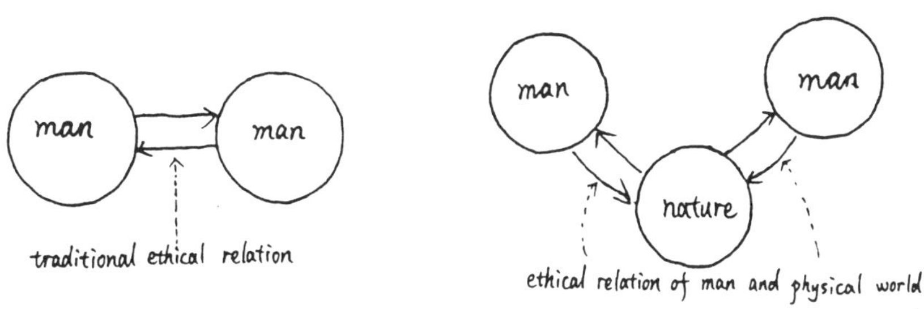

Fig. 2.

It is often asserted that nature should certainly serve man, which brings about many global problems: the overexploiting of nature, the exhaustion of natural resources, the destruction of ecological balance, environmental pollution, etc. The blind exploitation of nature by man leads to the serious punishment of man by nature. The traditional fit and proper exploitation of nature now has to exhibit some ethical complexion. Knowledge of this kind of ethical relationship, perhaps most embodied in ecology, the science of environment, etc. may be value-laden, that is, has some value judgements and implications which tell us what to do or what must not be done something for survival, for the health and welfare of human beings, including future generations.

Department of Philosophy,
Zhongshan University,
Guangzhou, China.

BIBLIOGRAPHY

1. *Introduction Reading in the Philosophy of Science* (1980). Prometheus Books.
2. Loren R. Graham (1981). *Between Science and Values*, Columbia University Press, New York.
3. Steven and Hilary Rose (1986). 'The Myth of the Neutrality of Science', in *Science and Liberation*, Black Rose Books, Montreal.
4. Gerard Radnitzky (1977). 'Science and Values: The Cultural Importance of the Is/Ought Distinction', in *The Search for Absolute Values: Harmony Among the Sciences*, Volume II, International Cultural Foundation Press, New York.
5. J. Bronowski (1956). *Science and Human Values*, Julian Messner, New York.
6. *The Ethics of Science: Issues and Controversies* (1989). English translation of the revised Russian text, Progress Publishers, Moscow.

YIN ZHENGKUN

TRUTH AND FICTION IN SCIENTIFIC THEORY

The debate between scientific realism and anti-realism deals mainly with the problem whether the terms in scientific theory genuinely refer or not, that is, whether the objects referred to by those terms genuinely exist in the objective world. In a sense, this problem may be understood as "Are those objects real entities or fictions?"

Some scientific realists insist that the referents of the terms such as "electron", "gene" are not artificial fictions but real existence. They regard the idea as a basic point of view of their truth theory. As L. Laudan has pointed out,

> a realist would never want to say that a theory was approximately true if its central theoretical terms failed to refer. If there were no entities similar to atoms, no atomic theory could be approximately true; if there were no subatomic particles, no quantum theory of chemistry could be approximately true. In short, a necessary condition – especially for a scientific realist – for a theory being close to the truth is that its central explanatory terms genuinely refer.[1]

On the contrary, anti-realists generally doubt the existence of the referents of those terms. From the logical empiricism to the recent constructive empiricism of van Fraassen, all of them regard the observability as the reason for the theoretical entity is existence, and the unobservable objects referred to by the scientific theoretical terms as "useful fictions" or "convenient theoretical symbols" which the cognitive subject has to introduce to explain the empirical phenomenon. As an empiricist, Quine regards "the conceptual scheme of science as a tool, ultimately, for predicting future experience in the light of past experience." He maintains that

> Physical objects are conceptually imported into the situation as convenient intermediaries – not by definition in term of experience, but simply as irreducible posits comparable, epistemologically, to the gods of Homer . . . the physical objects and the gods differ only in degree and not in kind. Both sorts of entities enter our conception only as cultural posits. The myth of physical objects is epistemologically superior to most in that it has proved more efficacious than other myths as a device for working a manageable structure into the flux of experience.[2]

R. S. Cohen, R. Hilpinen and Qiu Renzong (eds.), Realism and Anti-Realism in the Philosophy of Science, 265–277.

Some post-modernists went so far as to make on attempt to dismantle the line between truth and fiction. According to J. Derrida, there are fictions in the central part of truth because the contradiction between truth and fiction is deconstructed. Another post-modernist, Hilary Lawson, claims. "All our truth are, in a sense fictions – they are stories we choose to believe."[3] But "although truth may be a fiction, it is the most powerful fiction we have."[4]

In order to clarify the controversies listed above, this paper attempts to make further research about the following problems, in the scientific theories, what is fiction? are there fictions or not? what is the function of fiction? what is the relation between truth and fiction?

1. REFERENCE AND FICTION

If a creative activity is the process in which people can produce a new form in which something exists then fiction should also be one of the creative activities of a human being's thinking (when it refers to a process of thinking) In some fields such as in literature and art, fiction is even an important creative activity. But by means of fiction, people only construct some things or events existing not in the real world (or the objective world) but in the ideas (man's mind and language). In other words, the concepts and statements from fiction do not refer to the actual objects and express the actual events. In general, there are fictions in myths and novels. Characters such as Adam, Eve, Prometheus, Othello, Don Quixote, Sherlock Holmes and the stories about them are all fictitious. They were not like the historical characters and events that had existed as the material entities in the real world. But in the history of science there were also many concepts such as 'phlogiston', 'caloric', 'aether' and so on in the scientific theories. Now we know that they do not genuinely refer in the real world, namely, they are fictitious.

However, the meaning of 'refer' is equivocal. As Richard Rorty says:

> The term can mean either (a) a factual relation which holds between an expression and some other portion of reality whether anybody knows it holds or not, or (b) a purely 'intentional' relation which can hold between an expression and a nonexistent object.[5]

He calls the former 'reference' and the latter 'talking about'. He says that we cannot 'refer to' Sherlock Holmes but we can 'talk about' him, and similarly for phlogiston. 'Talking about' is a notion of common sense, and 'reference' is a term of philosophical art. We cannot refer to fictions. Some of the scientific philosophers have remarked the difference too. They call the former 'genuinely refer' so as to avoid confusing it with the other. Evidently Rorty's use of the term 'refer' is clearer than the other. I tend to accept his usage in this paper. But it make no difference whether we use 'refer' or 'genuinely refer', if no one knows whether the object expressed by a term actually exists (the material existence) or not. In this case how can we decide it does not refer to fiction? Therefore I consider that we may claim 'refer' only when it means a relation between a term and a matter that actually exists in the real world; and 'talking about' when it means the intentional relation between a term and a matter that is considered existing, but in fact it may not exist at all in the real world. Thus we can draw a clear distinction between 'refer' and 'talking about', and avoid referring to fiction, when talking about reality and fiction as well.

Nevertheless, the clear use of terms cannot solve the puzzle whether the scientific theoretical terms refer or not. According to Quine's view of 'Ontological commitment', when a person talks about a thing, he has received an ontological judgement. Any scientific theory takes an ontological position, or possesses an ontological premise that admits or refuses the existence of something. But "ontological commitment" does not tell us what there is. And as Quine says,

> but in order to know what a given remark or doctrine, ours or someone else's *says* there is; and this much is quite properly a problem involving language. But what there is is another question.[6]

Hence, "ontological commitment" only commits "the ideal existence" of the theoretical entities talked about in scientific theoretical terms, but not "the material existence", namely, it does not guarantee that the terms refer, or the entities have homologues in the real world. It is probable that a certain theoretical entity is fiction or the relevant theoretical terms refer to fiction before we have sufficient reason to determine the entity "materially existent".

Then how can we determine a scientific theoretical term refers? The anti-realistic empiricists claim that we can do so only by means of 'observability'. Bas van Fraassen says:

> Science aims to give us theories which are empirically adequate; and acceptance of a theory involves a belief only that it is empirically adequate . . . a theory is empirically adequate exactly if what it says about the observable things and events in this world is true.[7]

He especially emphasizes what is observable is what can be directly observed by means of our naked organs of sense. If we can not measure an entity (as a special micro-entity) without any apparatus, or say we must use a cloud chamber or any other apparatus to measure the paths of micro-particles, then the process can not be counted observing but detecting by means of apparatus.

Van Fraassen's point of view really is rather limited. In fact, scientists do not accept his view at work. As Dudley Shapere argues, when citing example of the solar neutrino experiment, "although the central core of the sun lies buried under 400,000 miles of dense, hot opaque material, astrophysicists nevertheless universally speak of the experiment as providing 'direct observation' of that central core."[8] In fact, they only observed the effect produced by the neutrino emitted from the central core of the sun in the apparatus. Most of the scientists tend to consider that, if we can explicitly infer the particular effect produced by a theoretical entity, such as atom, electron, neutrino, etc., under certain experimental conditions from the theory, and really and truly observe that effect under above conditions, especially when we can use the effect to expand our cognition about any other entities, even to make some artificial products in practice (for example, we use the wave character of electrons to make the electron-microscope and observe the structure of a virus with the help of it), then it is reasonable to say the theoretical entity is by no means a fiction, but actually exists in the real material world, and the theoretical term talking about it has reference as well. But if we cannot do so, then however much the theory seems coherent and convincing, scientists would hardly accept it, and they doubt the existence of the entity involved in it. For instance, some of the quantum physicists are suspicious of any 'hidden variables' models because there are no observable effects related to these variables which, if ever measured, could justify/ interdict their introduction. The 'hidden variables' may be fictitious, at least under existing conditions.

If after a series of changes in theory, especially the scientific revolution, there is no better reasons to owe the observable effects to anything else, then the reason that these entities producing the above effects actually exist, i.e. their corresponding terms still refer, is more sufficient and satisfying. For instance, with the continual change of theory, such terms as 'atom', 'electron' could keep their reference in a new one, though the meanings of the terms have undergone a great change, because of the past known effects, and could be inferred from them in the new theory. That may strengthen our faith in the existence of these entities.

According to the above, whether a theoretical term refer or not is determine to a great degree not only in virtue of the empirical test but also the evaluation of the theory system (or the conceptual scheme), even in virtue of its whole background knowledge, What is fictitious in science is not something given once and for all, but evolves along with scientific knowledge. It is possible that a theoretical entity which had been generally regarded as having actual existence for a long time would be fictitious. For example, in the 1780s, when J. Priestley said 'dephlogisticated air' and Lavoiser said 'oxygen', both of them almost depended on the same experiment. 'Dephlogisticated air' did 'refer' in the light of the phlogiston theory in fashion at that time, and was entirely observable. But according to the oxygen theory, oxygen refers and phlogiston is fictitious. Although 'dephlogisticated air' and oxygen in fact talk about the same thing, we cannot think 'dephlogisticated air' has reference because that means phlogiston also exists, which is contrary to the oxygen theory accepted by us. Phlogiston theory was given up not only due to the discovery of oxygen, but also, probably the main reason, due to the consequence of phlogiston having negative weight, which contradicts the mechanistic background knowledge.

Let's see another example; in the 19th c. no one doubted the existence of aether as the aether theory was a commonly accepted background knowledge in science. People thought that a lot of observable effects were caused by aether. Maxwell hypothesized his electromagnetic wave in aether. He regarded the wave as the transmission of aethereal vibration. H. Hertz confirmed the aether by demonstrating the existence of radio waves. A. A. Michelson figured out a way to interact with aether. He thought his experiment confirmed G. G. Stoke's aether drag theory, which was not to deny the existence of aether at that time,[9] and so on. In fact, scientists regarded aether as fictitious only after they had accepted Einstein's theory of relativity as the background knowledge, and all of the above effects had the new explanation.

2. FICTION IN SCIENTIFIC THEORY

Many scientists unconsciously have faith in realism that the scientific theory constructed or accepted by them should be the real description or explanation of matters objectively existing, and all concepts in scientific theory should have one-to-one correspondence with the objective world, instead of being fictions. This point of view is so-called 'metaphysical realism' (Putnam) or 'dogmatic realism' (Heisenberg on Einstein's position). But when we open a book of the history of science, we can find out with surprise that science attempts at any time to actually reflect the present world facing us and is surely to have described and explained the connection between/among the phenomena and their rules in an increasingly extensive, deep, exact and efficient way. On the other hand, there are fictional pictures of the world here and there in the history of science. The scientists unceasingly discard the old fictions and construct new ones. Even in modern science, we cannot get rid of the components which the future scientists will perhaps regard as fictitious and they may need them more and more, including the conjecture of unknown fields, and idealized entities or states to reduce experimental conditions, and mathematical treatment.

Any scientific theory emerges in the form of hypothesis. That is an essential character of science as a creative enterprise. But the hypothesis itself means an incomplete theoretical form, whose fundamental concepts and ideas are to be tested. It is probable that hypothesis involves some theoretical entities and laws which have not undergone experimental test but only exist in the idea, temporarily. Thus, fictions are unavoidable in the hypothesis. When Newton established his theory, he claimed: "Hypotheses non fingo", and made great efforts to establish his theory on hard data, but his theoretical central core involves at least two fundamental concepts: absolute space and absolute time, which modern scientists take for obvious fictions.

It is more important that scientists often have to postulate some unobservable theoretical entities to illustrate, explain or predict the observable empirical phenomena. This is a conjectural and research process, so called "Free creation of thinking". It is reasonable to doubt whether they exist or not, before we have sufficient evidences to confirm them. For example, in modern physics scientists propose a number of new hypotheses involving many new unconfirmed micro-particles so as to unite the known natural fundamental forces. There are various kinds of these micro-particles such as x-particle, super-electron, Higgs boson, etc., and there is a dazzling collection of names only for particles that make up leptons and quarks on the subquark level: subquark, substraton, prequark, preon, rishon, chromon, flavon, familon, alphon, beton, gleak, omegon, even maon for commemorating the great Chinese Mao Zedong who believed that matter can be divided infinitely. Who can assure us none of them is a fiction?

Moreover, some theoretical entities whose existence had been generally accepted in a rather successful theory might ultimately be proved to be fictions by new theory. Scientists predicted a new planet between the sun and Mercury to explain the precession of Mercury's orbit according to Newton's gravitation theory. Owing to the great success of Newton's theory in predicting new planets, such as Nepture (and then they in fact discovered Pluto in 1930 in this manner), they had no doubt of its existence though they had not found it for 85 years. They did not know it is an unnecessary fiction until Einstein proposed his general relativity. Another example is the magnetic monopole. There is up to now no known reliable direct experimental evident for its existence. Is it possibly a fiction? We need more and new experiments, perhaps a new theory, to answer the question.

It is a misunderstanding to think that scientists make fictions as the result of their boundless thinking, straying far from the reality as scientific fancy novelists. But in fact most of known fictions in scientific theories are proposed not due to their brave fancy but to, at least it seems to me, scientists' rather conservative outlook. In 1917, De Sitter proposed a simple cosmological model of an expanding universe, but it was not accepted because it lacked sufficient evidence. Einstein, almost simultaneously, added a special cosmic term to his field equations of general relativity, so as to maintain a picture of a static universe. Later a few scientists discovered the cosmic term was an entirely surplus fiction. It seems that any scientific theory must be established on certain background knowledge including background theory and known empirical data. Scientists, even such great scientists as Newton and Einstein with super-imaginative minds, are still often handicapped by people's cognitive and practical levels at their times. During the 1920s, A. Friedmann and G. Lemaitre discovered expanding universe models based on the field equations of general relativity, but they attracted little attention until E. Hubble's observational evidence was generally accepted in 1929.

There are many similar instances in the history of science. Although chemical revolutionist Lavoisier overturned the phlogiston theory, he could not entirely get rid of the restrictions of the old tradition. There were still fictions in his oxygen theory. He not only retained some alchemical symbols, but also put oxygen in the place of principle, as the acid principle (*Principle oxygine*); he even maintained 'caloric' without weight.

In addition to the above mentioned fictions, scientific theories need some theoretical terms which have evidently no actual correspondents. They discuss the theoretical 'entities' such as 'mass point', 'rigid body', 'point charge', 'ideal gas' and 'states' such as the thermally isolated system, which only exist in some limiting cases and which could not actually appear at all. Shapere calls these terms 'idealization terms' (or idealization concepts) which clearly have fictional characters and are used as tools in the theory, in order to make them different from the so-called 'existence concepts' or 'entity concepts'. After studies of the three cases about the use of 'rigid body', 'point charge' and 'mass point', he points out

> There are often good *scientific* reasons for distinguishing between the way in which a certain entity is asserted to be (or not to be) and the way in which it is *treated* (although, again for scientific reasons, it could not really be that kind of thing) for the sake of convenience in dealing with certain scientific problems.[10]

According to our definition in this paper, 'idealization' is, in fact, a kind of fiction and according to Shapere a fiction with sufficient scientific reasons. Let us look at what are the reasons through Shapere's instance of 'point mass'. Shapere says:

> . . . the rationale for supposing their employment to be both useful and possible is clear. (1) There are certain problems to be solved-problems relating to the positions, velocities, masses, and forces of bodies. (2) Mathematical techniques exist for dealing with such problems if the masses are considered to be concentrated at geometrical points (namely, the geometrical techniques of Newton and, for later scientists, the methods of the calculus). (3) It is, as Newton showed, possible to treat spherically symmetrical bodies as if their masses were concentrated at their centers (and, incidentally, Newton conceived the elementary mass-particles to be spherically symmetrical). And, as for bodies not spherically symmetrical, they could be considered in the same way, provided the distances between their centers was large in comparison to their radii – a condition fulfilled, happily, by the earth-moon system. More generally, bodies could be considered as if their masses were concentrated at their centers of gravity.[11]

Therefore, a term such as 'mass point' does not refer to any entity which actually exists in nature. And it even contradicts the other principles of Newton's mechanics, because Newton explicitly claimed, "We conclude the least particles of all bodies to be also all extended, and hard and impenetrable, and endowed with their proper inertia,"[12] but not the geometrical point. In classical mechanics, an infinite gravitational potential would result from the localization of gravitational mass in a dimensionless point. Hence, such concepts as 'mass point' and similarly 'rigid body', 'point charge' are only artificial abstract tools or reasonable fictions in order to enhance the exact degree of mathematical calculation or to reduce the physical conditions in solving some concrete scientific problems.

From the above argument we can see that scientific theory possesses the characters of abstractive generality and conjectures. Science, as a highly creative enterprise, can not avoid fictions in any time, but especially needs reasonable fictions, i.e. to use the abstractive, idealizative concepts as tools to make our cognition purer and more profound.

3. TRUTH AND FICTION

To many people, fiction is opposite to truth, and equal to falsity or falsehood. It seems that truth could not be established in a case in which fiction appeared simultaneously, and truth should rule out fiction. This is a point of view of metaphysical realism. In fact, fiction may be married with falsity as well as allied with truth.

Fiction does not make up something out of nothing, and is not wild imagination either. Fictions, both in literature and in science, are the results of creative thinking aiming at some rational goals, based on certain present models or empirical data, even on certain theories in science. Fiction in science is the fiction 'Possessing scientific reason'. In literature, the more typical the character the more successful the fiction is. From Hamlet to Holmes, from Don Quixote to Ah Q,[13] all of them demonstrate that. In science, fiction is one of the important ways in which hypotheses should be postulated as well as one of the important means of scientific research. It is because scientists presuppose some entities and their attributes and interrelations (i.e. the laws of motion and interaction among these entities), which have not yet been confirmed in practice, that scientists can be guided to make further experiments and observations, hence the new discovery. Even if these imaginations are indeed false in the final analysis, their anticipations are so successful that we are often justified in accepting them, and make science progress. Sometimes with the development of science, fiction, at present, may yet be reality in the future. The history of the aether's rising and falling offers us a typical case.

The concept of 'aether' appeared almost as early as 'atom'. It had been a mere philosophical fiction with a mysterious colour for a long time. If the concept 'atom' expressed the ancient Greek's imagination of material discreteness, then 'aether' expressed a view of material continuity. 'Aether' entered modern science as 'atom' did with the European scientific renaissance in 17th century. Gilbert, Kepler, Harvey, Descartes and others once introduced it into their own doctrines. But only after Huygens, Hooke and others employed it in the wave theory of light did 'aether' become a quite necessary concept. To people at that time, any wave transmission must be helped by a certain medium vibrating. Since light can go through the boundless intervening space, the space must be full of the medium – aether in every place. They thought aether existed. On the contrary, it seemed quite fantastic that a sort of wave could transmit itself without any medium. Scientists proposed many predictions that could be tested in virtue of aether, which developed not only the wave theory of light, but also the electro-magnetic theory between the 18th century and 19th century. For example, an influential set of non-mechanical aethers, designed to mediate electrical interaction, was proposed by several physicists, including J. Canton and G. B. Beccaria, after 1750. Such a concept assisted H. C. Oersted in the discovery of the magnetic effect of an electrical current. M. Faraday also owed something to this tradition.

But Maxwell and Hertz used aether to advance the electromagnetic theory of light, and on the other hand, their theory required that aether had to be rare enough not to impede the planets and rigid enough to transmit prodigiously rapid transverse vibrations. These conditions were hardly consistent with other background knowledge For example, H. A. Lorentz explicitly freed the aether from obedience to Newton's third law of motion, which placed aether in difficulty. Although aether no longer plays the roles it did in the 19th century due to the theory of relativity now, it remains in physical theory in various guises. There are some similar ideas in Dirac's "vacuum" theory. Dirac said: "If one reexamines the question in the light of present-day knowledge, one finds that the aether is no longer ruled out by relativity, and good reasons can now be advanced for postulating an aether."[14]

From the evolution of aether we can find out that a fictional metaphysical concept could enter science under certain conditions, and it was not only rationally accepted, but also advanced scientific theory in a certain period, and people could obtain some true knowledge with the help of the concept. It is just because of the development of the theory, with the concept as its core, that it's fictional nature was gradually revealed. People can treat it as a completely unnecessary fiction in a new theory (such as relativity), and equate it to falsity. As a result neither can we neglect its value during its long existence in history, nor can we give up research concerning what it talks about in a new light even while aether has lost its meaning in the new theory.

According to the structure of scientific theory, the more mature a theory is, the closer to an axiomatized deductive system it will be. The axiom, the central part of a theory, always consists of some concepts and statements which are self-evident or cannot be directly tested, and it must include some indirectly observable theoretical entities in a non-phenomenalistic theory. We can indirectly test them only by means of the observable effects which we assume to be produced by them. In other words, as long as we discover some laws expressed with mathematical relations between physical concepts and use them to correctly describe, explain and predict some observable phenomena, we can consider this is the result of the interaction between these entities assumed by us. But we assume and name them in virtue of the experience within the limits of the times. In the limits (including the background knowledge and experimental technique), as long as the theory constructed by us is logically self-consistent and not obviously contrary to the background knowledge and, can offer, which is more important, satisfying explanations and successful predictions of empirical phenomena, and even can be used in practical production, we can not find out whether these entities assumed in the theory are fictitious or not. For example, people could use caloric theory in explaining the heat transfer phenomena very well. If they were limited to the realm of heat transfer phenomena, people could hardly find caloric to be fictitious with their background knowledge between the 18th century and the early part of the 19th century. But once knowledge breaks through, they can find out that the natures of the entities assumed by us, and the names, are errors. And they can use other entities which obtain their natures and names in new theories to explain the above mentioned phenomena. At this time the entities past named by us show their fictitious natures. But this does not mean that we have no reason to insist on ontological realism, namely, insist that aforementioned phenomena must be the result of the interaction between certain unobservable entities. Nevertheless, although some 'entities' such as phlogiston, caloric, have exposed their fictitious natures, this is a change of our ideas, a new idea instead of the old one.

Any truth is not abstract, ultimate absolute truth, but a concrete truth. Truth is not the theory which covers and contains everything, but the correct theory which is defined and used in a certain limited realm and under some concrete conditions. No matter what is used, the tested criteria of truth theory are the correspondence theory, the coherence theory, the pragmatic theory or the Marxist practice theory based on the correspondence theory, etc. It can only test or define the concrete truth, or relative truth or approximate truth. We have no way to rid the concrete truth of what the later generations of scientists may regard as fictions. Moreover, I claim that scientific truth should contain not only knowledge of the material structure of entities, but also, more importantly, that of the regularity of the interactions between entities which we may not know yet. Even if someone discovers fiction in a theory so as to claim it erroneous, he can still retain some theorem or laws of the repudiated theory in the form of mathematical functional relations in the new one. In this sense, we can say that the truth content passed on and on to a certain degree.

Moreover, another kind of fiction such as 'mass point', 'rigid body' obviously not only helps us to seek truth, but also is a component of truth with profound epistemological significance.

Therefore if we had the idea that only scientific theory free from fictions is truth, then neither could we have truth in the history of science, nor could we affirm that the present theory is true. This plausible view for maintaining the purity of truth is, indeed, the view of an extreme relativism of denying truth. On the other hand, if one asserts that our present theory has no fiction, he expresses a God's Eye point of view, with dogmatism.

Provided that science is a creative enterprise searching for the unknown, it permits man's making fictions in it; As long as truth must be realized by means of the form of scientific theory, truth and fiction may coexist. But it is because scientific theory involves fictions that it is falsifiable. No scientific theory can be the eternal, ultimate truth.

Department of Philosophy,
Huazhong University of Science and Technology,
Wuhan, China.

NOTES

[1] Laudan, L. (1984). *Science and Values*, Univ. of California Press, Berkeley, pp. 120–121.
[2] Quine, W. V. (1961). *From a Logical Point of View* (2nd edn), Harvard Univ. Press, Cambridge, p. 44.
[3] Lawson, H. and Appignanesi, L. (eds) (1989). *Dismantling Truth*, St. Martin's Press, New York, pp. xxviii.
[4] *Ibid.*, p. 130.
[5] Rorty, R. (1979). *Philosophy and the Mirror of Nature*, Princeton Univ. Press, Oxford, p. 289.
[6] *Ibid.*, p. 16.
[7] Fraassen, B. van (1980). *The Scientific Image*, Oxford Univ. Press, Oxford, p. 12.
[8] Shapere, D. (1984). *Reason and the Search for Knowledge, Boston Studies in the Philosophy of Science*, vol. 78, D. Reidel Publishing Company, Dordrecht, p. 342.
[9] See Lakatos, I. (1978). *The Methodology of Scientific Research Programmes*, Cambridge Univ. Press, London, pp. 73–78.
[10] *Ibid.*, p. 365.
[11] *Ibid.*, pp. 364–365.
[12] Newton, I. (1946). *Mathematical Principles of Natural Philosophy* (Principia), Univ, of California Press, Berkeley, p. 899.
[13] Ah Q is the principal character in a famous novel of a Chinese modern author, Lu Xun.
[14] Dirac P. (1951). Is There an Aether?, *Nature* **168**, 906–907.

FAN DAINIAN

NIELS BOHR AND REALISM

Niels Bohr (1885–1962) was one of the greatest physicists of the 20th century and a rival of Albert Einstein. As an atomic physicist, Niels Bohr was concerned about the function of language as an unambiguous means of communicating experience related to the atomic world, and about the epistemological lessons arising from atomic physics. His main points of view are as follows:

There is "a fundamental limitation in the classical physical ideas when applied to atomic phenomena" (Bohr, 1934, p. 53).

> The quantum postulate attributes to any atomic process an essential discontinuity or rather individuality, completely foreign to the classical theories and symbolized by Planck's quantum of action (Bohr, 1934, p. 53)
>
> From the very nature of the matter, we can only employ probability considerations to predict the occurrence of the individual processes (Bohr, 1934, p. 109).

But Einstein disliked it and said: "I, at any rate, am convinced that He is not playing at dice" (Einstein, 1926).

So, Bohr developed the notion of "complementarity" and wrote:

> The very nature of the quantum theory thus forces us to regard the space-time coordination and the claim of causality, the union of which characterizes the classical theories as complementary but exclusive features of the description, symbolizing the idealization of observation and definition respectively (Bohr, 1934, pp. 54–55).

But, in 1958, he revised his idea and said: "The very fact quantum regularities exclude analysis on classical lines necessitate . . . in the account of experience a logical distinction between measuring instruments and atomic objects, which in principle prevents comprehensive deterministic description." He stressed, in the wider frame of complementarity, he did not claim "any arbitrary renunciation of the ideal of causality" (Bohr, 1963, p. 6).

He also expressed his notion of complementarity as follows:

> We are apparently forced to choose between two mutually contradictory conception of the propagation of light: one, the idea of light waves, the other, the corpuscular view of the theory of light quanta, each conception expressing fundamental aspects of our experience (Bohr, 1934, p. 107).

R. S. Cohen, R. Hilpinen and Qiu Renzong (eds.), Realism and Anti-Realism in the Philosophy of Science, 279–287.

> The problem of the nature of the constituents of matter presents us with an analogous situation. The individuality of the elementary electrical corpuscles is forced upon us by general evidence. Nevertheless, recent experience, above all the discovery of the selective reflection of electron from metal crystals, requires the use of the wave theory superposition principle . . . Just as in the case of light, we have consequently in the question of the nature of matter, so far as we adhere to classical concepts, to face an inevitable dilemma which has to be regarded as the very expression of experimental evidence. In fact, here again we are not dealing with contradictory but with complementary pictures of phenomena, which only together offer a natural generalization of the classical mode of description (Bohr, 1934, p. 56).

Einstein did not like Bohr's idea of complementarity. He called it "the Heisenberg and Bohr soothing philosophy or religion" (Einstein, 1928).

At the 5th and 6th Solvay Congresses held in 1927 and 1930 respectfully, the debate between Einstein and Bohr was developed. Einstein attempted to negate the uncertainty principle and the notion of complementarity with the thought experiments of the double-slit and the photon box, but both were refuted by Bohr (Bohr, 1958, pp. 32–58, Jammer, 1974, pp. 109–158).

In 1935, A. Einstein, B. Podolsky and N. Rosen wrote their paper 'Can quantum-mechanical description of physical reality be considered complete?' (Einstein, 1935) proposing the famous EPR argument, and challenged Bohr's interpretation of quantum mechanics once again. Einstein and his collaborators started from the locality (or separability) assumption and demonstrated the incompleteness of the quantum-mechanical description of physical reality. On the contrary, if the completeness of the quantum-mechanical description were insisted upon, then the locality assumption would be violated and special action at a distance (or telepathy as Einstein called it) would be permitted.

Bohr replied quickly in the *Physical Review* with the same title (Bohr, 1935). Based on the feature of wholeness (or individuality) typical of proper quantum phenomena, in Bohr's view, the completeness of quantum-mechanical description, challenged by Einstein and his collaborators, can be saved.

The EPR incompleteness argument was a big impulse to the revival of hidden variable theories in quantum mechanics. The advocators of local hidden variable theories denied the completeness of the quantum-mechanical description of the state of micro-objects in space-time. In 1964, J. S. Bell, in his paper entitled 'On the Einstein-Podolsky-Rosen paradox' proposed Bell's famous inequality which showed that any local hidden variable theory cannot reproduce all statistical predictions of quantum mechanics. From the early 1970's to the 1980's, about a dozen experiments were carried out and the results obtained seemed to confirm the predictions of quantum mechanics and not those of local hidden variable theories.

From the early 1950's, under the influence of the Soviet ideology, philosophers and physicists in mainland China believed that Bohr is an idealist or a positivist. The situation has changed since the end of the 1970's. In recent years, during the debate about the subjectivity of cognition, one scholar still believes that Niels Bohr is the main representative of the international idealist trend of thought in physics owing to the fact that Bohr had said: in quantum mechanics, "no sharp distinction can be made between the behavior of the objects themselves and their interaction with the measuring instruments" (Bohr, 1958, p. 61) and "the new situation in physics has so forcibly reminded us of the old truth that we are both onlookers and actors in the great drama of existence" (Bohr, 1934, p. 119, He, 1990). He also thinks, "in respect to the interpretation of quantum mechanics, Bohr is wiser than Einstein, but, regretfully, he cannot get the materialist conclusion from the correct physical interpretation (He, 1991). Another scholar believes Bohr is a subjective idealist; and owing to the experimental tests of Bell's inequalities in the 1970's which demonstrated that the Copenhagen Interpretation is correct, that thereby materialism is falsified (Jin, 1988). These two scholars, in spite of the fact that one adheres to materialism and the other does not, both believe Bohr is an idealist or a positivist, believe that Bohr claimed '*esse est percipi*' and "the moon (or the electron) does not exist when nobody looks at (observes) it". And some others think Bohr is a phenomenologist who recognizes only the phenomenal world and not the atomic world.

But, according to Bohr's philosophical writings, I think Bohr is basically a realist, instead of being an idealist or positivist. The reasons are as follows.

First, Bohr did not deny the reality of the external world including atomic objects. He recognized their existence and their objectivity independent of the subject.

For example, early in 1929, he wrote:

In physics, where our problem consists in the coordination of our experience of the external world (Bohr, 1934, p. 1).

Every doubt regarding the reality of atoms has been removed and as we have gained a detailed knowledge even of the inner structure of atoms, we have been reminded in an instructive manner of the natural limitation of our forms of perception (Bohr, 1934, p. 103).

In 1958, 4 years before his death, he pointed out clearly:

It is also essential to remember that all unambiguous information concerning atomic objects is derived from the permanent works – such as a spot of a photographic plate, caused by the impact of an electron – left on the bodies which define the experimental conditions . . . The description of atomic phenomena has in these respect a perfectly objective character, in the sense that no explicit reference is made to any individual observer (Bohr, 1963, p. 3).

According to an article written by the former Soviet academician M. A. Markov, Bohr had said to Professor Sliv clearly: the micro-world, micro-objects, exist independent of our consciousness (Markov, 1990).

According to another Soviet physicist, Fock, in his recollection, during a series of conversations on the philosophic significance of quantum mechanics in the early 1957, Bohr told Fock he was not a positivist and attempted simply to consider nature exactly as it is and he completely recognized the objectivity of atoms and their properties (Graham, 1987, pp. 337–338).

Therefore, imposing viewpoints such as *esse est percipi*, "the moon (or the electron) does not exist when nobody looks at (or observers) it" on Bohr, is groundless. And in my point of view the confirmation of the Copenhagen interpretation of quantum mechanics does not falsify materialism or realism either.

Second, Bohr never denied the distinction between object and subject. He only opposes the mechanical conception of nature which fixed the object-subject separation between the object and the measuring instrument. He only claimed the different placing of such a separation.

For example, Bohr wrote:

While, in the mechanical conception of nature, the subject-object distinction was fixed, room is provided for a wider description through the recognition that the consequent use of our concepts requires different placing of such a separation (Bohr, 1958, pp. 91–92).

He also wrote:

In view of the influence of the mechanical conception of nature on philosophical thinking, it is understandable that one has sometimes seen in the notion of complementarity a reference to the subjective observer, incompatible with the objectivity of scientific description. Of course, in every field of experience we must retain a sharp distinction between the observer and the content of observations, . . . In quantum physics, as we have seen, an account of the functioning of the measuring instruments is indispensable to the definition of phenomena and we must, so-to-speak, distinguish between subject and object in such a way that each single case secures the unambiguous application of the elementary physical concepts used in the description. Far from containing any mysticism foreign to the spirit of science, the notion of complementarity points to the logical conditions for description and comprehension of experience in atomic physics (Bohr, 1958, pp. 90–91).

Bohr's notion of complementarity indeed claimed:

no sharp distinction can be made between the behavior of the [atomic] objects themselves and their interaction with the measuring instruments (Bohr, 1958, p. 61).

But, he pointed out clearly:

The notion of complementarity does in no way involve a departure from our position as detached observers of nature, but must be regarded as the logical expression of our situation as regards objective description in this field of experience (Bohr, 1958, p. 74).

Here, he placed the object-subject separation between the detached observer and the atomic phenomenon which was produced by the interaction between the atomic object and measuring instruments.

Bohr expressed his viewpoint more clearly in his letter to Pauli on March 2, 1955.

A phrase like 'detached observer' . . . used in connection with the phrase 'objective description' . . . had to me a very definite meaning. In all unambiguous accounts it is indeed a primary demand that the separation between the observing subject and the object content of communication (i.e., the phenomenon to be described) is clearly defined and agreed upon . . . this condition is indispensable in all scientific knowledge . . . It appears that what we have really learned in physics is how to eliminate subjective elements in the account of experience, and it is rather this recognition which in turn offers guidance as regards objective description in other fields of science. To my mind this situation is well described by the phrase 'detached observer' (Folse, 1955, p. 215).

Therefore, Bohr used the phrase 'detached observer' and clearly distinguished the object and subject, meanwhile, noticing the dependence of atomic phenomena upon the measuring instruments, or the condition of observation. There is no trend of idealism or positivism. What Bohr had done was to opposed the mechanical conception of nature which improperly extended the subject-object separation in classical physics into atomic physics.

Indeed, according to Bohr's viewpoint, in the account of conscious life, psychical experiences, mental activities, the subject-object separation, is more flexible and the sharp distinction between subject and object is impossible. As Bohr wrote:

For describing our mental activity, we require, on one hand, an objectively given content to be placed to a perceiving subject, while, on the other hand, . . . no sharp separation between object and subject can be maintained, since the perceiving subject also belongs to our mental content. . . . Indeed, strictly speaking, the conscious analysis of any concept stands in a relation of exclusion to its immediate application. The necessity of taking recourse to a complementary . . . mode of description is perhaps most familiar to us from psychological problems (Bohr, 1934, p. 96).

In introspection it is clearly impossible to distinguish sharply between the phenomena themselves and their conscious perception (Bohr, 1958, p. 27).

It must be emphasized that the distinction between subject and object, necessary for unambiguous description, is retained in the way that in every communication containing a reference to ourselves we, so to speak, introduce a new subject which does not appear as part of the content of the communication. It need hardly be stressed that it is just this freedom of choosing the subject-object distinction which provides room for the multifariousness of conscious phenomena and the richness of human life (Bohr, 1958, p. 101).

It is evident, however, that all search for an ultimate subject is at variance with the aim of objective description, which demands the contraposition of subject and object (Bohr, 1963, p. 14).

But, Bohr still insisted:

Of course, in objective physical description no reference is made to the observing subject, while in speaking of conscious experience we say 'we think' or 'we feel' (Bohr, 1963, p. 28).

To the question whether complementarity has introduced a subjective element in physics or not, Bohr answered in his last published paper very succinctly:

This development has essentially clarified the conditions for an objective account in atomic physics, involving the elimination of all subjective judgement (Bohr, 1963, p. 24).

As to Bohr's statement that: "We are both onlookers and actors in the great drama of existence" (Bohr, 1934, p. 119), he just meant that the role of the subjective practice of human beings in changing the objective world should not be neglected. What he opposed was also the mechanical conception of nature.

Third, Bohr is not a phenomenologist who recognized only the phenomenal world and denied the atomic world.

In quantum physics, Bohr believed, the interaction between atomic objects and apparatus

forms an inseparable part of a phenomenon. Accordingly, the unambiguous account of proper quantum phenomena must, in principle, include a description of all relevant features of the experimental arrangement (Bohr, 1963, p. 4).

Bohr really claimed that only these atomic phenomena (i.e., the phenomena produced by the interaction between atomic objects and measuring instruments) can be visualized, can be accounted by the classical terms. But he also claimed, the atomic object can be described by the mathematical formalism of quantum mechanics (Psi-function). This is a kind of theoretical abstraction. He also insisted we use the concepts of radiation in free space and free material particles having no interaction with any apparatus.

And, Bohr wrote:

> On the whole, it would scarcely seem justifiable, in the case of the interaction problem, to demand a visualization by means of an ordinary space-time picture. In fact, all our knowledge concerning the internal properties of atoms is derived from experiments on their reactions or collision reactions, such that the interpretation of experimental facts ultimately depends on the abstractions of radiation in free space, and free material particles. Hence, our whole space-time view of physical phenomena, as well as the definition of energy and momentum, depends ultimately upon these abstractions (Bohr, 1934, p. 77).

Concerning these atomic objects, according to quantum mechanics, we cannot visualize them to form a space-time coordination, but can make probabilistic predictions of the possibilities of the phenomenal appearances of atomic objects under certain measuring instruments (such as corpuscular or wave pictures, space-time coordination or energy-momentum descriptions [claim of deterministic causality]). In view of these, Bohr did not confine himself to the domain of phenomena, he recognized theoretical abstraction, the quantum description of the atomic objects.

In 1953, Max Born had written to Bohr that he "dislikes thoroughly" the instrumentalist standpoint of those physicists and philosophers who hold that the goal of science is a purely observational description in which one does not inquire what there is behind the phenomena." And Bohr "agrees entirely" with Born's position that the mathematical symbols of the formalism which allow us to predict the phenomenal appearances of an atomic object express what we know about this reality behind the phenomena (Folse, 1985, pp. 247–249).

Therefore, the Einstein-Bohr debate is not, as someone thought, a debate between a realist and an idealist, positivist, or phenomenologist, but a debate between a classical, mechanical, deterministic realist with a modern, sophisticated, dialectical realist, a debate between two scholars with different epistemic or conceptual, framework within the realism. Someone thought that Einstein upheld realism and disliked the quantum mechanical description, while Bohr upheld quantum mechanics and abandoned realism. Actually, Bohr upheld realism and quantum mechanics, and insisted on wholeness (or individuality) instead of the separability of the quantum systems.

In 1937, during Bohr's visit to China, Professor Shu Xingbei of the Zhejiang University bluntly asked Bohr's own opinion of the Einstein-Bohr debate. Bohr answered candidly that he was right and Einstein was not (Wang, 1985). From 1970 to the early 1980's, Bohr's viewpoint was supported by most experiments testing the Bell inequalities.

Just as Leon Rosenfeld said,

> Untrammelled by formal schooling, guided only by the sure intuition of the investigator of nature, Bohr rediscovered the dialectical process of cognition which had so long been obscured by unilateral development of epistemology on the basis of Aristotelian logic and Platonic idealism (Rosenfeld, 1979, p. 522).

The epistemic framework of complementarity is a new form of realism with some elements of dialectics and someone think it may have some similarities with some ancient Chinese philosophical doctrines.

A leader of the School of Names, Gong-sun Long (c320B.C.–c250B.C.) expressed his doctrine of the separation of hardness and whiteness, stated that: "There is a hard, white stone. It is hard when you touch it with your hands. It is white when you look at it with your eyes. Whiteness cannot be felt by hands, while hardness cannot be felt by eyes. So, whiteness and hardness can be separated. The properties of matter that can be separated are hidden in the matter. Whether we can discover it or not, depends upon which sense organ we are using." Someone think this doctrine and the Taoist Tao of Yin and Yang seem similar to Bohr's notion of complementarity.

However, hardness and whiteness of a stone, Yin and Yang, coexist and can be observed at the same time, while the two aspects of complementarity do not. This is the essential distinction between Bohr's notion of complementarity and these two ancient Chinese doctrines.

Zhongguancun,
Beijing, 100080,
China.

BIBLIOGRAPHY

Bohr, Niels (1934, 1987). *Atomic Theory and the Description of Nature*, Ox Bow Press, Woodbridge, Conn.

Bohr, Niels (1935). 'Can Quantum-Mechanical Description of Physical Reality be Considered Complete?', *Physical Review* **48**, pp. 696–702.

Bohr, Niels (1955). Letter from Niels Bohr to Wolfgang Pauli, 2 March, 1955, BSC, NBA.

Bohr, Niels (1958). *Atomic Physics and Human Knowledge*, John Wiley, N.Y.

Bohr, Niels (1963). *Essays (1958–1962) on Atomic Physics and Human Knowledge*, Richard Clarn, Bungay, Suffolk.

Einstein, Albert (1926). Letter to Max Born, 4 December, 1926, *The Born-Einstein Letters*, Walker, N.Y., p. 91.

Einstein, Albert (1928). Einstein to Schroedinger, 31 Mai, 1928, *Schroedinger. Planck. Einstein. Lorentz. Briefe zur Wellenmechanik*, Springer, Wien, 1963, p. 29.

Einstein, Albert (1935). 'Can Quantum-Mechanical Description of Physical Reality be Considered Complete?', *Physical Review* **47**, pp. 777–780.

Einstein, Albert (1949). 'Remarks to the Essays Appearing in this Cooperative Volume', *Albert Einstein, Philosopher-Scientist*, Harper & Brothers, N.Y., p. 669.

Folse, Henry J. (1985). *The Philosophy of Niels Bohr – The Framework of Complementarity*, North-Holland, Amsterdam.

Graham, Loren R. (1987). *Science, Philosophy and Human Behavior in the Soviet Union*, Columbia University Press, N.Y.

He, Zuoxiu (1990). *Zhongguo Shehui Kexue* **2**, pp. 69–83.

He, Zuoxiu (1991). *Studies in Dialectic of Nature* **7, 3**, pp. 28–38.

Jammer, Max (1974). *The Philosophy of Quantum Mechanics*, Wiley, N.Y.

Jin, Guantao (1988). *My Philosophical Exploration*, People's Press, Shanghai.

Markov, M. A. (1990). 'Glazami Ochevidcha, *Priroda* **5**, p. 100.

Rosenfeld, Leon (1979). *Selected Papers of Leon Rosenfeld* (Ed. R. S. Cohen and J. Stachel (*Boston Studies Phil. Sci.* vol. 21), Reidel, Dordrecht.

Wang, Gan-chang (1985). 'A Profound Friendship and an Unforgettable Meeting', *Impact of Science on Society* **35, 1**, pp. 55–61.

HENRY J. FOLSE

THE BOHR–EINSTEIN DEBATE AND THE PHILOSOPHERS' DEBATE OVER REALISM VERSUS ANTI-REALISM

Although debates surrounding the interpretation of quantum theory are generally thought to have some bearing on the question of realism, the exact nature of the connection has remained controversial. Thus it is common to find the anti-realist arguing that the success of quantum theory has dealt the death blow to realism, while simultaneously some realists herald the quantum revolution as demanding a reconstruction of the conceptual framework of physics in order to attain a fuller understanding of nature's structure. At the center of this debate on physical reality and the quantum description of nature is the monumental encounter between Bohr and Einstein, certainly by any standard one of the most significant intellectual confrontations in the history of science. Yet the precise relation of this particular debate to the philosopher's discussion of realism remains not surprisingly as obscure and ill-understood as the relationship between quantum theory and realism generally.

In recent years many philosophers have complained that the opposing views in the traditional formulation of the issue of realism versus anti-realism are ill drawn. This suspicion is confirmed by the fact that the attempt to apply the philosophers' preconceived schema to the debate between Bohr and Einstein on the description of microsystems in quantum mechanics distorts that debate. This paper attempts to 'turn the question around' by taking what really was at issue in the discussion between Bohr and Einstein and applying it to the philosophers' debate over realism as a way of illuminating and reformulating what really is at issue.

1. THE REALISM DEBATE AND THE BOHR-EINSTEIN DEBATE

The question of realism was present at the birth of modern science in the Copernican revolution. In his famous anonymous "Introduction to *de Revolutionibus*" Andreas Osiander strikes three distinguishable chords in favor of his anti-realist reading of the "hypotheses" of Copernicus's work. First is the familiar "*epistemological*" theme which disengages

R. S. Cohen, R. Hilpinen and Qiu Renzong (eds.), Realism and Anti-Realism in the Philosophy of Science, 289–298.

"truth or probability" from acceptance.[1] Second, there is the equally familiar "*ontological*" theme that theoretical constructions are computational fictions and not descriptions of real processes.[2] Finally there is the somewhat less discussed "*axiological*" theme that the correct goal of science must be only "to present a correct basis for calculation" and not to "persuade anyone it is so" by constructing pictures or models of "reality."[3] In subsequent episodes throughout the history of science these anti-realist themes have recurred, generally with some selective emphasis on epistemological or ontological issues.

On one level the debate between Einstein and Bohr does reflect these familiar themes. After all Bohr 'accepted' quantum theory and Einstein 'rejected' it. Isn't it the *epistemic* question of the acceptability of quantum theory which was at stake? Bohr is easily cast as arguing an anti-realist line for the acceptability of the quantum description on its ability to save the phenomena. He did indeed argue that we must abandon the attempt to visualize atomic processes and rest content with an abstract 'symbolic' formalism for predicting the probability distribution for a range of possible experimental outcomes. At the same time, Einstein is easily portrayed as defending the realist demand for a description which would represent the quantum mechanical object in terms of properties possessed apart from any observation of them.

Moreover the *ontological* theme of the realism debate surfaces when the disputants disagree over what physical 'reality' can be accorded to the properties of the systems whose behavior quantum mechanics purports to describe. Again Bohr appears on the anti-realist side for he rules out any attempt to describe atomic systems apart from the circumstances in which they are said to be 'observed'. Indeed Bohr does insist that the interpretation of experimental phenomena as measurements of the values of various properties of *observed* objects makes use of stories that picture trajectories of particles moving through space or wave disturbances propagating across a field, but Bohr denies these pictures are pictures of observer independent microsystems and their real behavior in space and time. Einstein appears as the classical realist precisely because he wants to accord an "element of physical reality" to just those *unobserved* properties to which Bohr denies reality.

Philosophers debating realism speak about these issues in terms of 'metaphysical truth' versus mere 'pragmatic utility' or in terms of 'real' existing entities versus purely constructive 'fictions'. This way of posing the issues arises from the philosopher's tendency to see the scientists' acceptance of theories in their ability to lawfully connect groups of empirical phenomena. However, in providing such lawful connections an abstract theory can be interpreted by a model which may enable one to tell a 'story' about how empirical phenomena are produced by the behavior of unobservable 'theoretical' entities, their properties, and interactions. Thus the issue of realism is transformed into the issue of the truth of these stories and the reality of the players in them. On the one hand, for the anti-realist philosopher the criteria for theory acceptance warrant asserting as true only those statements which can be confirmed by empirical phenomena. Whatever role the stories might have in theory acceptance is purely pragmatic and has nothing to do with their possible or probable truth in the metaphysical sense of describing the properties of an objective reality. The realist philosopher, on the other hand, considers the criteria for theory acceptance as providing rational warrant for holding that at least some theoretical statements do truly refer to an objective world which possesses the properties we predicate of it independently of its relationship to empirical phenomena.

However, when we inspect more closely the way in which Bohr and Einstein debated the issues of "theory acceptance" and "physical reality", what we find is that they did not speak about these issues in this typically philosophical way.[4] We find absolutely no direct reference to the 'truth' of theoretical claims, nor would it be helpful to try to 'translate' their debate into a question over the truth conditions for theoretical statements, for like most scientific debates 'truth' is a parameter irrelevant to acceptability. Moreover, philosophers who speak of the realism debate as concerned with the ontological question of the existence of "theoretical entities", are surely not concerned with the same issue that arises in the opposition between Bohr and Einstein, both of whom were clearly convinced realists with respect to the independently real existence of atomic systems.[5] These facts suggest that if we attempt to stage this confrontation on the philosopher's battle plain of realism versus anti-realism we will inevitably misconstrue the problem with which it deals. Therefore, here I propose not to translate the Bohr-Einstein debate into the preconceived terms of the philosophers' debate, but to reflect on what Bohr and Einstein actually debated as a way of formulating a question which philosophers concerned with realism might find of interest. So we turn now to consider what was in fact actually maintained by each disputant.

2. RECENT CHANGES IN THE PERCEPTION OF BOHR AND EINSTEIN

The traditional perception of the Bohr Einstein debate presents Bohr as a phenomenalist who banishes the objective description of 'physical reality' from physics and preaches the instrumentalist message that the goal of physics is to "save the phenomena". Furthermore, the basis of Einstein's objection is often taken to be the indeterminism of quantum mechanics. Nevertheless, if we ignore this anecdotal history of the textbooks and follow the more recent research on Bohr and Einstein, it becomes obvious that Bohr is miscast when made to play the typical philosophical anti-realist's role and that, at least by the time of EPR, determinism was not the basis of Einstein's objection.

In previous publications I have argued that Bohr should be termed a 'realist' because his arguments are based upon what he considered to be a discovery about the real nature of microsystems, namely that they had to be described as changing their classical mechanical states discontinuously.[6] His objection against predicating the classical empirical properties to objects independently of their observation in experimental phenomena is derived from this physical discovery, not from any instrumentalist line that the criteria of theory acceptance warrants accepting only what can be empirically verified as true. Because Bohr speaks of different experiments as revealing complementary phenomenal appearances of *the same physical system*, the framework of complementarity requires we accord some transphenomenal reality to the atomic systems whose complementary phenomenal appearances the theory predicts so well.

As the tense debates between Bohr and Heisenberg in 1927 reveal, the opposing views were the young Heisenberg's truly anti-realist hope to dispense with classical models and rest content with a predictive formalism versus Bohr's view that *understanding* the experimental phenomena which confirm the formalism required using the classical descriptive models to 'interpret' these phenomena as observations of the properties of the systems thereby described.[7] The lawful connections between phenomena are made through the use of particle and wave stories that are expressed in terms of the properties to which the classical concepts refer and which provide the only way we have for interpreting the phenomena as observations in which one determines the measured value of such properties. Bohr tells us we cannot read these stories 'realistically' as visualizable pictures of transphenomenal objects. The physical discovery that microsystems change state discontinuously makes it "ambiguous" to attribute such properties to objects apart from their phenomenal manifestations in observational interactions.

Nevertheless, Bohr nowhere denies reality to the microsystem as the same entity which appears in a 'preparation' phenomenon and a 'detection' phenomenon and is the basis of the story by which we understand (not merely predict) the lawful connection between these phenomena. Moreover he must presuppose the independent reality of the microsystem when he speaks about that which could appear phenomenally as manifesting either complementary wave-like or particle-like properties, depending on the nature of the apparatus with which it interacts to create the experimental phenomenon described as an "observation" of such properties.[8] While this argument for calling Bohr a 'realist' is my own, I would venture to claim that other recent interpretations of Bohr at least agree to the extent that it is scholarly anachronistic to characterize Bohr's outlook as a rejection of the concept of physical reality.

New research on Einstein has also changed the traditional perception of his rejection of quantum theory. The work of Don Howard and Arthur Fine has shown that at least by the time of EPR, the source of Einstein's discontent with the quantum description had little to do with determinism and a lot to do with what is now called "separability".[9] The assumption of the principle of separability, that space-like separated systems can always be assigned separate mechanical states, was for Einstein an ontological presupposition so necessary for individuating distinct systems that he could not conceive abandoning it. Assuming separability, the EPR argument leads to the conclusion that physical reality contains elements which quantum theory, on Bohr's interpretation, prohibits attributing to the system, thus justifying rejection of quantum theory's 'completeness'. But as Bell's work has shown us, separability all by itself has empirical consequences which differ from quantum theory, and here the evidence seems to tell against separability in favor of the quantum description.[10]

Thus in the perception of more recent research, the difference between Bohr and Einstein consists in this: for Einstein, an ontological presupposition, essential for individuation, constrains what can be accepted as a 'complete' description of physical systems, leading to his rejection of quantum mechanics as complete. For Bohr, the discontinuous change of state of a microsystem in an interaction implies that predicating the classical state defining properties to an object independently of observation is 'ambiguous' and hence cannot be the basis of an objective description. For Einstein ontological constraints determine the acceptability of empirically based theory; while for Bohr, the conditions governing the well-defined predication of empirically defined properties constrain permissible ontological presuppositions regarding the objects to which such properties are predicated.

3. THE AXIOLOGY OF INQUIRY AND THE BOHR EINSTEIN DEBATE

The Bohr Einstein debate may be seen as concerned with the axiology of atomic physics.[11] Though their discussion *prima facie* concerned the proper description of the familiar *gedankenexperiments*, their different views reflected a deep division over whether or not the aim of atomic physics had been achieved, expressed in terms of the question "Can Quantum Mechanical Description of Physical Reality be Considered Complete?" Their opposing answers were based on a disagreement over the necessary conditions for predicating properties to the systems described by atomic physics. We can understand why this disagreement arose by considering how the physicist's philosophy of nature, *i.e.*, one's view of the relationship between ontology and empirical science, affects the choice of goals and values. Though they finally agreed that the theory was 'successful' in attaining the goal of empirical adequacy, their different views in the philosophy of nature led them to disagree over its final acceptability.

Here I want to alter our perception of the debate by stressing that they held in common the typically realist goal of securing 'harmony', 'unity', and 'consistency' in the conceptual framework physicists employ for describing phenomena. An anti-realist needs pursue no goal other than empirical adequacy. Both of our disputants of course valued predictive success, but for neither was this value paramount. Einstein's confidence in general relativity apart from its successful predictions is well known, and Bohr's opposition to Heisenberg reveals he accorded much higher value to *understanding the phenomena* than merely getting the numbers right. Bohr and Einstein were both involved in the philosophy of nature, and accepted the typically realist goals of *understanding*, as opposed to merely *predicting*, phenomena. In order to secure the goal of a harmonious understanding of natural phenomena both were prepared to reweave the conceptual nets by which the natural philosopher working in the empirical tradition of physics hopes to capture nature. Indeed if Bohr and Einstein were as far apart in their goals as are the realist and the anti-realist, it becomes astonishing that Bohr insists on arguing that the conceptual revolution required in quantum physics is analogous to the one Einstein precipitated with relativity.[12]

Perhaps more remarkable than the fact that both physicists held values that typify a realist axiology is the fact that *both men also shared another goal which typically motivates the anti-realist*. In seeking to fashion new perspectives for the description of nature Bohr and Einstein both followed the mathematician's drive to reduce fundamental assumptions to the simplest set by means of pruning claims to the truth of excess metaphysical assumptions. For Einstein this was the presupposition of absolute space-time as the locus of the ether; for Bohr it was the assumption that an independent physical reality can be described in terms of properties which can be predicated unambiguously only of systems observed in physical interactions.

Consequently, while it is certainly correct to say that each physicist held to a different standard for a complete description, and hence had different goals for atomic physics, it is misleading to say that one's goals were 'realist' goals and the other's those of an 'anti-realist'. A more revealing description would point to their different positions regarding the relation of ontology to empirical knowledge. The primacy Einstein accorded separability was a result of his conviction that only in this way could physics individuate and accord separate identity to the physical systems it describes. Thus in Einstein's philosophy of nature, ontological commitments constrain the goals of empirical knowledge, and failure to attain those goals reflects negatively on the final acceptability of any theory which aspires to empirical knowledge. For Bohr the value accorded accepting an empirically adequate theory capable of describing all possible phenomena forces even our most deeply held ontological commitments to submit to revision.[13] But while empirical adequacy may well be the ultimate anti-realist value, Bohr does not take the pursuit of this aim to imply abandoning all ontological commitments. Instead, for him it clearly implies the need to *revise* ontological commitments in order to accord with what attaining an empirically adequate theory requires us to assume about the phenomena, namely that they arise in the interactions between observing instruments and independently real, physical microsystems which possess some identity between their phenomenal appearances.

However, even though complementarity must make reference to the microsystem apart from its phenomenal appearances, Bohr fails to explain how we are to individuate such systems after abandoning separability. Indeed, while Bohr certainly wants to accord reality to atomic systems apart from our observations of them, his reluctance to engage in ontological reconstruction delimits an unfinished sector of the framework of complementarity. Einstein's values make his ontological commitments clear, but he is left without any acceptable theory, even though he has to admit reluctantly that an empirically adequate one is at hand. Bohr can lay claim to an acceptable theory, but the ontological implications of complementarity have yet to be made clear.

The Bohr Einstein debate arises from the need of physics to attain two goals: one is the acceptability of a theory which, as it appears, is in accord with all possible empirical evidence; the other is the demand for a description which is 'complete' against a standard posed by prior ontological presuppositions. Different physicists may prioritize the value of attaining these goals differently, and thus be said to adhere to different axiologies, but this is not the same as debating realist versus anti-realist axiologies, or at least not the way philosophers have debated it. How are we to decide, when these goals conflict, which is to take priority, and which is to be refashioned? What reason do physicists give for the preferences they display? Here perhaps philosophers can find an issue which could be discussed more profitably than the way in which realism and anti-realism traditionally have been counterposed.

Department of Philosophy,
Loyola University,
New Orleans,
U.S.A.

NOTES

[1] Osiander warns the reader that "... it is not necessary that his [Copernicus's] hypotheses be true, nor even probable. This alone suffices that they provide a computation that tallies with the observations" (Gingerich (1975), p. 301).

[2] Osiander addresses specifically those "ignorant of geometry and optics" who might "perchance" think the "epicycle of Venus as likely". He also actually argues that other

observations than those the epicycle is conjectured to predict rule out the reality of any such motion. *Ibid.*, pp. 301–304.

[3] Osiander makes it clear that anyone who mistakenly "takes models – with their own purpose – to stand for reality . . . leaves the discipline more ignorant than before he entered it". *Ibid.*, p. 304.

[4] When reading Bohr's account of these discussions, one cannot help feeling that this bears little or no resemblance to the philosopher's debate over realism; *cf.* Bohr (1949).

[5] No one doubts Einstein's realism, but many seem to forget that Bohr clearly proclaimed that "every doubt regarding the reality of atoms has been removed", and that "we have gained a detailed knowledge of the inner structure of atoms", including not only the behavior of aggregates of atoms, but also of single atoms (Bohr (1958), p. 103). For further analysis of this theme see Folse (1985), pp. 224–227; and (1987), pp. 161–167.

[6] See Folse (1985), (1986a), (1986b), (1987), and (1989b).

[7] Heisenberg (1949), pp. 103–107; and (1963), pp. 13–17.

[8] As he does, for example, at Bohr (1958), p. 26.

[9] Howard (1985), and Fine (1986), pp. 26–39. I have discussed the connection between separability and Bohr's response to EPR in Folse (1988) and (1989a).

[10] This theme appears in the papers of Howard, Wessels, Shimony, and Teller in Cushing and McMullin (eds.) (1989).

[11] I have analyzed the debate from this point of view in Folse (1990), for which I am much indebted to work in the NEH Summer Seminar conducted by Larry Laudan in 1989; see Laudan (1984) for his views on the axiology of inquiry.

[12] One might object that this is only a strategic move Bohr makes *ad hominem* for the sake of convincing Einstein; however, Bohr strikes this theme from the beginning of his arguments for complementarity, well before Einstein's opposition could have been considered 'entrenched'. The fact that Bohr clearly believed that his reasoning in many ways paralleled Einstein's made his failure to convert Einstein all the more frustrating, and this was felt as especially frustrating just because they held many goals in common.

[13] Bohr makes his values obvious when, commenting on their disagreement, he notes that "In my opinion, there could be no other way to deem a logically consistent formalism as inadequate than by demonstrating the departure of its consequences from experience or by proving that its predictions did not exhaust the possibilities of observation, and Einstein's argumentation could be directed to neither of these ends". Bohr (1949), p. 229.

BIBLIOGRAPHY

Bohr, Niels (1935). 'Can Quantum Mechanical Description of Physical Reality Be Considered Complete?', *Physical Review* **38**, pp. 696–702.

Bohr, Niels (1949). 'Discussion with Einstein on Epistemological Problems in Atomic Physics', in P. A. Schilpp (ed.), *Albert Einstein: Philosopher-Scientist*, Open Court, LaSalle, Illinois, pp. 199–241.

Bohr, Niels (1958). *Atomic Physics and Human Knowledge*, J. Wiley, New York.

Cushing, James T. and McMullin, Ernan (eds.) (1989). *Philosophical Consequences of Quantum Theory*, University of Notre Dame Press, Notre Dame.

Fine, Arthur (1986). *The Shaky Game: Einstein, Realism and the Quantum Theory*, The University of Chicago Press, Chicago.

Folse, Henry (1985). *The Philosophy of Niels Bohr: The Framework of Complementarity*, North Holland Physics Publishing, Amsterdam.

Folse, Henry (1986a). 'Niels Bohr, Complementarity, and Realism', in A. Fine and P. Machamer (eds.), *PSA 1986: Proceedings of the Biennial Meeting of the Philosophy of Science Association, Vol. I*, Philosophy of Science Association, East Lansing, Michigan, pp. 96–104.

Folse, Henry (1986b). 'Complementarity and Scientific Realism', in P. Weingartner and G. Dorn (eds.), *Foundations of Physics*, Holder-Pichler-Tempsky, Vienna, pp. 93–101.

Folse, Henry (1987). 'Realism and the Quantum Revolution', in *Abstracts of the 8th International Congress of Logic, Methodology, and Philosophy of Science, Vol. 4, Part I*, Institute of Philosophy of the Academy of Sciences of the USSR, Moscow, pp. 199–200.

Folse, Henry (1988). 'Niels Bohr's Concept of Reality', in P. J. Lahti and P. Mittelstaedt (eds.), *Proceedings of the Symposium on the Foundations of Modern Physics: The Copenhagen Interpretation 60 Years After the Como Lecture – Joensuu, Finland, August 6–8, 1987*, World Scientific Publishing, Singapore, pp. 161–179.

Folse, Henry (1989a). 'Bohr on Bell', in Cushing and McMullin (eds.) (1989), pp. 254–271.

Folse, Henry (1989b). 'Complementarity and Space-Time Description', in Menas Kafatos (ed.), *Bell's Theorem, Quantum Theory and Conceptions of the Universe*, Kluwer Academic Publishers, Dordrecht, pp. 251–259.

Folse, Henry (1990). 'Laudan's Model of Axiological Change and the Bohr-Einstein Debate', in Fine, A. *et al.* (eds.), *PSA 90, Vol. 1*, The Philosophy of Science Association, East Lansing, pp. 77–88.

Folse, Henry (1992). 'Complementarity and our Knowledge of Nature', in M. Carvallo (ed.), *Nature, Cognition, and System, Vol. 2*, Kluwer Academic Publishers, Dordrecht, pp. 51–66.

Gingerich, Owen (ed.) (1975). *The Nature of Scientific Discovery*, Smithsonian Institution Press, Washington, D.C.

Heisenberg, Werner (1949). 'Quantum Theory and Its Interpretation', in S. Rozental (ed.), *Niels Bohr: His Life and Work as Seen by his Friends and Colleagues*, North-Holland, Amsterdam, pp. 94–108.

Heisenberg, Werner (1963). Interview with Thomas Kuhn, Munich, 25 February, 1963, in *Archive for the History of Quantum Physics*, Tape 52a.

Howard, Don (1985). 'Einstein on Locality and Separability', *Studies in History and Philosophy of Science* **16**, 171–201.

Laudan, Larry (1984). *Science and Values*, University of California Press, Berkeley.

HE ZUOXIU

ON THE EINSTEIN, PODOLSKY AND ROSEN PARADOX AND THE RELEVANT PHILOSOPHICAL PROBLEMS

I. INTRODUCTION

The controversy on "Can the Quantum Mechanical Description of Physical Reality be Considered Complete?" was everlasting for almost sixty years [1, 2]. A clarification was made by J. Bell [3], who showed that any local, deterministic theory leads to a relation called the Bell's inequality, whereas the quantum mechanics violates the inequality. From 1972–1982, different kinds of experiments performed by experimentalists, gave results that indicated the violation of the Bell's inequality and were in agreement with quantum mechanics [4]. That the Bell inequality is violated now is generally accepted.

In 1979, B. d'Espagnat gave a detailed analysis of the philosophical problems pertinent to this topic. Finally, he drew the conclusion, the doctrine that the world is made up of objects whose existence is independent of human consciousness turns out to be in conflict with quantum mechanics and with facts established by experiment [5]. Similarly N. D. Mermin gave an alternative representation of this point of view, "we now know that the moon is demonstrably not there when nobody looks" [6].

Are their conclusions wrong or correct?

II. D'ESPAGNAT'S ARGUMENTS

B. d'Espagnat's arguments can be expressed by the following sketch, which is taken from "The Quantum Theory and Reality" [5].

He showed that local realistic theories originate from three premises: "One is realism, the doctrine that regularities in observed phenomena are caused by some physical reality whose existence is independent of human observers. The second premise holds that inductive inference is a valid mode of reasoning and can be applied freely, so that legitimate conclusions can be drawn from consistent observations. The third premise is called Einstein separability or Einstein locality and it states

R. S. Cohen, R. Hilpinen and Qiu Renzong (eds.), Realism and Anti-Realism in the Philosophy of Science, 299–305.

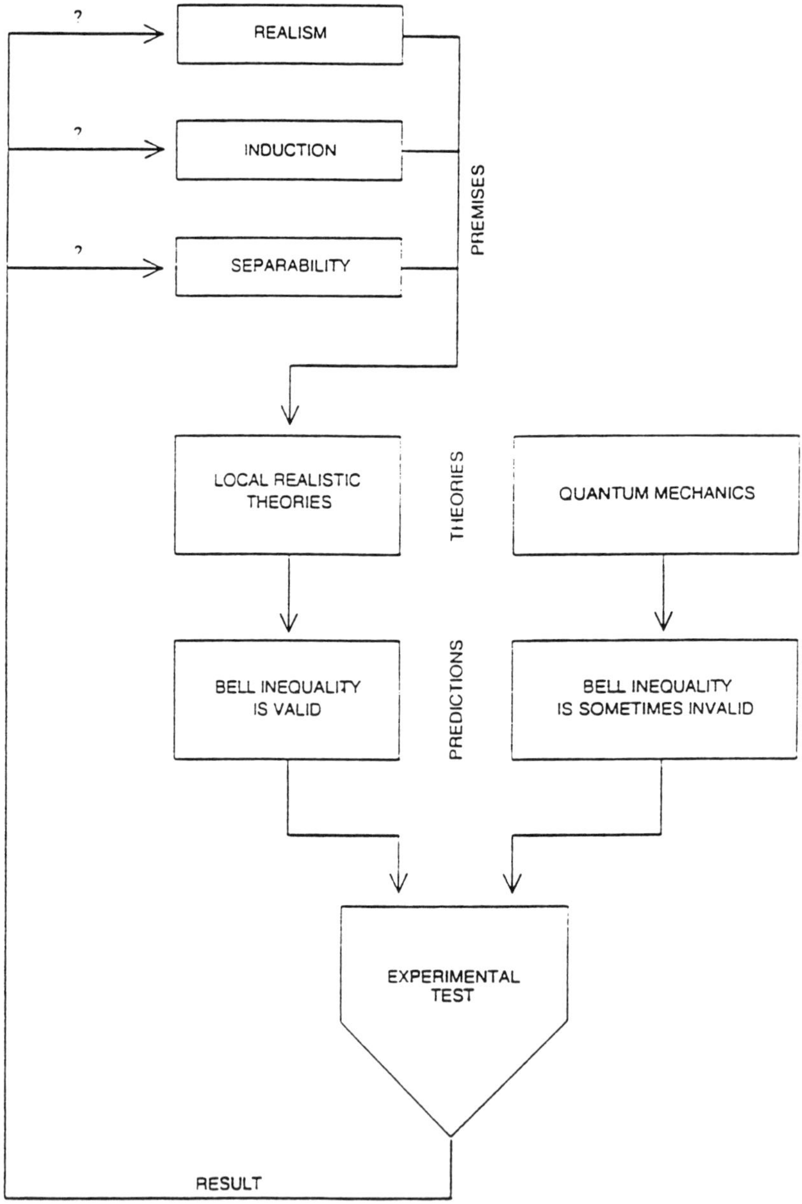
?
REALISM
?
INDUCTION
?
SEPARABILITY
PREMISES
LOCAL REALISTIC THEORIES
THEORIES
QUANTUM MECHANICS
BELL INEQUALITY IS VALID
PREDICTIONS
BELL INEQUALITY IS SOMETIMES INVALID
EXPERIMENTAL TEST
RESULT

Fig. 1.

that no influence of any kind can propagate faster than the speed of light" [5].

In short, these premises can be identified as: (1) realism, (2) the free use of induction and (3) separability.

Since the logical consequence of the local realistic theory is the Bell inequality, and there is strong experimental evidence that the inequality is violated, thus, at least one of the three premises of local realistic theories must be false.

d'Espagnat argued, that (1) "in a physics experiment, the separability assumption expressed the intuitively reasonable idea that the spin components of one proton have no influence over these of the other proton, provided the two particles are far apart, . . . this assumption must now be regarded as highly questionable". (2) If an unbiased and large amount of samples were tested, "the confidence of these assertions approaches certainty as the size of the sample increases". Hence, he leads to the conclusion: the "quantum mechanics curiously disagrees with the doctrine that the world is independent of mind".

III. D'ESPAGNAT'S LOOPHOLE

However, if we examine the proof of Bell's inequality more carefully, the assumption of realism really is one of the premises of a local realistic theory, but this premise is only a *special form* of realism, the deterministic realism, i.e. the existence of a hidden parameter. So that the violation of Bell's inequality can not be regarded as a violation of realism *in general*, e.g. a general statement, such as "disagreeing with the doctrine that the world is independent of mind"!

IV. REALISM IS ONE OF THE PREMISES OF QUANTUM MECHANICS

In this history of philosophy, there were different definitions about realism. What we shall adopt, is

Realism = the existence of world is independent of mind.

In order to show that the realism is one of the premises of the quantum mechanics also, we shall analyze the following example, the spin correlation of a pair of electrons or protons, in detail.

(i) Assume we have some composite system, composed of the electrons or protons A and B, with total angular momentum equal to zero.

(ii) The spin correlation function $E(\vec{a},\vec{b})$ is defined as

$$(1) \qquad E(\vec{a},\vec{b}) \equiv \overline{4A_a \cdot B_b},$$

where A_a and B_b are the spin component of particles A and B along the direction of unit vectors $\vec{a}$ and $\vec{b}$.

(iii) According to quantum mechanics, one easily shows, that

$$(2) \qquad E(\vec{a},\vec{b}) = \langle 0^+|\vec{\sigma}_A \cdot \vec{a}\ \vec{\sigma}_B \cdot \vec{b}|0^+\rangle,$$

in which $\vec{\sigma}_A$ and $\vec{\sigma}_B$ · are the Pauli spin operators of the particules A and B respectively, and the wave function $|0^+\rangle$ is

$$(3) \qquad |0^+\rangle = \frac{1}{\sqrt{2}}\left(\psi_{+\frac{1}{2}}(A)\psi_{-\frac{1}{2}}(B) - \psi_{-\frac{1}{2}}(A)\psi_{+\frac{1}{2}}(B)\right)$$

and the ψ's are the spin wave functions.

(iv) Direct calculation shows

$$(4) \qquad E(\vec{a},\vec{b}) = -\vec{a}\cdot\vec{b}$$

However, J. Bell started from local and deterministic theory, i.e., existence of some hidden parameter, he obtained the relation

$$(5) \qquad |E(\vec{a},\vec{b}) - E(\vec{a},\vec{c})| \le 1 + E(\vec{b},\vec{c})$$

in somewhat different form from quantum mechanics.

(i) If one lets $\vec{a} = \vec{b} = \vec{c} = \vec{n}$ = unit vector, then

$$(6) \qquad E(\vec{n},\vec{n}) \ge -1,$$

which is compatible with quantum mechanics.

(ii) If the angles lie between $\vec{a}$ and $\vec{b}$, $\vec{b}$ and $\vec{c}$ is 60°, and angle lies between $\vec{a}$ and $\vec{c}$ is 120°, then according to quantum mechanics, one has $E(\vec{a},\vec{c}) = -\frac{1}{2}$, $E(\vec{a},\vec{b}) = E(\vec{b},\vec{c}) = \frac{1}{2}$, substitute them to the Bell's inequality, one has

$$(7) \qquad \left|\frac{1}{2}+\frac{1}{2}\right| \le 1 - \frac{1}{2} \Rightarrow 1 \le \frac{1}{2}\ !?$$

Now we trace back to the procedure of the derivation of the spin correlation. One can find that the spin correlation $E(\vec{a},\vec{b}) = -\vec{a}\cdot\vec{b}$ is essentially originated from the special form of spin wave function

$$(8) \qquad |0^+\rangle = \frac{1}{\sqrt{2}}\left(\psi_{+\frac{1}{2}}(A)\psi_{-\frac{1}{2}}(B) - \psi_{-\frac{1}{2}}(A)\psi_{+\frac{1}{2}}(B)\right)$$

where the spin component $+\frac{1}{2}$ of particle A is always combined with the spin component $-\frac{1}{2}$ of the particle B and vice versa, i.e. the $(+\frac{1}{2}) + (-\frac{1}{2}) = (-\frac{1}{2}) + (+\frac{1}{2}) \equiv 0$, because of law of conservation of angular momentum.

All these are objective, independent on the human mind, it could exist, even far before the appearance of the mankind.

V. THE ROLE OF THE APPARATUS

It should be clarified, whether the spin correlation is arised or originated from the interaction between the particles and apparatus or not. This is related to the roles of apparatus in the theory of quantum mechanics.

As the theorem of measurement shows: the physical quantity $\bar{L}$ of certain particle system is equal to

$$\bar{L} = \sum_i |c_i|^2 L_i, \tag{9}$$

where L is the corresponding operator of the physical quantity, L_i is the eigenvalue of the operator L, with the eigen function $\psi_i(x)$, i.e.,

$$L\psi_i(x) = L_i\psi_i(x), \tag{10}$$

the c_i is the coefficient of expansion of the wave packet $\psi(x)$ in the basis of the eigenfunction $\psi_i(x)$, i.e.

$$\psi(x) = \sum_i c_i\psi_i(x), \tag{11}$$

The value of L_i can be measured and obtained by some apparatus, while the $|c_i|^2$ is the transition probability, i.e. the wave packet collapse to

$$\psi(x) \Rightarrow \begin{cases} \psi_1(x) & & |c_1|^2 \\ \psi_2(x) & \text{with the transition} & |c_2|^2 \\ \vdots & \text{probability as} & \vdots \\ \psi_n(x) & & |c_n|^2, \end{cases} \tag{12}$$

which can be measured by experiments also. This theorem of measurement clearly answers: even if the apparatus interferes with the wave packet of the particles system, this theorem still guarantees how the *real value* of physical quantity L of the *original quantum system* could be found!

It seems difficult to understand what is the essentiality of the collapse or the reduction of the wave packet. Actually, the collapse of the wave packet is a general phenomenon occurring in any statistical measurement. For example, a "die" will collapse to certain number, eg. the "red 4", when the "die" hits on the table, while before this acting, all the number 1, 2, 3 . . . , 6 is indefinite, each of them has a probability of 1/6. Although the table *disturbs* the *original* state of the "die", it does not prevent the *objective* of studying the probability of the appearance of the "red 4".

When we apply the same theorem to the process of the spin correlation measurement, the wave packet of this composite system

$$|0^+\rangle = \frac{1}{\sqrt{2}}\left(\psi_{+\frac{1}{2}}(A)\psi_{-\frac{1}{2}}(B) - \psi_{-\frac{1}{2}}(A)\psi_{+\frac{1}{2}}(B)\right)$$

will collapse to

$$\text{(13)} \qquad \Rightarrow \begin{cases} \psi_{+\frac{1}{2}}(A)\psi_{-\frac{1}{2}}(B) & \text{with the transition} & \left(\frac{1}{\sqrt{2}}\right)^2 = \frac{1}{2} \\ \psi_{-\frac{1}{2}}(A)\psi_{+\frac{1}{2}}(B) & \text{probability as} & \left(-\frac{1}{\sqrt{2}}\right)^2 = \frac{1}{2} \end{cases}$$

The peculiarity of the collapse of the wave function of this composite system, is that the eigenfunctions should be taken as the products of $\psi_{+\frac{1}{2}}(A)\psi_{-\frac{1}{2}}(B)$ and $\psi_{-\frac{1}{2}}(A)\psi_{+\frac{1}{2}}(B)$.

This process of measurement is like a special "die" in the following form.

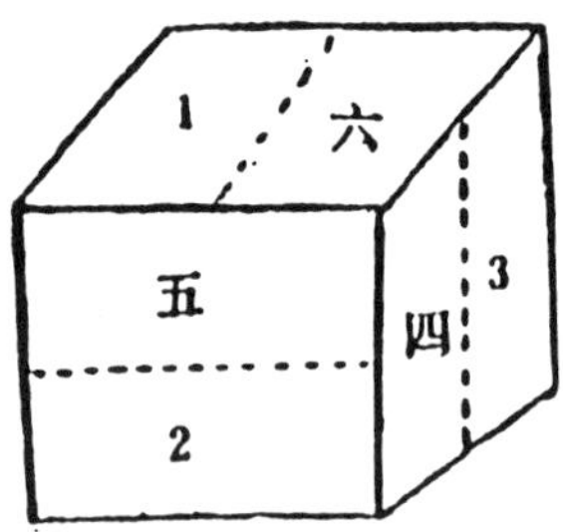

Fig. 2.

with the following law of conservation. i.e.

(14) Arabic No + Chinese No. = constant = 7

VI. ON THE MECHANISM OF THE COLLAPSE OF THE WAVE PACKET

Of course, the collapse of the wave function in quantum mechanics is different from the classical "dice", since the former is always accompanied with "the disappearance of the interference". This had been explained, for instance, by a successful theory of measurement, developed by A. Daneri, A. Loinger and G. M. Prosperi in 1962 [7]. The essential of this theory is that the detectors are made from certain thermodynamical quasi-stable systems, the collapse of the wave functions occurs during the interaction between the particles and the apparatus. It can be shown that the inference terms are multiplied by some numerical factors practically zero, while the non-inference terms are multiplied by a factor 1, both of which originate from the variables, which describe the apparatus.

That is the role the apparatus plays, i.e., the confirmation of the spin correlation of the quantum mechanical system, a reflection of the physical reality which is independent of the human consciousness.

VII. CONCLUSIONS

(1) Quantum mechanics is compatible which objective realism i.e., the world is made up of objects, which are independent of human consciousness.

(2) Deterministic realism is not supported by the experiments done in quantum mechanics. The realism of quantum mechanics should be interpreted as local and stochastical or statistical realism.

Zhongguancun, Huangzhuang,
Beijing, China.

BIBLIOGRAPHY

1. Einstein, A., Podolsky, B. and Rosen, N. (1935). *Phys. Rev.* **47**, p. 777.
2. Bohr, N. (1935). *Phys. Rev.* **48**, p. 696.
3. Bell, J. (1964). *Physics* **1**, p. 195; (1966). *Rev. Mod. Phys.* **28**, p. 447.
4. For example, Kasday, L., Ulman, J. and Wu, C. S. (1975). *Nuovo Cimento Lett.* **B25**, p. 633.
5. d'Espagnat, B. (1979). *Scientific American* **241**(5), p. 128.
6. Mermin, N. D. (1981). *J. Philos.* **7**, p. 397.
7. Daneri, A., Loinger, A. and Prosperi, G. M. (1962). *Nucl. Phys.* **23**, p. 297.

HONG DINGGUO

ON THE NEUTRAL STATUS OF QM IN THE DISPUTE OF REALISM VS. ANTI-REALISM

There is a widely spread impression that realism has been swept away from the reign of physical science once and for all since quantum mechanics came into being. This implies that QM provides a decisive support to anti-realism and is not compatible with any realism, whichever form it takes. However, such an opinion is unable to hold its position. In fact, QM is neither a menace to realism (or anti-realism) nor a decisive support to anti-realism (or realism). That is to say: the status of QM in the dispute of realism vs. anti-realism is neutral.

ON THE ONTIC REALITY AND THE EMPIRICAL REALITY

The word 'reality' frequently appears in philosophy literature. Its primitive meaning is 'existing being'. Owing to its various implications in different contexts, this word is in fact of multiple meanings. To make the discussion proceed clearly, we clarify two different notions about the ontic reality and the empirical reality.

The ontic reality is the reality which is external and independent of human beings, having strong objectivity. One of its essential characteristics is its transcendentality. Naturally, those philosophers who think that the only source of knowledge is experience will hold that the ontic reality is unrecognizable, and so deny that the notion of ontic reality possesses any epistemological significance. In contrast, holders of realism (referring to ontic realism) think that the notion of the ontic reality possesses an epistemological implication. They insist that although the ontic reality is beyond reach of experience, it is within reach of mind. Human beings can explore ontic reality by means of intuition and insight produced by the free creation of the mind, and grasp it in the form of scientific concepts. As Einstein said [1], "physics is an attempt conceptually to grasp reality as it is thought independently of its being observed." No doubt, the exploration and grasp of ontic reality can be convincing only when it is in concord with experience. In fact, the notion of ontic reality is continually sublimated and deepened in the interactions between the free creation of mind and the verification of experience. Trying to explore

R. S. Cohen, R. Hilpinen and Qiu Renzong (eds.), Realism and Anti-Realism in the Philosophy of Science, 307–316.

the ontic reality is in the nature of mind. Since the ontic reality transcends the sum of all experience and knowledge of mankind at any time, such a process will never end.

The empirical reality is the reality which depends on subjects and cannot be separated from experience so as to have the weak objectivity characterized by intersubjectivity. Empiricism (i.e. empirical realism) insists that science (and ordinary knowledge as well) is indissolubly linked with human experience, so that the task of science is but to describe the phenomena constructed by the collective experience of human beings.

Obviously, empirical realism contains a thick atmosphere of instrumentalism (or operationalism). However, d'Espagnat correctly points out that there is a difference between empiricism and strict instrumentalism. To offer an image, empiricism is a boat that is sailing between ontic realism and pure instrumentalism, and completes the instrumentalism with ideas about empirical reality as natural as possible.

We suggest an open-minded philosophy under which the ontic reality and empirical reality are two complementary notions existing side by side with their own implications and functions respectively. On the one hand, we confess that intersubjectivity is the foundation of all empirical sciences, and the notion of empirical reality is of paramount importance to practical sciences. On the other hand, we believe that the ontic reality potentially dominates the activities of mankind and the progress of sciences. Although the ontic reality is always hidden, its various surface structures can be perceived by human beings in the form of empirical reality. And its strong temptation for human beings is the eternal motivation of the development of sciences.

THE MINIMAL STRUCTURE OF QM AND ITS EXTENSION

Any complete physical theory is a system with four dimensions: its mathematical formalism, the operational definitions of its basic concepts, its interpretation, and its unfolding [2]. The first two dimensions constitute the minimal structure of a physical theory. It is a formal but operational algorithmic system, which simply and effectively describes the relevant physical phenomena. The last two dimensions constitute a philosophical extension to the first structure to manifest any physical meaning it may have. Such an extension is a self-contained system of metaphysical ideas which is compatible with the first structure.

What on earth is the minimal structure of QM? That is the quantum algorithmic system centered around the concept of probability amplitude. In Bohr's words [3], "the whole formalism (i.e. QM) should be regarded as an instrument, by which the predictions with definite or statistical information could be deduced under certain experimental conditions described in classical terms."

The minimal structure of QM has the following features: (1) it has been deprived of all physical meanings, being a complete theory only in the operational sense; (2) it has so far been one of the most successful theories in relating experimental information; (3) the phenomena it applies to, all possess the characteristic of the irreducible wholeness of experimental context; (4) it can apply only to the ensembles of quantum measurements described in classical terms.

Instrumentalism insists that human knowledge is but the set of all the effective 'recipes'. In the view of instrumentalists, the minimal structure of QM is all of QM. However, most physicists are not willing to follow instrumentalism. In a sense, even the proponents of the Copenhagen School oscillated between a *de facto* instrumentalism and a declared but undeveloped realism. It is inconceivable for a physicist to have a formalized physical theory without any physical meaning. However, to explore an interpretation of QM is to explore a philosophical extension of it. How is such a metaphysical business relevant to physical science? What is the value of such a business? Let us examine these questions from two sides.

From the passive side, we notice that metaphysics is inevitable, and the point is how to take a correct position. On this Bohm said [4]:

> We don't know the nature of reality ultimately. What is the point of describing it or talking about it? Well, one view is to say there is no point, and a lot of modern philosophers and scientists say we shouldn't do it. And that's a reasonable approach up to a point. But everybody has tacit metaphysical ideas even though he has explicitly disavowed them. And therefore ultimately that way of doing it is only going to put you under control of whatever metaphysical ideas you happen to have adopted as presuppositions, perhaps in early childhood. So, I think it is valuable to explore your metaphysical ideas, to question them, and to propose new metaphysical ideas, and so on.

From the active side, we notice that metaphysics is necessary. This point is suggested by the so-called Gödel theorem about the incompleteness of a formal system in mathematics. The theorem tells us that in any formal mathematical system some true propositions can be constructed, which cannot be proved nor negated within the framework of the system. That is to say: the truth of such propositions must be judged by means outside the formal system. No doubt, Gödel's theorem can be applied to formalized physical theories. In other words, such a formalized physical theory contains some true propositions which cannot be proved or negated, and the truth of such propositions should be judged by means and notions outside the formalized physical theory. Therefore, to complete the meaning of a formalized physical theory, it is necessary to make some philosophical extension of it. For creative theoretical work, such a philosophical extension is indispensible. The only point is to make such an extension reasonable and consistent.

Moreover, any formalized physical theory results from the creation and evolution of some physical insight, while the creation of any new physical insight and the formation of any new physical notions is not logical and algorithmic but meta-physical and non-algorithmic. In this sense, the minimal structure of QM can be thought of as the result of sublimation and crystallization of the metaphysical ideas which have nourished the growth of QM.

The above analysis indicates that the birth of QM is by no means to deny meta-physics but put more severe requirements to the possible forms of the concrete notions of metaphysics.

THE FIRST TYPE OF PHILOSOPHICAL EXTENSION OF QM

Therefore, when we talk about the philosophical extension of QM, we speak of a consistent system of metaphysical ideas which cannot be deduced from nor refuted by the minimal structure of QM. If such a system has the notion of empirical reality as its essential content, we call it the first type of extension of QM. According to D'Espagnat's analysis [5], the notion of empirical reality includes the following ideas:

(1) At any time, the center of mass of macroscopic bodies has the property of locality.
(2) A property of a system must be defined operationally but counterfactually.
(3) The time evolution of the properties of a system in an ensemble is determined by the computation rules of a theory.
(4) It is impossible to influence the past.
(5) A theory of empirical reality is obtained by postulating (a) that replacing very large times by infinite times and/or very large particle numbers by infinite numbers is a valid abstraction; and (b) that on

the other hand the possibility of measuring observables exceeding a certain degree of complexity is to be considered as nonexistent, even in matters of principle, even though this non-measurability does not follow from the theory and even the only way we have for making it compatible with QM seems to be to ascribe it to some basic inaptitude of men.

It is not difficult to find that the first four ideas come from the notion of reality in common sense and the essential requisite of the minimal structure of a physical theory; while the last is the product of efforts to coordinate the first four ideas with the minimal structure of QM. To elucidate the last point, let us look at some important progress in the field of fundamental researches of QM.

At first, we must say that the conventional (or orthodox) theory of QM has been a somewhat philosophical extension of QM because it contains some concepts implying physical meanings such as 'particle', 'wave', 'complementarity principle', 'uncertainty principle', etc. However, such extension is unsuccessful since if it does not regress to strict instumentalism, various conceptual difficulties cannot be exempted. A prominent example is that it cannot explain the collapse of the wave function in a single quantum mechanical measurement. Many quantum physicists have claimed that many conceptual troubles in the conventional QM may be avoided by resorting to the more elaborate theory known as algebraic QM. This means that by extending QM in the algebraic framework, and endowing the observable algebra with a primitive status, it may be possible to speak of both classical reality and quantum reality in a united epistemological framework.

Here, we would like only to mention some works of signpost significance. First, S. Machida and M. Namiki were able to prove that [6] the disappearance of the cross terms in the measurement process (corresponding to the collapse of wave function) results in the macroscopic nature of instruments. They used the continuous direct sum of Hilbert spaces H_{n_i} labeled by the continuous different number of particles $n_i \in (n_o, \Delta n)$ to describe the instrument involved, when n_o is the mean of particle numbers of the instrument. In their theory, observables are described by self-adjoint operators defined within each H_{n_i}, and the following limits were taken: $n_o \to \infty$, $\Delta n \to 0$. This means that the continual superselection rule is an essential component of their theory.

Next, H. Araki has further proved that [7] (1) A continual superselection rule can result in the collapse of wave packets in the separation procedure of QM in the infinite time limit. (The so-called separation procedure differs from a measurement procedure in that it doesn't bring instruments into different states.) (2) In the case of a discrete selection (or no selection at all), it is impossible for the collapse of wave packets to take place. (3) Even for the case of a continual superselection rule, the collapse of wave packets and the measurement procedure must be carried out in two separate steps (i.e. step by step).

Finally, based on the above results, especially on the work of Araki and K. Hepp [8], C. M. Lockhart and B. Misra proposed an interesting algebraic quantum measurement theory not long before [9]. In this theory, when the instrument dynamics is complex enough, the states of 'system plus instrument' will spontaneously evolve from a pure state to a mixture (after an infinite time) via an irreversible process. Thus, a well-known essential prerequisite in quantum measurement problems had been logically deduced.

From the above brief introduction, we could see that it is based on his investigation of the above theoretical progress that D'Espagnat has designed his fifth idea. It is reasonable to say that such an idea is an embodyment of an empirical realism compatible with QM. Thus, it is not strange that he called it the axiom of empirical reality.

THE SECOND TYPE OF PHILOSOPHICAL EXTENSION OF QM

The second type of philosophical extension of QM is a system of QM that includes the notion of ontic reality. It is goal pursued by those quantum physicists who persist in their ontological research. What we are mainly concerned with here is the total new program of quantum mechanical ontology developed in the gap between the ontologies of Einstein and Bohr.

This research line was at first proposed by L. de Broglie in the name of 'the pilot-wave theory' [10], and was given up soon because it was attacked intensively and had no support then. Quite later, Bohm proposed similar ideas independently [11], introducing particle variables (local beables) into non-relativistic dynamics and endowing the wave function with the meaning of a realistic field (global beables), so as to transfer a non-determinant QM into a determinant QM without losing experimental information. Moreover, in his reformation of QM, all quantum paradoxes are resolved in a consistent way, with all quantum processes being explained 'intuitively'. Since Bohm has cleverly answered all reasonable criticisms, his theory stands and thus not only makes de Broglie return to his original position but also attracts J. S. Bell to join their efforts; and Bell later proposed his famous and elegant 'Beables for quantum field theory' [12]. Although there are many differences between the individuals, we call their common thought BBB theory in short.

In BBB theory, 'observer' is no longer at a fundamental level. Observables are replaced by beables. The so-called beables are referred to those quantities corresponding to realistic elements (i.e. existed beings). Their existence is not dependent on 'observation'. On the contrary, observing instrument, observation behaviour, and even observers all come from beables. Naturally, the crossed influence between the state of global beables and the behaviours of local beables are determined by the equations of QM. The totally new ontic relationship between the whole and parts appears running through all quantum mechanical problems (both the measurement problems and non-measurement ones). This is a fundamental characteristic which distinguishes the quantum ontology from the classical one. It is simply called the wholeness of quantum reality, or the unseparability of the quantum system, or the EPR non-local correlation of quantum system. However, as long as the influence of global beables of QM to local beables is far less than the effects of classical interactions, a quantum mechanical system is dissolved, and quantum reality is thus retrograded to a classical one which can be described by classical physics. So, in this type of extension model of QM, a smooth transference between the quantum reality and the classical reality can be realized, without a non-transcendent gap between them.

People may be anxious that the causality paradox may be brought by the confrontation between the non-local correlation of quantum mechanical ontology and the principle of special relativity. You should not worry about that for two reasons. First, the non-local correlations of QM are very fragile. Once one tries to draw information from some location of a quantum system, the correlation is dissolved immediately. So, one can never utilize it to transfer a distant signal simultaneously. Second, the quantum measurements carried out so far are only of statistical significance. The principle of special relativity thus possesses only statistical significance. That is to say: the non-locality refused by experiments is the one contradicted by the principle of special relativity in the statistical sense but not one in a presumably basic ontic process.

Here it is probably suitable to mention the ontological interpretation of the Dirac equation, proposed by Bohm and B. J. Hiley recently [13]. In their ontological model, the equations satisfied by global beables (i.e. the Dirac equation and the continuity equation) is Lorentz covariant, while the guidance equation of local beables (i.e. velocity equation of particles) is of a non-locality characteristic. That is to say: no matter how far the distance between the particles, the displacement of any particle may generally mean the corresponding instant variations of the velocities of all the particles. However, this does not mean that the ontological interpretation of QM cannot be extended toward the situation of relativity, since the theoretical statistical prediction relies only on the covariance of the equation of global beables. So, it may be suggested that there exists a preference-based reference frame, in which the non-locality correlations are simultaneous, and there exists a non-covariant basic stochastic process, which makes whichever distribution of the initial velocities tend dramatically to the equilibrium distribution determined by globe beables. To all practical purposes, the equilibrium probability distribution is the sole distribution. The statistical results for all existent quantum mechanical experiments computed by means of the Bohm-Hiley theory is therefore covariant, as the results computed in the conventional QM. It is obvious that at the moment there is no experiment to prove their theory wrong.

In fact, Bohm and Hiley have brought us to the vicinity of a new physics. The statistical process on the sub-relativity level proposed by them cannot be revealed in any quantum mechanical experiment. This implies that it is sub-QM as well. The existent quantum theory and relativity come only from the statistical results of experiments. They are not sensitive to the details of basic ontological models, just as the laws of thermodynamics are not sensitive to the basic atomic models. So, in this period, we have a wide range of probabilities within which to choose various models of sub-relativity and sub-QM.

Last, we would like to mention the relationship between the algebraic QM and the extension of QM. It is remarkable that H. Primas has made a concise and systematic research study of it [14]. He emphasized that the early Hilbert space formalism of QM is limited to locally compact phase space, so as to rule out the proper descriptions of genuinely irreversible stochastic processes, spontaneous symmetry breaking, phase transition and the classical behaviour of quantum world. It is therefore unsuitable to take it as a universally valid theory. However, the structure of algebraic QM contains all physical systems, not only finite systems (with a local compact phase space), but also infinite systems (whose phase space is not locally compact). Infinite systems have infinitely many physically non-equivalent W*-representations which account for the stupendous complexity of observable phenomena in nature. For example, it may be proved that an open quantum system generally experiences symmetry breakings, and possesses classical observables, while contextual classical observables are not the function of intrinsic observables but the product of the intrinsic algebra combined with a contextual topology.

Basing on the algebraic QM, the philosophical extension of QM constructed by Primas is a consistent system of ideas with the following distinguished statements: (1) A strictly closed physical system without any concept of an observer is called an endosystem. (2) If the endosystem is divided into an observing and an observed part, we speak of an exophysical description. (3) The world of observers with their communication tools is called an exosystem. (4) The empirical interpretation of a physical theory refer to our knowledge of the properties or modes of reactions of the system "as we perceive them". (5) The ontic interpretation of a physical theory refers to the nature of 'object itself'. It is irrelevant with respect to our knowing it or not, and it is independent of any perturbation by observing actions. In the view of Primas, the ontic interpretation of QM refers only to a fictitious theoretically immanent reality, and not to the ultimate reality. But, under the working hypothesis that QM is a universally valid theory, an ontic interpretation allows us a consistent way of speaking as if we would refer to reality.

CONCLUSION

In sum, we might construct various consistent extended structures of QM on the minimal structure of QM with different physical commitments, which constitute an interpretation spectrum of QM in accord with a series of different attitudes toward reality, from extreme anti-realism to ontological realism. The work of extending QM is not only a philosophical undertaking but also a physical one. It is an unfinished business. We hope that more philosophers familiar with theoretical physics and more physicists with philosophical quality will take care of, and be engaged in, such a fundamental project concerning the future of physical science and that of modern philosophy as well.

Department of Physics,
Hunan Nonmal University,
Changsha, China.

BIBLIOGRAPHY

1. Einstein, A. (1937). *Mein Weltbild*. English version by C. Seelig, *Ideas and Opinions*, Dell Publishing Co., New York, p. 292.
2. Hong Dingguo (1988). *The Structures of Physical Theories and Their Unfolding* , Science Press, Beijin.
3. Bohr, N. (1949). *Dialectica* **2**, pp. 321–329.
4. Bohm, D. (1985). 'A Talk on the Value of Metaphysical Discussion', in D. Facctor (ed.), *Unfolding Meaning – A Weekend of Dialogue with David Bohm* , Foundation House Publication.
5. D'Espagnat, B. (1989). *Reality and the Physicist*, Cambridge Uni. Press, pp. 232–253.
6. Machida, S. and Namiki, M. (1980). *Prog. Theor. Phys.* **63**, pp. 1457, 1833.
7. Araki, H. (1980). *Prog. Theor. Phys.* **64**, p. 719.
8. Hepp, K. (1970). *Helv. Physica Acta* **45**, p. 237.
9. Lockhart, C. M. and Misra, B. (1987). *Physica A* **136A**, p. 47.
10. Broglie, L. de (1927). *Journal de Physique* **5**, p. 225.
11. Bohm, D. (1952). *Phys. Rev.* **85**, pp. 166, 180.
12. Bell, J. S. (1987). 'Beables in Quantum Field Theory', in B. J. Hiley and F. D. Peats (eds.), *Quantum Implications* , Routledge and Kegan Paul, London.
13. Bohm, D. and Hiley, B. J. (1989). *Phys. Rep.* **173**(3), pp. 93–122.
14. Primas, H. (1991). 'Realism and Quantum Mechanics', preprint.

JIN WULUN

DOUBLE STRUCTURE OF TAO REALITY

The legitimacy of interpreting scientific theory in a realistic way has been increasingly accepted by physicists and philosophers of science. For centuries the results of relevant investigations in natural sciences have made it possible to use the hypothetico-deductive method fruitfully. This tremendously strengthens the rationality of, and the belief, in realistic interpretations.

Heisenberg (1958) used *potentia* to denote the independent reality. Shimony (1978, 1986) and Stapp (1979, 1985) followed Heisenberg, exploring further the realistic interpretation of the quantum formalism; Bell (1964) talked about physical reality by 'Beable' in quantum field theory; Bohm (1957, 1980) used the concepts 'quantum potential' and 'the enfolding and exfolding of implicate and explicate order' to describe the wholeness of reality; d'Espagnat (1983) put forward 'veiled reality' to describe reality independent of all observers. And beside these there are other interpretations, such as the many-worlds interpretation of Everett and others. These inquiries are fruitful, but still there is much room for a more detailed account of reality.

In this paper I'll try to discuss the structure and properties of reality with Tao and Yin-Yang in ancient Chinese Philosophy, and to explain some quantum paradoxes. In doing so, I prefer d'Espagnat's opinion: to solve the difficulties a realist faces we shall establish that "a link with some philosophical or cultural tradition is necessary".[1]

THE THEORY OF TAO-REALITY

Tao is the ontology of the universe and also the laws of the universe. As an ontology, just as Professor Jin Yue-lin says: "Tao, as the most lofty concept, as the most fundamentally primal power, is not absolutely empty. Tao must be real . . . , Tao can be spoken both integrally and separately."[2]

First, Tao is real, so we can call it 'Tao-reality', and therefore we can call the theory about it the 'theory of Tao-reality'.

Second, integrally speaking, it can be called Tao. Tao is "the beginning of heaven and earth", "the mother os *Wanwu* (ten thousand things)".

R. S. Cohen, R. Hilpinen and Qiu Renzong (eds.), Realism and Anti-Realism in the Philosophy of Science, 317–328.

It is the unity of having (有) and no having (无), [being and non-being, nothing] these two spring from the same source but differ in name. Separately speaking, it can be divided into "the eternal Tao" and "not the eternal Tao", that is "having and no having".

Third, the unity formed by having and no having is nonobservable. It "differs from the observable concrete things": "look, it cannot be seen – it is beyond form. Listen, it cannot be heard – it is beyond sound. Grasp, it cannot be held – it is intangible. These three are indefinable; Therefore, they are joined in One." (Tao Te Ching, p. 14). However, it is not absolutely empty. On the contrary, "Tao is elusive and intangible. Oh, it is intangible and elusive, and yet within is image, Oh, it is elusive and intangible, and yet within is form." (Tao Te Ching,p. 21) It is "standing alone and unchanging, even present and in motion." (Tao Te Ching, p. 25)

Tao-reality has having and no having as two components, therefore we can say it has a double structure. Having and no having are two existential states of Tao in the cyclic movement. No having is the start-point and the end-point, whereas having is the middle-point or extreme-point in this cyclic movement. Therefore, although no having and having are different, they are to be used to refer to Tao. They are two indispensable aspects. Tao on the one side and having and no having on the other side have relations as follow.[3]

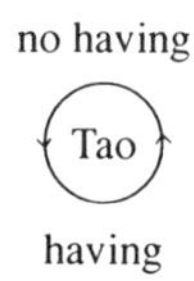

Fig. 1. The relation between Tao and having and no having.

Here Tao is the ultimate reality. It is the united body of having and no having. It is the primal source of *WanWu*. Everything emerges from it, and returns to it, but Tao itself is eternal existence. *WanWu* have the variant principles, but Tao comprises the principle of *WanWu*. The unity and mutual interrelation of all things and events, the experiences of all phenomena in the world are manifestations of this basic oneness. All things are seen as interdependent and inseparable parts of this cosmic whole; and as different manifestations of the same ultimate reality.

From this, we can know that Tao-reality has many more implications than Heisenberg's *potentia*. Heisenberg's *potentia* is simply tendencies or possibilities of Being. It has no internal structure, and it only expresses a process from *potentia* to actualization by measurement. Tao-reality has a double structure. It is an unity of having and no having, but *potentia* is only equivalent to 'having' in the Tao-reality; however 'having' cannot exist without 'no having'. More importantly, Tao-reality cannot be directly measured, only *WanWu* generated from Tao-reality can be measured. That is to say, only the observable things generated from Tao-reality can interact with a measurement. Tao-reality and its emergence are all outside our consciousness; to use d'Espagnat's words, Tao-reality is a far reality,[4] far realism is the theory about it.

THE THEORY OF TAO-BECOMING

Tao is real, it has a double structure.

Tao can be born and destroyed. It has the mechanism of double-directions movement.

Tao is changeable. All things arise from it. (The Tao begot one, one begot two, Two begot three, and three begot *WanWu*) This begetting is natural, there is no artificial force in it, and it involves only the force of Yin and Yang, as Harvard University astronomer Professor David Layzer says:

> In Chinese Philosophy, it recalls the doctrine of Yin and Yang; Yang, the integrative force, corresponds to the growth of order; Yin, the disruptive force, to the entropic decay of order.[5]

Becoming is an important feature of the unity of having and no having. Becoming is the unity of having and no having. Becoming has generating and destroying, and the generating has the beginning. According to Hegel, when a thing is at its beginning and has not yet become actuality, it is not a mere nothing but already contains something, or existence, of it. Beginning itself is becoming, but beginning also contains the meaning of development.

In fact, a thing or an object that the objective probability expresses is the representation of the unity of having and no having. There is nothing in appearance, but there is the having in fact. Not only the having is real, but also no having is real. Having and no having are interchangeable, "something for nothing".[6] There is the 'becoming' and 'beginning' implicated in it. For example, consider radioactive material: before it spontaneously emitted a particle a radioactive atom has neither happened nor not happened. For the emitted particle, it will be in the unitary state of having and no having. No having means no radiation; Having means a tendency state of radiation. When this state will be destroyed, the particle will come out of it. This emerging particle is exactly observable, though the unity of having and no having here is nonobservable.

The vacuum fluctuation is the best example of Tao-becoming. Vacuum is certainly not absolute emptiness. The vacuum state of a unity of having and no having will spontaneously break the symmetry or lose the symmetry. The symmetry breaking requires a new field, the Higgs field, and the associated particles are also called Higgs. We now know that the Higgs field is not zero. So "the universe and everything in it may be no more and no less than one of those vacuum fluctuations that allow collections of particles to burst forth out of nothing, live for a while, and then be reabsorbed into the vacuum"[6] The vacuum symmetry-breaking, the vacuum phase changing, generate elementary matter and fundamental interactions. "A closed universe has zero energy overall, and it is not so difficult to make something with zero energy overall out of a vacuum fluctuation."[7]

Only by this time, *WanWu*, namely observable things, can emerge and the empirical phenomena can appear, and then we can observe, measure and grasp them. This suggests that the process of becoming from potentiality (the unity of having and no having) to actualization (empirical phenomena) is very complex.

An important feature of this generative process is an integrative genesis. For example, the generation of a particle is not isolated from the original potential state which existed in it, but the result of the integrity of all information in the net of the unity. It is not like a watermelon seed in the watermelon which is peeled off, but rather like a young shoot growing from the watermelon seed. This is an integrative genesis.

The view of integrative genesis emphasizes that the explication of becoming or changing would be made in terms of 'coming into being', 'passing away' or 'transformation'. This view can be called the *genesis theory*. It is plainly contrary to the *constructive theory* which is dominant in western thought, for the latter claims that the changing of things is the 'combining' and 'separating' of the unchangable elements, and the method is the method of analytical reconstruction.

As we known, the Feynman diagram represents the creation and disappearance of particles and their interactions. For example, a photon traveling through space and time spontaneously creates an electron/positron pair, and the positron meets another electron and disappears, and the like. However, the constructive theory cannot interpret these phenomena; only the genesis theory can interpret them.

The genesis theory specifically emphasizes that this generation is not a mechanistic separation, but an integrative genesis.

To see measurement from this view, we can say that measurement concerns only generative things, not Tao-reality itself. Reality is independent of measurement. This is the view of reality we need to defend. This view of reality differs from that of metaphysical reality. And the main difference consists in that the latter takes measured things not as generated ones, but as originally real existence there. Observation and measurement simply pick out the measured objects from the universe, and the measurement does not act on the rest of the universe, only on the measured object. The view of Tao-reality emphasizes that the measured object is a generative object of the unity of having and no having. This generative thing is the product of the integrative genesis of Tao-reality. The measurement in turn affects the whole reality.

The mechanism of the generation and interaction have been governed by Yin-Yang force. The doctrine of Yin-Yang tells us that the unity of having and no having is a combination of Yin and Yang which are unseparable. The generative things are loaded with both Yin and Yang. This is because "*WanWu* carry Yin and embrace Yang. They achieve harmony by combining these force" (Tao Te Ching, p. 42).

To speak with emphasis, for measurement the generative things only appear as Yin force, and the measurement represents Yang force. Therefore, the relation between the measured object and the measuring device is a relation of Yin-Yang complement. This situation is consistent with Bohr's framework of complementarity. To see this change of Yin-Yang roles, we might as well see the change of Yin-Yang roles in the relation between father and son, male and female. For a single male has both Yin and Yang, but for his wife, he has only a Yang force. The relations between *WanWu* in the universe are all like this. However Yin and Yang ever combine two into one, but separable.

Starting with this view, we can explain the EPR correlation, nonseparability and nonlocality.

When the system AB is separated into A and B, far from each other, any one of them under measurement will influence the other though there is not any physical interaction between them. The measured object, say A, will soon be transformed into a Yin force, while the measuring device appears as the Yang force as we said above; when A is transformed into Yin, then B is transformed into Yang. That is to say, if we measure A, then A appears as Yin and B as Yang; and vice versa (see figure 2). When observation or measurement begins, the measured object becomes a mediator linking reality with the measuring device. Since Yin-Yang is always co-existent, it integrates reality, observable objects and the device into an undivided whole. Thus, A and B are a whole connected by Yin-Yang, even if they are separated from each other and have no physical interactions. We can say that Yin-Yang's becoming is exactly the very essence of the EPR correlation mystery; that is the essence of the microparticle's entanglement.

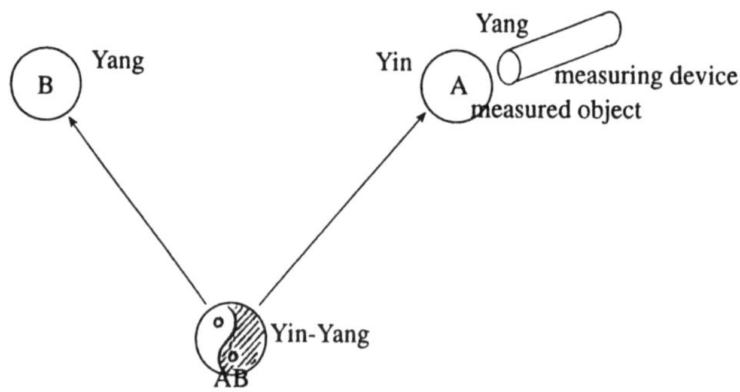

Fig. 2. Interpretation of EPR correlation with the Doctrine of Yin-Yang.

We can hold that the correlation-at-distance linked together by Yin-Yang force in the result of measurement is characterized as non-separability, nonlocality, world's wholeness, and general connection. It is characterized as 'passion-at-a-distance', as Professor Shimony says.[8] It is a kind of overall holism.

The double-slit experiment can also be interpreted with the doctrine of Yin-Yang. According to this doctrine, the combination of Yin and Yang produces everything in the world, which are composed of Yin and Yang.

Nothing can exist unless neither Yin nor Yang is absent. But two force, Yang and Yang or Yin and Yin, are in conflict, and they cannot co-exist within a single thing. This feature of Yin and Yang leads to showing the special pictures on the screen of the single-slit and double-slit experiments.

When combining Yin and Yang into one, particles are produced, so a peak appear on the screen. However, Yang and Yang would repel one another, and Yin and Yin would repel one another too, just as two like electric charges repel one another. Therefore, the wave valleys appear on the screen when Yang and Yang or Yin and Yin repel one another. See figure 3.

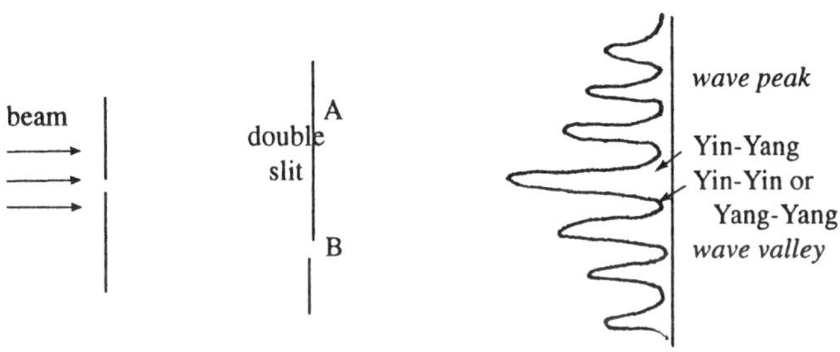

Fig. 3. Interpretation of Double-slit experiment with the Doctrine of Yin and Yang.

Some of other quantum phenomena can be also interpreted with the Doctrine of Yin and Yang. Since we have limited space we cannot offer a detailed discussion.

The Tao-becoming is a holistic process. To clarify this process further we need to make further inquiry for holism, including a search of the physical mechanism of Yin-Yang interaction. In the 5th section of this paper we will discuss this problem.

THE THEORY OF CONSTRUCTION

Scientists construct scientific theories to explain the observational characteristics of the physical world. They often construct hidden-structured models of the objects studied. This structure is used for a causal account of observable phenomena, and the theoretical models provide approximate explanations of phenomena.

Scientific realism argues that to a certain extent the theoretical structure is an insight into the structure of the world or of reality, but the scientific realist does not maintain that the constructed theoretical structure must be the true structure which we live in.

Generally speaking, this procedure is roughly this: scientists observe sensory phenomena, from which they apply their cognitive structure (including background theory) to advance a set of basic concepts or basic principles as the theoretical foundation from which they deduce a sequence of conclusions with the aid of logical deductive procedures. These conclusions can be compared with sensory phenomena, and are required to be compared with other well-known theories as well as the new observable phenomena. In the course of comparing, they will be continuously revised, substantiated and increasingly completed.

This construction is not, as a Chinese saying goes, to "draw calabash according to the original example", because there is no original calabash. The construction is rather like to "draw a tiger in the light of a cat". The tiger drawn consists simply in a cat's form and a painter's imagination. As a result the picture drawn is often like a dog rather than a tiger.

This constructive procedure is called the 'method of analytical reconstruction'; in other words, it attempts to represent the whole structure, or the whole of reality by constructing local phenomena. There are some apt metaphors in China: 'looking at the sky from the bottom of a well' or 'a blind person groping for the elephant'. Although a part of the sky is real and true, and a part of an elephant is also real and true, yet one part differs entirely from other parts. Every part falls under conditions and circumstances. Maybe they are really incommensurable in Kuhn's sense.

THE DOCTRINE OF EMBEDDING

Our present world picture is the result of constructing and embedding, just as a beautiful painting drawn by a famous painter. It is a product which has been made bit by bit, just as a painter works out the plot of a picture.

The scientific world picture is a bit similar. The scientists first construct a theory from local phenomena and then embed it within the general appearance of a picture; as Einstein says, "the real difficulty lies in the fact that physics is a kind of metaphysics; physics describes 'reality'. But we do not know what 'reality' is; we know it only by means of the physical description!"[9] That is to say, the reality we know is only the result of models constructed by scientific theories that are hopefully embedded in the nature.

The atomic pictures we know, from Democritus's atom to Dalton's atom to Bohr's atom, are all the products that scientists constructed. The electron was also constructed in this way. Therefore, atom, electron, elementary particle, and quarks too are all products constructed and then embedded in nature by scientists. Their existential features are closely linked to scientific theories; as the physicist Heinrich Hertz said, "from the view of humankind, light is a kind of electromagnetic wave." That is to say, "light" and "electromagnetic wave" both are something seen from the view of humankind. They are the descriptions and interpretations for natural phenomena in terms of language.

Physicist D. Bohm made this clearer. After having thought deeply, he said, "I suddenly realized that everything is all subjective." I am sure that the world picture provided by science has artificial properties. Of course, experience provides limiting constraints upon this world picture. But construction and embedding are not arbitrary, they must be subjected to many constraints so as to satisfy the weak objectivity in science. To make an embedding is to assign theoretical models constructed according to the objectivity principle in nature, to Tao-reality. In the words of Professor Jin Yue-lin, this process is "to receive from the given and to return to the given."[10]

However, the reality picture constructed by science is greatly different from picture of nature itself. It is in view of this that various difficulties arise for convergent realism just as Larry Laudan criticized;[11] And it is also in this situation that the transformation of incommensurable paradigms and the Gestalt shifts of world views take place, as Thomas Kuhn shows clearly.[12]

HOLISTIC METHOD FOR SEARCHING TAO-REALITY

Reality is a whole. It is correlated by Yin and Yang. Observable things spring from it with the aid of Yin and Yang. Observable things are only its generating products. Tao-reality itself cannot be measured.

Reality (or the world picture) as revealed by science is a constructed and embedded picture. Its method of revealing reality is the method of analytic reconstruction. For this method, "there continue to be influential dissenting voices, even among physicists."[13] The limitations of this method are as follows:

(1) The description of a whole would be achieved by knowing its parts plus a mechanical assembling method. If we take a man as example, then this constructed man is a robot, not a living man.

(2) Its philosophical basis is reductionism. According to reductionism, a whole is simply composed of its parts. One can achieve a completely adequate understanding of the behavior of the whole by analyzing it into its component parts and discovering the laws according to which these operate. This so-called "simplistic principles of the universe" as I. Prigogine wrote, is a myth which has been challenged by new sciences.

(3) It devotes itself to revealing the analysis of static structure but it neglects to search for the connections, processes, and becoming. It notes that the whole is constructed of its proper parts which relate differently from the sum of the various parts; rather they have a "supervening relation" as Paul Teller says.[14]

The method of analytic reconstruction has met refutation by holists. The holists claim that some individuals in the domain in question have properties that are not determined by the properties of their parts. "Now it is clear that historical or individual properties of a whole cannot be expected to supervene on a basis of properties and relations of their parts which fail to include their historical or individual properties and relations".[15]

There is also a similar view in Chinese philosophy, that is the so-called Yin Tuo Luo (因陀罗) net. According to this view, every intersecting point in this net has a bright pearl which illuminates other pearls and is also illuminated by others.[16] All relations in the net are based on the nonrelational properties of every bright pearl in itself (a single bright pearl's illumination). This, however, is not the dominant view of Tao-reality in Chinese philosophy.

We have distinguished the holistic method of Tao-reality and the method of the analytic reconstruction. The differences between Tao-reality and Heisenberg's theory of *Potentia* or Everett's many-worlds interpretation can be listed as following.

Comparive list of various views of reality.

Heisenberg Shimony, Stapp	Potentialities	actualisation *by measurement*
Everett	Many-worlds	real world *by measurement*
Whitehead	Potentialities	actualisation *by value*
Bohm	Implicate order	explicate order by *soma-significance*
d'Espagnat	Veiled reality ¦ empiricial reality	phenomenal world *by observation*
our model	no having ¦ observable things Tao 道 having →¦ generated →	phenomenal world *by observation*

The theory of Tao-reality maintains the holistic method of integrative genesis. It especially emphasizes the following ideas:

(1) Tao-reality is existence independent of our consciousness. It is a really independent reality. This reality is an indivisible whole. All sensible phenomena are generated from this independent reality.

Humankind itself is the product of natural evolution, according to Darwin's theory of Evolution; Humankind's consciousness is also a transforming process, from no having too. The independent reality does precede consciousness. Admitting the independent reality, such concepts as subject-object interaction, or the unity and separation of subject-object, might be significant concepts in epistemology.

(2) Tao-reality goes beyond the space-time framework. It cannot be described by existing scientific concepts; by *WanWu* are in the space-time framework, and can be described by the scientific concepts.

(3) The processes of construction and embedding show that our current picture of the universe is obtained through generalization of scientific knowledge. We have this picture thanks to the work that scientists do in understanding, describing and expressing scientific achievements. This picture itself is only a part of the real world. The independent reality is large than and precedes the picture constructed by the sciences.

(4) *WanWu* generated from Tao-reality are not mechanically cut off,

but organically and integratively generated from it. Everything in it carries information of the whole universe. Therefore, the method for understanding and grasping reality cannot be limited to analytic reconstruction and induction. We should turn to the holistic method of integrative genesis, and transform our method from static analysis to dynamic integration.

The theory of Tao-reality can be also called the genesis theory; its core is integrative genesis. The mechanism of the process of this integrative genesis remains to be explored further.

Institute of Philosophy,
Chinese Academy of Social Sciences,
Beijing, China.

NOTES

[1] d'Espagnat, B. (1983). *In Search of Reality*, Springer, N.Y., p. 97.
[2] Jin Yue-lin (1987). *On Tao*, The Commercial Press, p. 17.
[3] Wang Bo (1991). 'Exploring the Relation between Tao and Having, No Having in Chinese Philosophy', *Philosophical Research*, No. 8, pp. 44.
[4] d'Espagnant, B. (1990). *In Search of Reality*, p. 95. He said there, "The visions of the world of Buddha, of the Tao, . . . are examples of far reality".
[5] Layzar, David (1990). *Cosmogenesis*. Oxford University Press, Preface for the Chinese Edition.
[6] Gribbin, John (1984). *In Search of Schrödinger's Cat – Quantum Physics and Reality*, Bantam Books, pp. 194–202, p. 271.
[7] *Ibid.*
[8] Cushing, James T. and McMullin, Ernan (ed.) (1989). 'Reflections on Bell's Theorem', *Philosophical Consequences of Quantum Theory*, University of Notre Dame Press, p. 77.
[9] Einstein to Schrödinger, 19 June 1935.
[10] Jin Yue-lin (1986). *Theory of Knowledge*, The Commercial Press, p. 493.
[11] Laudan, Larry (1984). *Science and Values*, Chap. 5, University of California Press.
[12] Kuhn, Thomas S. (1970). *Structure of Scientific Revolutions*, University of Chicago Press, Second Edition.
[13] Healey, Richard A. (1991). 'Holism and Nonseparability', *Journal of Philosophy* **88**, No. 8, August, p. 398.
[14] Teller, Paul (1986). 'Relational Holism and Quantum Mechanics', *Brit. J. Phil. Sci.* **37**, pp. 71–81. 'Relativity, Relational Holism, and the Bell Inequalities', in Cushing and E. McMullin (eds.)
[15] Healey, Richard A. *Holism and Nonseparability*, pp. 401–402.
[16] Xu Zhou-yun (1991) (许倬云), *Chinese Culture and World Culture*, Chinese Edition, GuiZhou People's Press, p. 67.

LARS-GÖRAN JOHANSSON

REALISM AND WAVE-PARTICLE DUALITY[1]

Quantum mechanics is a touchstone for realism: everyone claiming to be a realist faces the challenge of giving a realist interpretation of quantum mechanics. There are two features of quantum mechanics which trouble the realist: one is wave-particle duality, the other is quantum theory of measurement. This paper discusses wave-particle duality only. The extent to which the ideas put forward will highlight the measurement problem is discussed elsewhere.[2]

Realism, as opposed to idealism, holds that the reality is independent of human minds. This independence consists of three aspects: (1) real entities exists whether or not there are minds in the world, (2) these real entities are external to human minds, and (3) real entities exist objectively, i.e. independent of anyone's opinions.

These views imply that any scientific theory candidate for being true must describe its subject matter in conformity with these demands. In other words, if a description of a real entity contains a (implicit or explicit) reference to a human perspective or point of view, any such reference must be possible to get rid of. A description which contains irreducible relations to human minds is not acceptable for a realist. Einstein, for example, felt the force of this demand when developing his relativity theory. The demand for a theory free of human references implies that the fundamental equations must be invariant under changes of coordinate systems and relativity theory meets this demand.

The dual nature of the objects in quantum mechanics appears to be a violation of these realistic demands, because it seems as if quantum objects show wave or particle properties depending on which experiment we perform. This dualism is by itself not contradicting realism, if strictly physical conditions which tell us when the objects are waves and when they are particles can be given. However, this seems impossible: one and the same physical experiment shows the wave or the particle character of the fundamental entities depending on which question we are interested in. Consider for example the well-known two-slit experiment: we can choose photons, electrons, neutrons or whatever when performing this experiment. The result of the experiment is an intensity distribu-

R. S. Cohen, R. Hilpinen and Qiu Renzong (eds.), Realism and Anti-Realism in the Philosophy of Science, 329–338.

tion on a photographic film which reflects the wave character of the objects having passed the double slit. By measuring (1) the distance between the interference fringes, (2) the distance between the photographic film and the double slit and (3) the distance between the two slits, we can calculate the wave-length of the particles using the well-known formula $n\lambda = d \sin \alpha$. This calculation takes for granted that each object has passed through both slits in the screen. In other words, when calculating the wave-length we assume that each object is at two places simultaneously, viz. when passing through the screen with the double-slit. (Young, who first made this type of experiment with visible light, was thus convinced that light must be waves and not corpuscles as Newton had said, and the experiment was considered the crucial experiment in favour of the wave theory of light.) On the other hand, each particle had a definite position when hitting the photographic film, because each particle had made a tiny spot on the film. The position is determined with an accuracy determined by the mean diameter of the grains in the film emulsion. This accuracy, i.e., the mean diameter of the grains of the emulsion is much smaller that the distance between the two slits in the screen. But if the object really was a particle when hitting the photographic film, how then is it possible for these objects to show an intensity distribution corresponding to waves spread out over the entire target region? The conclusion generally drawn from this and other experiments is that quantum objects show wave or particle properties depending on what questions we pose to nature and not depending on any objectively existing circumstances. This is not an acceptable stance for a realist; consequently, every realist faces the problem of accounting for wave-particle duality in a way which is both empirically adequate and philosophically acceptable. At the very least we need a criterion which in an observer-independent way tells us when an object is a wave and when it is a particle. Here is my proposal:

EVERY OBJECT IS A WAVE PHENOMENON DURING MOTION AND A PARTICLE WHEN TAKING PART IN IRREVERSIBLE INTERACTION WITH OTHER OBJECTS

The criterion should be understood so as to imply that if no irreversible interaction occurs, the object is moving and thus a wave.

This criterion uses three vague concepts, viz. (1) wave, (2) particle and (3) irreversible interaction.

1. *Waves.* The natural language word 'wave' has many connotations. In this context it is appropriate to focus attention on two properties common to all sorts of waves: they obey the superposition principle and they have a non-negligible extension in space. The last point means that no wave can be said to be in one point.

2. *Particle.* The word 'particle' hardly belongs to natural language, it is a theoretical concept in classical mechanics. One often thinks of a particle as some small object, but that is not correct. A planet is in some contexts a particle. The core of the particle concept is that we can disregard the extension of the object called a 'particle'. Furthermore, particles do not follow the superposition principle because that would mean taking into account the inner constitution and thus its extension. It is obvious that no object can be considered to be a particle and a wave simultaneously.

3. *Irreversibility.* Irreversibility is a conceptual problem in physics generally: how can it be the case that truly irreversible processes exist when all fundamental equations of motion are reversible? Without further argument I will take it for granted that irreversibility enters physics in systems made up of a great number of (at least partly) independent objects. A monoatomic gas (at a pressure higher than the air pressure) in a container is a simple example. Once the container is opened most of the gas leaves the container. The reverse process, i.e., a gas distributed in a big volume assembling into a small part of that volume, never occurs and molecular kinetics explains this as a statistical effect. The reverse process is not logically conflicting with the fundamental laws of motion, but it is extremely improbable. But considering only a few gas atoms, it not at all unlikely that they will be found inside the container long after the container has been opened. The irreversibility thus enters at the macroscopic level as an effect of the motions of many independent parts of the macroscopic system.

I assume that this line of reasoning is correct even in the framework of quantum theory. Hence, the phrase 'irreversible interaction with other objects' refers to interaction with objects big enough to have the irreversibility property, i.e., in which the probability of a reversal of state change can be neglected.

A measurement on a quantum object is undoubtedly an irreversible interaction, but is should be observed that I am not supposing that measurement interactions are the only irreversible processes. For example, in many experiments only a fraction of the prepared objects is measured, the rest being lost outside experimental control. All these objects will sooner and later hit some macroscopic object, for example some irrelevant part of the experimental equipment. These collisions are not measurements but nevertheless irreversible interactions. There is reason to believe that even those interactions can be irreversible because the hit part of laboratory can undergo an irreversible state change.

(The term 'measurement' here refers only to those measurements where the wave function collapses, but those cases are the problematic ones for the realist.)

This analysis describes the wave-particle duality as a mind-independent property of quantum objects, and so far the analysis does not conflict with any empirical evidence.

How then can this criterion be used as an explanation of the two-slit experiment? The answer is: every object is a wave from the emission from its source to its collision with some big object. When passing the screen the object passes both slits and does not interact irreversibly with the screen: the object neither exchanges energy nor momentum with the double-slit screen. Behind the screen the wave is separated into two parts whose total momentum is the same as it was before the screen. These two parts interfere with each other and the intensity distribution of the total wave shows an interference pattern. When hitting the photographic screen the wave collapses and gives away its kinetic energy to one of the grains. This collapse is a sudden disappearance of the wave; it 'condenses' into one small region. Which one? That is impossible to predict, the place of the collapse being an truly indeterministic event. If we simply assume that the probability for collapse at a certain point equals the intensity (the squared amplitude) of the wave, the interpretation fits observed data.

I think this interpretation resolves the conflict between realism and wave-particle duality, but it could be argued that that is not enough: the sudden collapse of the wave when hitting a macroscopic object is still a complete mystery. A philosopher could reply that the explanation of the wave collapse is now a task for the physicist, not the philosopher. I think this is a proper defence, because the collapse of the wave is a completely physical process without relevance for the realism debate. However, the wave collapse can in fact be given some sort of explanation without making any assumptions which are not already implicit in quantum mechanics.

When assuming that real objects are wave-like objects except in irreversible interactions we would like to know why these wave-like entities show particle behaviour in some circumstances only. The answer is: the particle behaviour follows from *quantization of action*.[3]

Although it belongs to the very foundation of quantum theory, quantization of action is not a very well understood concept. Quantization does not mean that physical quantities generally are made up of multiples of least possible portions of every quantity, such as the elementary charge *e*. It is rather the case that each interaction process is one indivisible whole: when two objects approach each other either one indivisible quantity of energy is exchanged or nothing at all happens. This is the meaning of quantization of action and it is the guiding idea behind the choice of quantum mechanical operators to represent measurable quantities. Bohr was of the same opinion:

> . . . its (quantum theory) essence may be expressed in the so-called quantum postulate, which attributes to any atomic process an essential discontinuity, or rather individuality, completely foreign to classical theories and symbolized by Planck's quantum of action.[4]

It is to be noted that quantum theory does not contain any general rule by which we can find the operator representing an observable. Specifically, it is generally *not* true that if the observable A is a function of the observables B, C, . . . , i.e., A = f (B, C . . .), a similar functional relationship **A** = f(**B, C . . .**) exists between the corresponding operators **A, B, C**, . . .[5]

Now, I conjecture that quantization of action is the root of the particle character of waves in interaction. In order to illustrate the argument, the following thought experiment is suitable. Imagine a possible world in which Planck's constant *h* has a macroscopic value, 1 Js say, instead of its actual world value $6.6 \cdot 10^{-34}$ Js. Let us further assume that we perform a wave experiment in a water tank, akin to those made in elementary physics courses in wave mechanics. A dipper in the form of a straight line touches the water surface and water waves expand on the water surface from the dipper. Let us further assume that at the opposite wall of the water tank we have placed a row of sensors capable of registering the incoming waves. Taking into account quantization of action, the interaction between the dipper and the water must be quantized. That implies that each wave is an indivisible portion of energy. When this energy portion approaches the row of sensors, at most one sensor can take up this energy. Even if the energy spacing in each sensor is small enough to allow for a distribution of the energy to all sensors

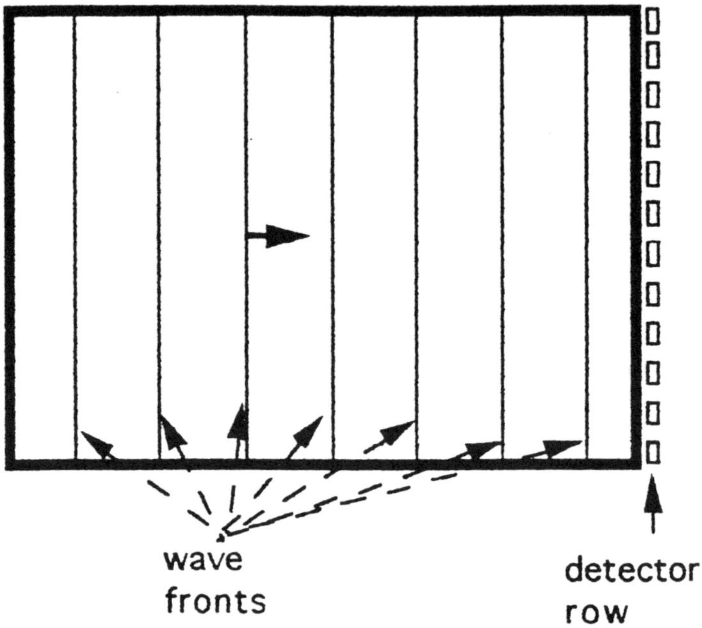

Fig. 1.

hit by the wave, this will not occur, because that would contradict quantization of action. Let us assume that the only observable results of this experiment are the records of the sensors. This record has the property that it never occurs that two sensors simultaneously are triggered. After a sufficiently large number of registrations the distribution of these registrations among the sensors maps the intensity distribution of the waves approaching the sensor row. But it seems as if the sensor row has been hit by a number of particles. There is nothing in the records telling us that waves were propagated on the water surface.

One could complain that this explanation does not show how the collapse of the wave happens: we have not been given any new insight into the details of the process. My answer to this argument is that this complaint confuses philosophical and physical explanation; asking for a detailed account of the collapse of the wave during interaction with certain other objects is to ask for more physics, it is not the philosopher's task. Moreover, the indivisibility of the energy exchange implies that it is impossible to describe further details. Hence it seems that a deeper physical explanation is not to be had. This is the limit where physical explanation must stop. But still I think this account has substantial explanatory value, because we have a deeper understanding of quantum phenomena. In my view, the salient feature which provides better understanding is that we no longer have to accept collapse of the wave function as a brute fact, but it is shown to be a consequence of a previously accepted trait of nature. Hence, the number of independently accepted statements in the theory is reduced. That this is a form of explanation is now a common view which is elaborated by Friedman[6] and somewhat differently by Kitcher.[7] Kitcher dubbed this form of explanation 'unification', and that term is now widely accepted, I think.

Schrödinger once proposed a wave interpretation of quantum mechanics, claiming that the intensity $\Psi^*(x, y, z)\Psi(x, y, z)$ of the wave function is a literal description of the object in question. In the case of the electron, the intensity of the wave function describes the density distribution of the charge of a single electron governed by the stated wave function. However, Schrödinger met severe criticism and was convinced that his ideas could not be correct. The objections raised against his wave interpretation,[8] could be summarized in three points:

(1) Wave functions are complex whereas real waves must be real entities.

(2) The wave function for a multi-particle system is multi-dimensional; the number of dimensions of the wave function is three times the number of particles in the object system. How should we interpret this wave function in physical three-dimensional space?

(3) The wave interpretation must take for granted that waves collapse during interaction with other objects. What is the mechanism and cause of this collapse?

1. The first problem can be solved if the real object is identified with the intensity of the wave function because this quantity is always real. Admittedly, the intensity, i.e., the squared amplitude, is not a wave function so one could criticize this stance by saying that real objects are not waves after all. However, the intensity has a number of features in common with real macroscopic wave phenomena and it is quite reasonable to extend the everyday word 'wave' to cover this type of objects. Specifically, the intensity of a wave packet made up of a number of monochromatic waves is a rather wave-looking phenomenon, and a wave packet is a much more plausible model of a real object than a monochromatic wave.

2. The second difficulty, the multi-dimensionality of many-particle

wave functions raises an interpretative problem only in EPR-type experiments, i.e., when objects partake in non-local interaction. In order to see this, consider first two non-interacting objects, *a* and *b*. The total wave function for this pair is $\Psi_{tot} = \phi_a + \phi_b$, i.e., a sum of two independent wave functions, one for each particle. We can ascribe to both objects a definite state independently of the other. The situation is not fundamentally different from a system of classical particles in this respect. However, if the particles are connected to each other non-locally, the situation is different. The well-known wave function for a spin zero pair,

$$\Psi_{tot} = \frac{1}{\sqrt{2}} (\phi_a(+)\phi_b(-) + \phi_a(-)\phi_b(+))$$

shows this clearly. It is not possible to ascribe a definite state to each particle independently of the other. So in this case the objection is valid. However, this is no objection specifically against the wave interpretation. The scientific community has settled on the view that non-local correlations are with us no matter what interpretation of quantum mechanics we choose. Non-locality seems to be an irreducible fact of the world and no philosophical interpretation can explain away this feature. The multi-dimensionality of the wave functions thus merely tells us, in the case of a correlated pair of particles, that the two waves are dependent on each other. It does not make up an argument against viewing objects as waves.

3. The objection that no explanation is provided for the wave collapse can be given an answer analogous to the answer to the second objection. No one else has provided a physical explanation of the mechanism of the collapse, so the wave interpretation fares no worse than its competitors. And if you accept that the proper form of explanation is unification, an explanation is in fact given. In any case, the advantage of the wave interpretation is that the projection postulate is superfluous. The collapse follows from the two assumptions made above, viz., that quantum objects are waves and that exchange of conserved quantities is an indivisible process.

Using this wave interpretation we can resolve not only a number of problems connected with wave-particle duality but some other problems as well. One such problem is the interpretation of probability formulas in quantum mechanics. The standard interpretation, due to Born, goes:

$P(x, y, z) = \Psi^*(x, y, z)\Psi(x, y, z)$ is the probability that we will *detect* the object at the point (x, y, z) when performing a measurement.

It is well-known that this formulation cannot consistently be translated so as to say that P(x, y, z) is the probability for the object to *be* in the point (x, y, z) irrespective of any measurements. That this latter more committing interpretation is impossible is a great obstacle for realism because this impossibility forces us to say that the objects have no definite properties before any measurements are made. However, this obstacle for realism can be removed if we adopt the wave interpretation. A wave with a non-negligible spatial distribution cannot reasonably be said to have a precise position. It is natural to propose that the probability $P(x, y, z) = \Psi^*(x, y, z)\Psi(x, y, z)$ is simply the *part* of the wave being in this point. If we further assume that the probability for interaction at a certain point equals the intensity of the real wave at that point, we have solved the interpretative problem. Thus, $P(x, y, z) = \Psi^*(x, y, z)\Psi(x, y, z)$ is not the probability for the entire object to be in (x, y, z), but the propensity for interaction, i.e. the propensity for exchange of energy at this point. This latter formulation fulfils realistic demands and if we assume that the proportion of the wave at a certain point equals the probability for interaction with measuring devices at this point, the interpretation is empirically equivalent to the Born interpretation.

But, it could be argued, using propensities in the explanation is a case of *obscurum per obscurus*. Propensities, and dispositions in general, are as much in need of clarification as are quantum probabilities. So we need a further analysis. The standard analysis of a disposition is in terms of inner constitution. Fragility, for example, is explained in terms of the strength of chemical bonding. Could we find an analogous explanation of the concept of propensity? Lawrence Sklar has argued that a general explanation of propensity is not to be had, because the propensity and hence the probability for an event depends on which class the event belongs to and there seems to be no objective way of classifying events. The propensity for an event depends on what we happen to know about the conditions, not on which conditions obtain. This argument we must accept, I think, but what if we know everything there is to know about the conditions and the future development still is indeterminate? What more can one say than there is a certain propensity for a change in a specified direction? Nothing, I think, and the request for an analysis of the concept of propensity when applied to irreducibly indeterministic events is to ask for too much. Hence, propensities could be acceptable in an analysis of indeterministic quantum phenomena. Lawrence Sklar hinted at the same conclusion. He rejected propensities as a general analysis of probability, but admitted that in the case of quantum mechanics, propensities might be acceptable because quantum events are irreducibly indeterministic.[9]

In conclusion: two important problems for a realistic interpretation of quantum mechanics can be solved by assuming that quantum objects are waves which exchange conserved quantities in discrete steps.

Department of Philosophy,
Stockholm University,
Sweden.

NOTES

[1] The ideas presented in this paper are part of my doctoral thesis *Understanding Quantum Mechanics; a Realist Interpretation without Hidden Variables*, Almquvist & Wiksell International, Stockholm, 1992.

[2] *Op. cit.*

[3] The concept of action was introduced by d'Alembert. It denotes the time integral of energy $\int Edt$.

[4] Cf. Bohr, N. (1934). *Atomic Theory and the Description of Nature*, Cambridge University Press, Cambridge, p. 53.

[5] Cf. Cohen, L. (1973). 'Joint Probability Distributions in Quantum Theory', in C. A. Hooker (ed.), *Contemporary Research in the Foundations and Philosophy of Quantum Theory*, Reidel, Dordrecht.

[6] Friedman, M. (1974). 'Explanation and Scientific Understanding', *J. Phil. 71*, no. 1, January 17.

[7] Kitcher, Philip (1981). 'Explanatory Unification', *Phil. Sci.* **48**, pp. 507–531.

[8] For an overview of this discussion see Jammer, M. (1974). *The Philosophy of Quantum Mechanics*, John Wiley & Sons, New York, pp. 31–33.

[9] Sklar, L. (1970). 'Is Probability a Dispositional Property?', *J. Phil.* **67**(11), June 11, pp. 355–366.

LARS-GÖRAN JOHANSSON

VAN FRAASSEN'S CONSTRUCTIVE EMPIRICISM – A CRITIQUE

Van Fraassen's Constructive Empiricism rests on the distinction between believing a theory to be true and believing it to be empirically adequate. This distinction in turn rests on an assumption which seems doubtful in the light of Goodman's paradox.

According to van Fraassen,[1] to believe a theory to be empirically adequate is to believe that the theory has a model, a part of which is isomorphic with all appearances. An appearance is a structure which can be described in a measurement or observation report. Van Fraassen takes pain to point out that this applies to all appearances, whether actually observed or not. In other words: a theory is empirically adequate if and only if a part of this theory is isomorphic with an infinite set of observation statements.

That means that to believe a theory to be empirically adequate is to use induction: having made a number of observations and measurements and found that a theory accounts for all observed phenomena, we might conclude that the theory is empirically adequate. Thus we are assuming that the theory also in the future will be in accord with our observations. Hence, van Fraassen claims that induction from observed to unobserved but observable phenomena sometimes can be justified.

On the other hand he claims that we are *never* justified in inferring the truth of a theory. To believe a theory to be true means to believe that it correctly represents the world and such a belief can never be justified according to van Fraassen. His reason for this rejection is underdetermination. If any amount of evidence is insufficient to distinguish between a number of incompatible theories, then none of the alternatives is more trustworthy than the rest. But then we must admit that theoretical statements cannot justifiably be believed to be true. This does not mean that they are void of meaning. Van Fraassen's position does not imply instrumentalism, he merely claims that we must stay forever agnostic about theoretical statements.

Obviously, van Fraassen's position rests on the distinction between observational and theoretical statements and the main criticism against

R. S. Cohen, R. Hilpinen and Qiu Renzong (eds.), Realism and Anti-Realism in the Philosophy of Science, 339–342.

him has focused on this issue. He has not been able to tell us exactly where to draw the distinction between theoretical and observational statements, i.e., the critics have argued he has exactly the same problem as the old positivists. His answer to this criticism could be summarized as follows:

1. Constructive empiricism does not face the same problem as logical empiricism because there is no semantic difference between theoretical and observational statements according to constructive empiricism. Both observational and theoretical statements are meaningful and have truth-values, whereas in logical empiricism theoretical statements lack truth-value. This implies that for the constructive empiricist it is acceptable to have a somewhat vague distinction observational/theoretical. It suffices if there are some clear cases on both sides.[2]

2. It is not the philosopher's task to draw a general distinction between observational and theoretical statements. Such a distinction is the business for scientists, because it is an empirical question, a question about us qua organisms in the world.[3]

3. The alternative of not using the distinction observational/theoretical is either to believe every accepted theory to be completely true, or to be a sceptic. Rejecting both these extreme positions, something in between must be found and if we want to say more about a scientific theory than it fits the facts so far, it is reasonable to say that it will continue to do so, i.e., that it is empirically adequate.[4]

It seems to me that these remarks make up a rather good defense of using a somewhat vague distinction observational/theoretical.

However, I believe that there are other grounds for disbelieving van Fraassen's distinction. I want to recall Goodman's discussion about induction in which he introduced the famous example of the predicate grue, which "applies to all things examined before t just in case they are green but to other things just in case they are blue".[5] It is clear that Goodman means that the time t lies in the future, hence it is plain that all hitherto available evidence is equally good evidence for emeralds being green as well as being grue. So simple induction cannot distinguish between projectible and non-projectible predicates and we can not distinguish between law-like and accidental generalisations.[6]

Goodman's conclusion was that the induction problem cannot be solved using any merely syntactical criterion. In other words, we cannot solve the induction problem by only considering the formal relations between the proposed general hypothesis and its evidence, i.e., its instances. Goodman proposed the concept of entrenchment as additional vehicle: among a number of rival hypotheses with equal evidence we shall choose the most entrenched one. This is admittedly not very clear. What sort of criteria do we use when deciding out of a number of rival hypotheses which one is the most entrenched?

In the present context there is a crucial question: 'is it possible to distinguish between rival hypotheses without using statements which van Fraassen would count as theoretical statements?' In other worlds, can we solve the induction problem solely within the observational realm? It seems impossible. Singular observational statements cannot do the job: either they are instances of all the rival hypotheses or they are irrelevant. Neither is it possible to rely on another general hypothesis about observable states of affairs: that move only triggers the ensuing question: how do we know that this new general hypothesis is a law-like generalisation and not merely an accidental generalisation? It is clear that we still have the same problem. These two alternatives, singular observational statements or generalisations about observable states of affairs, outrun our possibilities of finding distinguishing evidence in the domain of statements which according to van Fraassen can be known to be true. Hence, we must use theoretical statements when choosing the better entrenched hypothesis among a number of alternatives.

Now, the question is 'can van Fraassen use this way out of the induction problem?' No he cannot! Theoretical statements can according to constructive empiricism never be known to be true. But then they cannot be used as evidence for or against any hypothesis. Using an unsettled statement in an inductive inference to a statement which we claim to be true (a statement about a future observable event) is plainly irrational.

The conclusion to be drawn is that van Fraassen cannot accept Goodman's conclusion that we must go outside the realm of singular observational statements when making reliable inductive generalisations. Van Fraassen's constructive empiricism in fact presupposes exactly what is denied by Goodman and that throws serious doubt on van Fraassen's distinction between believing a theory to be true and believing it to be empirically adequate.

Of course, van Fraassen could reject Goodman's conclusion by providing a method for distinguishing between law-like and accidental generalisations without using theoretical statements. What he cannot do is to take for granted that this can be done. But that is precisely what he does when he makes the distinction between believing a theory to be true and believing it to be empirically adequate.

Department of Philosophy,
Stockholm University,
Sweden.

NOTES

[1] Cf. Fraassen, Bas van (1980). *The Scientific Image*, Clarendon Press, Oxford, p. 12.
[2] *Ibid.*, p. 16.
[3] *Ibid.*, pp. 57–58.
[4] Cf. Fraassen, B. van (1985). 'Empiricism in the Philosophy of Science', p. 254, in Churchland and Hooker (eds.), *Images of Science*, The University of Chicago Press, Chicago.
[5] Cf. Goodman, N. (1972). 'The New Riddle of Induction', p. 318 in his *Problems and Projects*, Bobbs-Merrill Inc., Indianapolis & New York.
[6] *Ibid.*, p. 388.

LI LI

ON COMPLEMENTARITY REALITY

Many interpretations have been given to quantum mechanics. Unfortunately, no one of them is completely successful and satisfactory. The most influential interpretation is the 'orthodox Copenhagen interpretation', which has been criticized as a positivistic and vague theory, and which always concludes with anti-realism. The realistic 'hidden variables theory', recently of greater and greater influence, faces the problems of nonlocality and unobservability, and it does not have more content than the orthodox theory. What I want to argue is that we should and can establish a realistic foundation for the anti-realistic Copenhagen interpretation – that is a foundation of complementarity reality to provide a realistic and reasonable interpretation, and to avoid the problems in interpreting quantum mechanics.

I. PROBLEMS IN THE COMPLEMENTARITY INTERPRETATION

Why do people feel that complementarity is difficult to understand? The enthusiastic advocate of complementarity, von Weizsacker, gets a definitely negative answer from Bohr after his most careful study of it. The famous opponent of complementarity, Einstein, complains that: "despite much effort which I have expended on it, I have been unable to achieve a sharp formulation of Bohr's principle of complementarity."[1] And in the current argument between realism and anti-realism, why is Bohr considered as a realist by some people and as an anti-realist by others? I think the reason is that there are some hidden contradictions in the complementarity principle, which seem to fit Bohr's aphorisms that "the opposite of a deep truth is also a deep truth"; "truth and clarity are complementary."[2] These contradictions give to complementarity a sense of vagueness, mystery and inconsistency.

1. *About the Reality of Atomic Systems*

Even though Bohr still uses the term 'atomic object', emphasizes the objectivity of the description of atomic phenomena, and tries to exclude

R. S. Cohen, R. Hilpinen and Qiu Renzong (eds.), Realism and Anti-Realism in the Philosophy of Science, 343–357.

the subject from the phenomena,[3] in fact, he does not consistently keep the reality of the atomic system in the complementarity interpretation. In fact, conversely, when he explains atomic phenomena by his complementary principle, he denies the independent reality of the atomic system. In his Como Lecture, he writes:

now the quantum postulate implies that any observation of the atomic phenomena will involve an interaction with the agency of observation not to be neglected. Accordingly, an independent reality in the ordinary physical sense can neither be ascribed to the phenomena nor to the agencies of observation.[4]

Obviously, he denies the independent reality of the atomic object and holds that the reality of the atomic object cannot be independent of observation. He notes that

the finite interaction . . . entails . . . the necessity of . . . a radical revision of our attitude towards the problem of physical reality,[5] of course there is in a case like that just considered no question of a mechanical disturbance of the system under investigation during the last critical stage of the measuring procedure . . . there is essentially the question of an influence on the very conditions which define the possible types of predictions regarding the future behaviour of the system, since these conditions constitute an inherent element of the description of any phenomena to which the term 'physical reality' can be properly attached.[6]

He denies the assumption of E-P-R that there is an independent reality of the object without disturbance by observation, and proposes that the term 'physical reality' must be revised so that it include not only the instrument which interacts with the object, but also the circumstances of the object, and these influence the conditions for predictions about the object. The dependent reality of the object results in the positivist interpretation of quantum mechanics. Bohr remarks

there is no quantum world. There is only an abstract quantum physical description,[7] quantum mechanical formalism which represents a purely symbolic scheme . . . ;[8] In this theory . . . in every stage of the computation only directly observable quantities enter . . . the new quantum mechanics does not deal with a space-time description of the motion of atomic particles;[9] the properties of atoms are always obtained by observing their reactions under collisions or under the influence of radiation;[10] radiation in free space as well as isolated material particles are abstractions, their properties on the quantum theory being definable and observable only through their interaction with other systems.[11]

These quotations show that Bohr denies that quantum theory describes the independent reality of the object and asserts that the properties of the atomic object can exist and be described and observed only under observation. He does not assume the independent reality of the atomic object and only admits the reality of the object with its circumstantial content under observation.

The denial of the independent reality of the object leads to the following problems in Bohr's interpretations: (A) If there is not an independent reality of the object, how can he say that the quantum theory describes the phenomena of the atomic object? What is the meaning of the "atomic object"and how can he explain the law which the quantum theory describes if the reality of the atomic object depends on the various and changeable circumstances? (B) If there is no independent reality of the object, what do the instruments which are used to observe the atomic phenomena act on, and how are the atomic phenomena created? Are the phenomena created by the instruments alone which have the ordinary meaning for their reality, and how can the object get its properties under observations? (C) If the reality of the object depends on our observations, the existence of the world would depend on our observations. But our practice in an experiment without an observer, and the existence of the world before our existence, both contradict this conclusion.

In fact, mutually exclusive phenomena under different observations reflect the fact that showing the properties of the object depends on its circumstances such as provided by the instruments but also properties both of the object and of the instrument are independent of any observer and his observation. Otherwise, there must be the problems mentioned above in the interpretation.

2. *About the Reality of Complementarity*

In the article 'Quantum Physics and Philosophy', Bohr states that the complementarity description is similar to relativity theory in keeping objectivity and avoiding subjectivity. In his 'Discussions with Einstein on Epistemological Problems in Atomic Physics', he notes that

> the term 'complementarity,' suited to embrace the characteristic features of the individuality of quantum phenomena . . . evidence obtained under different experimental conditions cannot be comprehended within a single picture but must be regarded as complementary in the sense that only the totality of the phenomena exhausts the possible information about the object.[12]

From these words, it seems he believes that the complementarity represents an objective attribute of the phenomena of the atomic objects, and it seems that it can be seen as representing the reality of the objects. But in fact he does not admit the reality of complementarity. He points out that the concept of complementarity represents ". . . the objective existence of phenomena and is related to the method of observing them."[13] That means that complementarity does not represent an objective attribute of the phenomena. It represents the dependence of phenomena on observation of them. It does not derive from the mutually exclusive properties of the reality of the objects. It derives from the different observations which create mutually exclusive phenomena of the objects. This meaning is also shown in these words of Bohr: ". . . the study of the complementary phenomena demands mutually exclusive experimental arrangements."[14] This says that the complementary phenomena result from the mutually exclusive experimental arrangements, not from the reality of the objects. But how do the instruments create mutually exclusive phenomena? He explains:

> According to the quantum theory, just the impossibility of neglecting the interaction with the agency of measurement means that every observation introduces a new uncontrollable element . . . the measurement of the positional co-ordinates of a particle is accompanied not only by a finite change in the dynamical variables, but also the fixation of its position means that a complete rupture in the causal description of its . . . momentum always implies a gap in the knowledge of its spatial propagation.[15]

That is to say, the interaction with the instruments destroys the foundation of a consistent description of the object and causes a gap in our knowledge of the object, which results in the complementarity of our knowledge about the objects. This idea is expressed in another style in the following words of Bohr:

> the occurrence of individual effects [permits] neither a choice of 'nature' nor a choice of 'observer', that only can be accounted as we are dealing with individual phenomena and that our possibilities of handling the measuring instruments allow us only to make a choice between the different complementary types of phenomena we want to study.[16]

In other words, the complementarity description does not describe the choice of nature; it represents our limited ability in handling the measuring instruments. That is why he always relates the complementarity to the failure of our forms of perception. For example, he declares,

> a close connection exists between the failure of our forms of perception, which is founded on the impossibility of a strict separation of phenomena and means of observation, and the general limits of man's capacity to create concepts, which have their roots in our differentiation between subject and object.[17]

These words show that complementarity, which indicates the impossibility of a strict separation of phenomena from means of observation, establishes the failure of our forms of perception, and has a close connection with the impossibility to differentiate subject and object. This results in the general limits of man's capacity to create concepts. In short, complementarity has a close relationship with the impossibility to differentiate subject and object. So it does not represent the reality of the objects.

His denying of the reality of complementarity leads to the following problems: (A) If complementarity results from the mutually exclusive instruments, why do different instruments react on the same object according to mutually exclusive styles? Why are the different instruments, which have the same physical reality in ordinary common sense usage, mutually exclusive? (In fact, they are mutually exclusive just because they show mutually exclusive phenomena of the atomic subjects.) (B) If complementarity shows the limitation of our perception and the capacity to create concepts, the reality of the independent world would be beyond our epistemic ability, and it would be unknowable. This conclusion contradicts the scientific object in its capacity to recognize the reality of nature.

Actually, both the objective existence of mutually exclusive phenomena of the objects, together with successful complementary description of the object prove the reality of the complementarity. The complementarity is the reality of the object.

3. *About the Reality of the Instruments*

On the one hand, Bohr does not consider the instrument as independent reality when he says that, because of the interaction between the instruments and the objects, "an independent reality in the ordinary physical sense can neither be ascribed to the phenomena nor to the agencies of the observation,"[18] which means the instruments have no independent reality, just as is said of the objects. So in Bohr's analysis of the experiments in "Discussions with Einstein on Epistemological Problems in Atomic Physics", the momentum of the instrument is also uncertain when its position is certain. But if the instrument is also dependent and uncertain, how can it interact with the object and how can we get definite phenomena from it? On the other hand, in order to get the complementary phenomena and descriptions, he considers the instruments as certain and an independent reality in the ordinary physical sense. He claims that in the analysis of the quantum phenomena, the essentially new character is the fundamental differentiation between the instruments and the objects, which is the result of the necessity that we must use classical concepts to explain the function of the instrument and we exclude any consideration about the quantum of action.[19]

The instrument is described as having double realities: independent and certain reality in the classical physical sense and dependent and uncertain reality in the quantum physical sense. How can it play double roles at the same time? And if the instruments are considered as independent reality and can be described by classical concepts, the atomic objects would belong to a different world from the instruments. This conclusion contradicts the consistency principle of science.

About the influence of the instrument on the object, Bohr again provides two explanations. On the one hand, it is considered as physical interaction, in which there is exchange of energy and momentum. On the other, in these analysis of the E-P-R experiment, it is considered as an influence upon the conditions of the types of predictions of the objects and there is no mechanical disturbance in the process (as note 6 notes). What the actual influence of the instrument on the object is remains unclear. In reality, in order to keep scientific consistency, we must consider the instruments to belong to the same world of the objects. Both of them must be considered as independent reality but their properties are uncertain according to quantum theory. When they exist in a certain relationship, such as when the instrument is used to measure a certain property of the object, both of them show a certain property and certain phenomena. What is special is that their showing of the properties depend on the relationship between them, but not the mysterious influence of the dual-roles-instruments.

4. *About the Reality of Physical Concepts in Quantum Theory*

We ask whether the physical concepts in quantum theory represent any reality and what reality do they represent? Bohr declares that the classical theory cannot describe atomic objects: "the quantum theory is characterized by the acknowledgement of a fundamental limitation in the classical physical ideas when applied to atomic phenomena,"[20] the essential discontinuity or rather individuality in any atomic process is "completely foreign to the classical theories."[21] Correspondingly, the classical concepts are also considered limited in describing atomic phenomena, which is shown in these words of Bohr:

the fundamental contrast between the quantum of action and the classical concepts is immediately apparent from the simple formulae . . . $ET = I\lambda = h$.[22] The Limitation in the classical concepts expressed through relation (2) ($\Delta t \Delta E = \Delta x \Delta I_x = \Delta Y \Delta I_y = \Delta Z \Delta I_z = h$), is, besides, closely connected with the limited validity of classical mechanics.[23]

But he contends that only the classical concepts can be used to definitely explain the result of the experiment. He asserts:

however far the phenomena transcend the scope of classical physical explanation, the account of all evidence must be expressed in classical terms.[24]

Even though the classical concepts are limited in describing atomic phenomena, the phenomena must be expressed in classical terms. How can we do that? His analysis:

on one hand, the definition of the state of a physical system, as ordinarily understood, claims the elimination of all external disturbances. But in that case, according to the quantum postulate, any observation will be impossible, and, above all, the concepts of space and time lose their immediate sense. On the other hand, if in order to make observation possible we permit certain interactions with suitable agencies of measurement, not belonging to the system, an unambiguous definition of the state of the system is naturally no longer possible, and there can be no question of causality in the ordinary sense of the word.[25]

According to Bohr, the definition of state claims the elimination of all observation including the observation of the state. The definition of state and the observation of it are mutually exclusive. If so, what is the meaning of the 'observables' in quantum mechanics and how can we get a definite description of the state of the object from an experiment on it? According to Bohr's words above, the definition of state excludes observations of the state and the space and time as well as the meaning or definition of space and time. But in the analysis of the ideal experiments in 'Discussion with Einstein on Epistemology Problems in Atomic Physics', the observation of the energy and the momentum make it possible to get the definite description of the energy and the momentum, which implies the definite definition of them, and bring uncontrollable changes of time and space making both the definitions and the observations of them impossible. And the observation of space and time make it possible to describe the space and the time and bring uncontrollable changes of momentum and energy, making both the definitions and the observations of them impossible. The observation of the state both excludes and infers the definite definition of it. The relations among the observations and definitions of the energy, the momentum, the space and the time of the object are ambiguous and entangled. How these concepts can properly be used in quantum mechanics to become the 'observables' is unclear.

Bohr declares:

an adequate tool for a complementary way of description is offered precisely by the quantum-mechanical formalism which represents a purely symbolic scheme permitting only predictions, on lines of the correspondence principle, as to results obtainable under conditions specified by means of classical concepts.[26]

This means the quantum theory is merely a symbolic scheme which can only predict the result under conditions specified by means of classical concepts. But how can the conditions be specified by means of classical concepts? Bohr claims:

In the quantum-mechanical description our freedom of constructing and handling the experimental arrangement finds its proper expression in the possibility of choosing the classically defined parameters entering in any proper application of the formalism.[27]

This means that the proper use of classically defined parameters is conditioned by the proper constructing and handling of the experimental arrangement. But how can the proper experimental arrangement make the limited classical parameters suitable to the atomic phenomena if the observation in atomic phenomena is essentially different from classical observation? Bohr does not give a clear explanation. He just declares ambiguously:

the very nature of the quantum theory thus forces us to regard the space-time co-ordination and the claim of causality, the union of which characterizes the classical theories, as complementary but exclusive features of the description, symbolizing the idealisation of observation and definition respectively . . . In the description of atom phenomena, the quantum postulate present us with the task of developing a complementarity theory the consistency of which can be judged only by weighing the possibilities of definition and observation.[28]

That is to say, the space-time co-ordination and the claim of causality are exclusive. We get definite space-time co-ordination only under the condition of the impossibility of the claim of causality. The limitation of classical concepts is overcome by using the two exclusive descriptions without combining them in one picture. The complementary descriptions can describe the atomic phenomena but the complementarity does not come from the reality of the objects, it comes from the limitation of the classical concepts. The classical concepts can represent the reality of the atomic phenomena through the method of complementarity.

As a matter of fact, the de Broglie relation and the uncertainty relations show that the concepts of momentum p, position q, energy E and time t in quantum theory are no longer the classical concepts, because the classical p.q.E. t have not these special relationship. They have the classical meaning when the instrument and the objects are in certain relationships and they show their proper p.E. or q.t. So the physical parameters in quantum theory represent the reality of the atomic objects.

In brief, the presumption of the uncontrollable, interaction between the object and the instrument which is the core of the complementarity interpretation, is bound to the denial of the independent reality of the object and the instrument, and the limitation of the classical concepts, which cannot be observed and defined simultaneously. But in order to get definite phenomena and a finite description of them, the independent reality of the instrument and the proper application of the classical concepts under the certain using of the instrument must be introduced in the interpretation. The interaction between the object and the instrument actually presupposes the independent reality of the object and the instrument. These contradictions are all caused by the presumption of the uncontrollable interaction between the object and the instrument. In reality, no uncontrollable interaction between the object and the instrument happens. There are only the different and certain relationships between them, which determine the different properties and the different phenomena that they show.

II. THE VIEW OF COMPLEMENTARITY REALITY AND ITS INTERPRETATION OF QUANTUM THEORY

The view of complementarity reality can be concluded from the analysis in Part I: both the object and the instrument are independent complementarity realities which contain mutually exclusive properties (which are described by uncertainty relations); each of the properties is uncertain when the object has no certain relationship with its circumstantial content. When the object is in a certain relationship with its circumstance the object changes to show the certain property which corresponds to the certain relationship at the same time, the circumstance such as the measuring instrument also changes to show the same kind of property. Because of the quantum state of the object and the process of the change, the objects statistically show certain values of the property under certain relationship with its circumstance.

1. *The Reasonableness of the View of Complementarity Reality*

(A) The view of complementarity reality can resolve the problems in the complementarity interpretation. (a) in this view, both the object and instrument are independent complementarity reality, which can avoid the problem that the reality of the object depends on observation and that the instrument belong to a different world from the object. The dependence of the atomic phenomena on the observation does not mean the reality of the object and the instrument depend on the observation. It means that the certain presentation of reality depends on the certain relationship between the object and the instrument. (b) According to this view, the mutually exclusive phenomena directly derive from the complementarity reality of the object, which avoids the problem of attributing the complementary phenomena to different instruments and the limitation of man's epistemic ability, (c) In the light of this view, the influence between the object and the instrument is mutual. Their certain relationship determines their certain properties, which avoids the problem of the unclear and different interactions between the instruments and the objects. (d) On the basis of this view, quantum mechanics describes the complementarity reality of the object. The uncertain description of the state of the object without measurement, and the certain description of the state of it under measurement in quantum mechanics, reflect the uncertain state of the complementarity reality of the object without a certain relationship with its circumstance and a certain state of it under certain relationship. It avoids the problem of using the limited classical concepts to describe the atomic phenomena, in which the concepts have no clear meaning and do not describe the reality of the object but merely describe the phenomena through the complementary method.

(B) The complementarity reality is consistent with facts, theory, and practice in the atomic field. The complementarity reality is shown by the objective complementary phenomena of the atomic objects. The fact that our view of complementarity reality can reasonably interpret quantum mechanics (as this paper shows later) indicates that the complementarity reality is described by the successful quantum theory. In practice the processes of the atomic objects (without observers) obey the quantum theory which shows the independent reality of the objects and the objectivity of the quantum theory. In fact, in our ordinary common sense field, there are also some remarkable phenomena which show the complementarity reality. For example, the metaphor that Bohr always quotes, that we are both actor and audience in the stage of life, describes our complementary properties. When we are on the stage, relating to the audience, we are actors, but we also can suddenly become the audience when watch the actors on the stage. When there is no stage, or all people are on the stage, our properties are uncertain. We have the mutually exclusive properties of actor and audience, our property is certain when we have a certain relationship with our circumstance. We change from uncertain property to a certain property, or from one certain property to another one, when our relationship with our circumstances changes. But the ordinary object always show a certain property and it is difficulty to observe its complementary properties, because an ordinary object always exists in a certain relationship with its circumstances, which are the ordinary world in which we live. Therefore the complementarity reality is general, not suitable only to the microfield. The complementarity does not derive from our limitated ability to observe and recognize the atomic objects.

(C) The view of complementarity reality can keep the consistence between ontology and epistemology. It assumes that what we describe and observe are what exist.

(D) The view of complementarity reality is consistent with the scientific belief which is the foundation of all scientific researches. The scientific belief is definitely described by Einstein as "the belief in an external world independent of the perceiving subject is the basis of all natural science."[29] It is also expressed by Bohr, when he notes that the description of atomic phenomena is absolutely objective – it does not involve any observer.[30] The scientific belief that there must be reality outside of us and independent of our observation, and that our theories objectively describe it, is retained in quantum mechanics by the view of complementarity reality.

2. *The Reasonable Interpretation of Quantum Mechanics in the View of Complementarity Reality*

(A) About the Wave Function. According to the view of complementarity reality, the ψ function describes the uncertain state of the complementarity reality of the object. Because the complementarity reality contains mutually exclusive properties, its state is uncertain and each of the properties is uncertain, but the range of the properties obey the uncertainty relations. The properties of the energy E and the momentum P, and the properties of the position q and time t, are the essential mutually exclusive properties of the object. The object cannot be described by space-time description of its causal motion. The properties of E and P and the properties of q and t are not certain at the same time. The ψ function reflects the attributes of the uncertainty and mutual exclusion of the complementarity reality of the object. The Schrödinger equation ($H\psi = ih\, \partial\psi/\partial t$) reflects the law of complementarity reality. The $\psi(K)^2$ describes the probability of the change of the object to a certain k value of a certain k property under a certain k relationship with its situation. For example, when the position of the object is measured, the object has a certain position relationship with its circumstance (the instrument), and $|\psi(x)|^2$ describes the probability of the transition of the object to a certain positional value x. When the result of the measurement of the position of the object is λn, the state of the object is one of the eigen-functions ϕn (λn is the eigenvalue of position, ϕn is the eigenfunction of position).

(B) About the Observables. In the view of complementarity reality, the concept 'observables' should be changed to the mutually exclusive properties of the object, because the existence of these properties is independent of observations, even though the certain appearance of the properties depends on observations. The certain extents of the appearances of the mutually exclusive properties are determined by the uncertainty relations [$\Delta p \Delta q \geq (\sim h)$, $\Delta E \Delta t \geq (\sim h)$], which reflect the mutually exclusive relations between the extents of the appearances of the complementary properties of the object by the reciprocal relations. The conventional axiom about the observables in quantum mechanics is:

the probability of observables (A, B, . . .)
if observed at time t
being observed to be (a, b, . . .)
is

$$\Sigma|\langle a, b, \ldots q|t\rangle|^2$$

where q denotes additional quantum numbers which together
with the eigenvalues (a, b, . . .)
form a complete set.

should be replaced by

> the probability of the properties of complementarity reality (A, B, . . .).
> under the (A, B, . . .) relationship with its circumstance at time t
> being (a, b, . . .)
> is
>
> $$\Sigma|\langle a, b, \ldots q|t\rangle|^2$$
>
> where q denotes additional quantum numbers which together with the eigenvalues (a, b, . . .)
> form a complete set.

(C) About Measurement. In the light of the view of complementarity reality, the measurement does not mean an uncontrollable interaction between the object and the instrument; it means the changes of both the instrument and the object to show certain properties under certain relationship between them in the measurement. Because both the object and the instrument are the independent complementarity reality, the appearances of their properties are uncertain before the measurement. When they are given in the measurement, their certain relationship determines their changes to show the same certain property. When the appearances of their properties are certain, the situation in the measurement is similar to that in the classical measurement, i.e. the instrument reflects the property of the object. Then the classical concept which describes the property in the measurement can reasonably be used, even though at the same time the other concept (which is mutually exclusive to this concept) can not be used since the object does not show the other property under this particular certain relationship of the object with its circumstance. The wave packet reduction in quantum mechanics reflects the change of the object from an uncertain state to a certain value of a certain property under the certain relationship with its circumstance.

(D) About the E-P-R Paradox. On the basis of the view of complementarity reality, the E-P-R paradox reflects the relative attribute of the complementarity reality, which shows a certain property under a certain relationship with its circumstance. In the E-P-R experiment, when object A interacts with object B, there is a relationship between them. When they do not interact to each other, the relationship still exists.

This relationship determines that both of them show the same property of their complementarity reality. When object A is measured and changes to show its certain momentum. P_A (or position q_A), the object B also changes to show its certain momentum P_B (or position q_B) because the object A and the instrument acting on object A, which are the circumstance of object B, determine the certain appearances of the properties of object B. Therefore, the E-P-R paradox does not mean we get the complementary descriptions of object B simultaneously without disturbing it. It just shows the radical relativity of the appearances of the properties of the complementarity reality of the object to its circumstance.

Institute of Philosophy,
Hebei Academy of Social Sciences,
Shijiazhuang, China.

NOTES

1 Jammer, M. (1974). *The Philosophy of Quantum Mechanics*, John Wiley & sons, pp. 90–91.
2 Bell, J. S. (1987). *Speakable and Unspeakable in Quantum Mechanics*, Camb. Univ. Press, p. 190.
3 Bohr, N. (1958). *Quantum Physics and Philosophy – Causality and Complementarity.*
4 Bohr, N. (1934). *Atomic Theory and the Description of Nature*, Camb. Univ. Press, p. 54.
5 Schilpp, P. A. (ed.) (1949). *Albert Einstein: Philosophyer-Scientist*, Tudar Pub. Co., p. 233.
6 *Ibid.*, p. 234.
7 Jammer, M. *The Philosophy of Quantum Mechanics*, p. 204.
8 Schilpp, P. A. (ed.), *ibid.*, p. 210.
9 Bohr, N., *ibid.*, p. 48.
10 Bohr, N., *ibid.*, p. 95.
11 Bohr, N., *ibid.*, pp. 56–57.
12 Schilpp, P. A. (ed.), *ibid.*, pp. 209–210.
13 Bohr, N. (1958). *Quantum Physics and Philsophy – Causality and Complementarity.*
14 Schilpp, P. A. (ed.), *ibid.*, p. 211.
15 Bohr, N., *Atomic Theory and the Description of Nature*, p. 68.
16 Schilpp, P. A. (ed.), *ibid.*, p. 223.
17 Bohr, N., *ibid.*, p. 96.
18 Bohr, N., *ibid.*, p. 54.
19 Bohr, N. (1958). *Quantum physics and philosophy – Causality and Complementarity.*
20 Bohr, N., *Atomic Theory and the Description of Nature*, p. 53.
21 Bohr, N., *ibid.*, p. 53.
22 Bohr, N., *ibid.*, p. 57.

[23] Bohr, N., *ibid.*, p. 60.
[24] Schilpp, P. A. (ed.), *ibid.*, p. 209.
[25] Bohr, N., *ibid.*, p. 54.
[26] Schilpp, P. A. (ed.), *ibid.*, pp. 210–211.
[27] Schilpp, P. A. (ed.), *ibid.*, pp. 230.
[28] Bohr, N., *ibid.*, pp. 54–55.
[29] Schilpp, P. A. (ed.), *ibid.*, p. 248.
[30] Bohr, N. (1958). *Quantum Physics and Philosophy – Causality and Complementarity.*
[31] Jammer, M. (1966). *The Conceptual Development of Quantum Mechanics*, McGraw-Hill Book Comp.
[32] Cohen, R. S. and Stachel, J. (ed.) (1979). *Selected Paper of Leön Rosenfeld*, D. Reidel Pub. Comp., 1979.
[33] Bohm, David (1954). *Quantum Theory*, Prentice-Hall Inc., New York.
[34] Ge Ge (1985). *Niel's Bohr – His Life, Research, and Ideas*, Shanghai People's Pub House.
[35] Zhou, Shi-xun (ed.) (1979). *A Course of Quantum Mechanics*, People's Educational Pub. House, Beijing.
[36] Lu, He-Fu (1984). *Textual Interpretation of the Quantum Theory of the Copenhagen School*, Fudan Univ. Pub. House, Shanghai.
[37] Guo, Gui-chun (1991). *Contemporary Scientific Realisms.* Scientific Pub. House, Beijing.
[38] Wu, Da-You (1984). *Quantum Mechanics*, Scientific Pub. House, Beijing.

LUO JIACHANG AND HU XINHE

RELATIONAL REALISM ON REFORM OF THE VIEW OF PHYSICAL REALITY AND ITS LOGICAL MANIFESTATION

The interpretation of quantum mechanics (QM) is one of the most controversial issues of 20th century science and philosophy of science. As some scientists and philosophers mentioned, many problems in it, such as wave collapse and the measurement problem, wave-particle duality and non-visualizable reality, the EPR argument and correlation-at-a-distance, etc., concerned both physical theory and philosophical concepts. Many quarrels came from the divergences of the disputers' philosophical views, especially of their view of reality. Then, a key question for interpretation of QM is whether we should and can hold on to the classical view of reality after such a radical change from classical physics (CP) to QM, and if not, how should we modify it? By examining the classical view of reality, the conceptual features of QM, and some affirmative interpretations of QM, we try to give a somewhat new view of reality, its logical form, and some applications of it. We hope that our tentative answer to this question may be helpful to provide a new perspective in understanding physical reality.

1. THE CLASSICAL VIEW OF REALITY

The classical view of reality could be called "the view of substance (orientities)", or "the view of the primary quality". The main ideas are: The objects of our recognition, especially in scientific research, are the entities in the world. These entities and their attributes are realistic and objective and by these terms the classical view stresses their independence and invariance. An entity is both qualitatively and quantitatively definite, independent of human observation, existent in itself, as well as independent of other entities or environment, existent separably or locally.

But as a fact even found in ancient times, we know that some physical phenomena, like the colour of light, sound, smell and contact feelings, etc., are related to human senses as well as entities themselves. To explain these phenomena, a strategy had been developed in traditionally phi-

R. S. Cohen, R. Hilpinen and Qiu Renzong (eds.), Realism and Anti-Realism in the Philosophy of Science, 359–379.

losophy, from the atomist and sceptical school of the ancient world via the Middle Ages to Kepler, Galileo and to its climax with Locke. The strategy was to make a clear distinction between that in the world which is absolute, objective, immutable, mathematical, and that which is relative, subjective, fluctuating and sensed. The former used to be called primary qualities which could be attributed to substance as its intrinsic, essential attributes, while the latter were called secondary qualities, declared to be effects of the action of primary qualities on human sense organs. So, the secondary qualities are binary (or plural) functions, neither reducible to the substantial entities, i.e., the primary qualities completely, nor ascribable to the state of human senses in and of themselves, namely

(1) $y = f(x_1, x_2, \ldots)$

where y are the secondary qualities, x_1 are entities and their essential attributes, x_2 are the states of human senses. Then, the y's are relative manifestations, or projections of x_1 or x_2, or relations between x_1 and x_2. But when the y's are the primary qualities.

(2) $y = g(x_1)$

they have nothing to do, in their existence, with surrounding variables. That is the very ideal of the view of an [objective] entity. Therefore it was based on the objectivity and invariance of the primary quality, called also "the view of primary qualities".

The classical view of reality is consistent with concepts and experiments of CP as well as our daily experiences. In fact, some basic constants, like mass, extension, time interval and some basic characteristics of matter, like particle and wave, in CP were often used as examples of primary qualities, while the view of entity [or substantial thing] constituted a part of the conceptual structure of CP. In Lakatos's terms, we can analyse the 'hard core' of a physical research program into components: a theoretical one and a metaphysical one. Referring to the former, we mean those basic postulates and laws of a theory which constitute the physical foundation of a research program. In CP, for example, Newton's three laws of motion and the law of universal gravitation are such components which are applicable to all the mechanical phenomena of objects. The latter refers to the basic conceptual framework of the world-picture of physicists, consisting of sediments accumulated by physical and philosophical thinking from time to time, and serving as criteria both in construction and evaluation of theories, in a sense called 'physical ontology'. Causality, continuity of motion, concepts of space and time, the view of reality, are all cases of it and its example in CP is just such a mechanism that all the various phenomena in the world could be explained in terms of continuous motion of the particles with definite, objective properties in absolute space and time, according to exact causality. In principle, we have no reason to hold on to either of these two sorts of components during a scientific revolution, although the latter is more resistant to variation. In practice, we can give the field form of matter in Maxwell's theory and the view of space and time in Einstein's relativity as historical cases of such reforms of metaphysical components.

2. THE BASIC FEATURES OF QM

With the development of physics, the ideals of the view of [substantial] entities seems to have gone to ruin. From the theory of relativity, it follows that the geometric-kinematic properties of position, time and velocity (as well as basic constants in CP like length, size, duration and mass), previously regarded as invariant features, depend upon reference frames. They have different values in different frames but all of the values are real and equivalent. So, like the secondary qualities, the primary qualities are also not intrinsic and invariant but relative to, or manifested in, the relations with observation. Concerning the primary qualities, physical phenomena are binary functions of both x_1 and x_2.

(3) $y = g(x_1, x_2)$

where now x_2 are the reference frames. There are still some philosophers, especially in China, who want to insist on the view of entities proper by giving other interpretations to the theory of relativity. One depends on the distinction between the effects of kinematics (due to velocity of motion) and dynamics (from acceleration), and argued that most effects of relativity belong to the former which are not real but apparent. So, measurement values = proper value + kinematical effects, in which proper value is measured in the rest frame. But this interpretation seems to contradict the basic idea of relativity that all measurement values are real and all frames have equal right. In other words, it seems to come back to Lorentz's idea of 'local times' in some extent.

Another interpretation tries to maintain objectivity of the primary qualities by appealing to invariant variables. With this aim, it defines physical reality in 4-dimension space in which the interval is invariant to Lorentz transformation. But it seems that to retain the definition of reality as invariant, you have to pay a very high price: giving up the reality of all physical properties in the 3-dimensional world, defining a set of new primary qualities in the 4-dimensional world which might be grasped only in ideas, not practices, and getting the 'block world' which is abstract, static, not evolving, all determined including all causal chains from the past to the future. Such a world picture will cause a lot of difficulties.

Let us go back to modern physics. We find that much more serious and fundamental clashes come from QM in which the whole framework of concepts seems to contradict the conceptual view of entities. In fact, QM, in contrast to CP, is an essentially new research program with a hard core whose theoretical and metaphysical components are both radically different from classical ones. So we shouldn't hold on at any price to an invariant concept of reality, but rather adjust it to be coherent with the whole conceptual structure. From the basic postulates of the QM formalism, including postulates of state-vectors, operators and eigenvalues, probability of prediction, superposition, Schrödinger equation, etc., to its inferences and interpretations such as the measurement problem, uncertainty relations, wave-particle duality, we can find a far-reaching change, not only in the content of physical knowledge, but also in the fundamental concepts in terms of which such knowledge is to be expressed. The three basic features of these concepts are (Bohm, Chapter 8): (i) Discontinuity of the quantum process, i.e., replacement of the notion of continuous trajectory by that of indivisible transitions; (ii) Probability or indeterminacy of quantum laws, i.e., replacement of complete determinism by that of causality as a statistical trend; (iii) Wave-particle duality of quantum objects, manifested in different experimental conditions, i.e., replacement of the assumption that the world can be analysed correctly into distinct parts, each having some fixed 'intrinsic' nature (for instance, wave or particle), by the idea that the nature of quantum objects depends on the external experimental conditions. That means a quantum phenomenon, even the world, is an indivisible whole in which parts appear as abstractions or approximations, valid only in the classical limit.

The quantum feature, namely the indivisibility of the quantum, implies the indivisibility of both the quantum of action and the quantum process. From the former, we cannot draw a clear line of demarcation between quantum object and measuring instrument; in principle, we cannot recognize the state of an object as thing in itself, but rather the integral phenomena as the results of interaction. From the latter, we can find a way to a probabilistic feature, i.e., deterministic laws in CP which were based on the principle of continuity are invalid in QM, and we can only give probabilistic predictions concerning changes of state.

So, causality becomes statistical trend, physical properties become the probabilistic spectra of eigen-values, and even the quantum objects, described completely by wave functions, become probabilistic. Just as was revealed in the wave-particle duality, quantum objects are unique in the sense that they are essentially different from those in CP. It is not completely identical with either a particle or a wave, nor both a particle and a wave, but instead it is something potentially capable of developing either one of these aspects of its behavior at the expense of the other. Which of these opposing potentialities will actually be realized in a given case depends as much on the nature of the system with which the quantum object interacts as on the object itself. Then, wave property and particle property, which were thought to be utterly different and absolutely opposite forms of matter in the classical framework, become relative manifestations of the same system under different experimental conditions in the quantum formalism.

Obviously, the above features of QM clash not only with the basic principles and world picture of CP, but also with the classical view of reality, because they reveal the quantum integrality, the probability and potential possibility of quantum objects, while denying the separability, locality, and definite, invariant properties, both qualitative and quantitative, of quantum objects. The alternative ways are: holding on to the view of entities, you must at least deny some of above features, as Einstein did, or give them compatible interpretation, such as local hidden variable theory; admitting and insisting on these features, it is necessary to develop a view of quantum reality which is consistent with them.

3. THREE VIEWS OF THE QUANTUM REALITY

Due to the aim of this paper, we will discuss only three views of reality in QM whose common features may be found after our sketch of them.

3.1. *Complementary Realism or Phenomenal Realism*

This is Bohr's view of reality. At first, the significance of the principle of complementarity is mainly in epistemology, in revealing the complementary limitation of the classical concepts (which are suitable to continuous processes) when applied to atomic phenomenon. Because "the quantum postulate implies that any observation of atomic phenomena will involve an interaction with the agency of observation not to be neglected" (Bohr, 1934, p. 54) and not to be analyzed, so two basic concepts of CP, the definition of state, and of observation or measurement, are incompatible. But as long as classical concepts are used, both of them are necessary for the complete description of quantum phenomena. Therefore, clear definition and precise observation (also space-time description and causal description) are mutually exclusive and complementary. From this reply to EPR, in which he claimed the ambiguity involved in not distinguishing the system as an observed phenomenal object from the system as an independent reality led to the invalidity of Einstein's argument, Bohr began to pay great attention to ontological meanings of complementarily (see Folse, pp. 154–161). He emphasized "it should be clear that the whole situation in atomic physics deprives of all meaning such inherent attributes as idealizations of CP [i.e., 'particle' and 'wave' pictures] would ascribe to the object" (1937, pp. 292–293). He used the word 'phenomena' to refer now not to independent reality but to the whole observational interaction, "to reserve the word 'phenomena' for the comprehension of the effects observed under given experimental conditions" (Bohr, 1939, p. 24). The "rational account of a phenomenon . . . must be taken to involve a complete description of the experimental arrangement as well as observed results" (Bohr, 1947, pp. 59–60). From this notion, we can get different phenomena, namely different observational interactions under different experimental conditions, and also get the complementarity of different phenomena as a logical consequence of the complementarity of experimental conditions.

We can thus draw the outline of Bohr's view of quantum reality: quantum objects cannot be described by classical concepts except through their interactions with observing instruments, the attributes of quantum object are real only in a context involving the experimental arrangements. These interactions and real quantum attributes are what we called 'phenomena'. It is the key word in Bohr's view of reality, at least since 1939. The characteristics of quantum phenomena are:

a. They are indivisible.
b. They are real.
c. They are complementary.

So all the information about the same object x_1 can be obtained by a set of phenomena y(i)

$$(4) \qquad \{y(i)\} = \{F(x_1, x_2(i))\}$$

where $x_2(i)$ refer to experimental arrangements. So the conclusion is that Bohr's phenomenal realism is characterized by the relations between quantum objects and experimental conditions, by stressing "These conditions constitute an inherent element of any phenomenon to which the term 'physical reality' can be properly attached" (Bohr, 1935, p. 700).

3.2. *Potential Realism and Latent Realism*

The first is what we call Heisenberg's view of reality. As also a main figure of the Copenhagen School, Heisenberg surely had many ideas which were identical or similar to Bohr's, but also he had his own unique ideas. As an eminent physicist, Heisenberg clearly recognized "that the results of modern physics do touch such fundamental concepts as reality, space and time, the confrontation may lead to entirely new developments which cannot yet be foreseen." "It is in quantum theory that the most fundamental changes with respect to the concept of reality have taken place" (Heisenberg, p. 28). In the analysing the relations between 'Language and Reality in Modern Physics' (*Ibid.*, chap. x), Heisenberg in fact distinguished two sorts of language (mathematical and ordinary) and two levels of object (reality, and facts as the results of measurement); his main ideas may be illustrated thus:

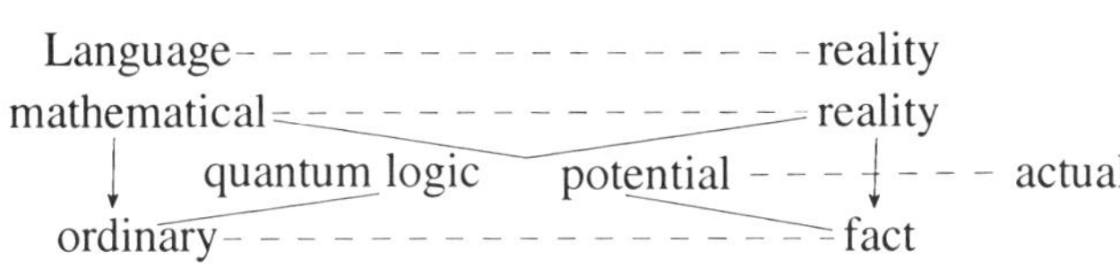

In CP, each language is applicable to both level of reality, but in QM, "we cannot speak about the atom in ordinary language. The analysis can now be carried further in two entirely different ways. We can either ask which language concerning the atoms has actually developed among the physicists in the thirty years that have elapsed since the formulation of QM. Or we can describe the attempts for defining a precise scientific language that corresponds to the mathematical scheme" (*Ibid.*, p. 179). The second way seems to choose resolution of the contradiction between ordinary language and quantum objects by replacing the former with a different precise language which follows definite logical patterns in complete conformity with the mathematical scheme of quantum theory. Some progress has been achieved in this direction (quantum logic), but a regression is always involved because "we have to speak at least about our eventual modification of logic in the natural language" (*Ibid.*, p. 181). The ontology that underlies this view is 'coexistent states' or 'coexistent potentialities'. The first way is the same as what Bohr had done in introducing complementarity, namely to use an ambiguous language to talk about the atomic world. Thus atoms and their attributes, when they appear in such a scheme, only acquire the same statistical significance as 'temperature of the atoms'. That is a modified concept of reality. "One might perhaps call it an objective tendency or possibility, a '*potentia*' in the sense of Aristotelian philosophy. In fact, I believe that the language actually used by physicists when they speak about atom events produces in their minds similar notions as the concept '*potentia*'. So the physicists have gradually become accustomed to considering the electronic orbits, etc., not as reality but rather as a kind of '*potentia*' " (*Ibid.*, p. 181). Thus, according to Heisenberg, quantum reality, when it appears inevitably in the framework of classical concepts and ordinary language, will be described as statistical expectations, objective tendency, or '*potentia*'. Only after interacting with measuring instruments, does it become as certain as in classical state, namely a well defined fact. Reality tends from 'potential' to 'actual' during a measurement.

Similarly, Margenau held a view of reality which we call 'latent' realism. He distinguished properties in QM to be attributed to objects conditionally (latent attributes) with those in CP attributed unconditionally (possessed attributes) (H. R. Holcomb III). "In QM the problem of reality seems to center in the meaning of states. Here it became clear for the first time, and not without a shock to the physicist, that even physical state, previously regarded as a nice and tidy collection of possessed observables geared together and running like wheels in a clock, is in fact a composite of latent observables loosely coupled to nature by rules of correspondence of a statistical sort" (Margenau, p. 452). Only a measurement of an observable projects it from latent to actual. So both Heisenberg and Margenau described quantum reality as potentiality which may be actualized in measurements.

3.3. *Dispositional Realism*

Similar to our 3.2, this describes quantum properties as dispositions of objects which manifest themselves as observable actions under certain conditions because of common groundings (R. Harré, 1985, 1991). Together with conditions, or more precisely in our case, with apparatus, as one of us argued (see Xinhe Hu, 1991), a dispositional account of physical properties can give us a satisfactory resolution to some problems associated with it. It can be sketched by the formula:

(5) Propensity + Apparatus $\rightarrow$ Physical Properties

among which propensity is a probabilistic & potential micro-world. Taking the left side as a whole and giving the apparatus the meaning of 'open to development', we can get a self-sufficient theory to answer such problems as regressions in both directions, namely ungrounded basic dispositions and an unobservable and unmaniputable cosmos, etc. But it seems we still can say nothing about disposition in itself. What we can say is just that it is discovered and conformed statistically on the experimental setup. We cannot talk about its physical content and give it an objective description except as it is bonded to experimental arrangement. But taking disposition in that sense, we know that even thinkers of the 17th or 18th century like Descartes, Hobbes and Boyle, viewed powers as not being inherent in matter; that means 'powers' are not ascribable to bodies in and of themselves. Rather, powers are manifested only when bodies are in relation to specifiable extrinsic circumstances. So it projects the common features of all these interpretations as follows:

(1) Distinguish two levels of reality: independent and phenomenal. The former can only be described by mathematical formalism as a catalogue of potential interactions or as a propensity to produce various possible results on measuring instruments, while the latter, as actual results of interaction, as realized possibilities, can give us all the information we can get by different instruments.

(2) Emphasize the role of instruments in the concept of quantum reality.

(3) Project the relational character of quantum reality. Described in

quantum phenomena, the attributes of quantum objects are manifested as results of interactions, as conditional properties, and as relations between objects and instruments. In a word, it seems relations take a much important place in our description of Nature, one which was once occupied by properties. What we are describing is not the isolated properties of objects but their extrinsic relations with circumstances. This is very thesis of this paper: relational realism.

4. RELATIONAL REALISM

Relational realism emphasizes the relational characters of reality. It regards reality as potential possibility, provided and emergent in a set of relational structures, and aims to explore these relations, and the relations between them.

The emphasis on the relational characters and relational structures of reality is not a fresh point of view in the history of both philosophy and science. As early as in his *Parmenides*, Plato had discussed the relational structure of existence and constructed entities as combinations of relations, although those relations were abstract, prior and about ideas (see Sternfeld, R., Chapter IV, V, VI).

Following Plato, Leibniz also developed an ontology of relations and the ontological structure of relational facts (see Hector-Neri Castaneda). Leibniz thought there are universal relations between the elements of the universe. Every individual, whether person or thing, in some way reflects everything that is happening in the universe, each from its own point of view.

We can cite many more historical examples (see Jammer [1966] 'Concluding Remarks'), but the most impressive progress of relational realism comes from modern physics. Besides the three interpretations of QM described above, we can find the most complete and convincing statements from the two well known books by Max Jammer. In the first one, he indicated those statements of relational realism (though not with such a title) "found their full corroboration in modern physics. At first, the theory of relativity revealed that the geometric-kinematic properties of position, time, and velocity (as well as length, size, duration, and mass), previously regarded as objective features, depend upon the frame of reference. Subsequently, QM showed in addition that these properties are relative also to the means of observation" (*Ibid.*, p. 404). And in the second, he analysed "the relational conception of quantum states" (Jammer, 1974, pp. 197–211). Its main points were:

(1) Bohr "found the solution of the emerging difficulties in the adoption of a relational and holistic conception of the state of a physical system" (*Ibid.*, p. 199). "In other words, the descriptions of the state of a system, rather than being restricted to the particle (or system of particles) under observation, expresses a relation between the particle (or particles) and all the measuring devices involved" (*Ibid.*, pp. 197–198).

(2) "Because every measurement involves both the entity to be observed and the measuring experimental setup, no attributes of physical reality can be ascribed to the former alone. In particular, a particle, even after having ceased to interact with another particle, can by no means be considered as an independent seat of 'physically real attributes' and the EPR thought-experiment loses its paradoxical character" (*Ibid.*, p. 198).

(3) "One may also regard the connection between Bohr's relational conception of state and Einstein's theory of relativity from the general point of view that in the historical development of physics attributes were gradually replaced by relations" (*Ibid.*, p. 201). The finite size of the velocity of light removed the discrepancies originating in the theory of relativity and led to the choice of different frames of reference, and the relational conception of quantum states reconciles the contradictions in QM. In this sense, Fock said: "A QM description of an object by means of a wave function corresponds to the relativity requirement with respect to the means of observation. This extends the concept of relativity with respect to the reference system familiar in CP" (*Ibid.*, p. 202).

Another more recent example is Paul Teller's relational holism. By this term, he means "the claim that objects which in at least some circumstances we can identify as separate individuals have inherent relations, that is, relations which do not supervene on the non-relational properties of the distinct individuals" (Teller, p. 73). Applying this idea to QM, he thought that inherent relations are important to interpreting QM, because all these cases involving superpositions manifest the inherent relations. According to this doctrine, we can see the EPR argument as revealing inherent relations of two particle systems and see the key to the measurement problem as the relations which hold between measuring devices and measured objects.

We have outlined relational realism and given it a sufficient argument in the micro-recognition field. Now, let us elaborate our relational realism further as following:

(1) Reality is relational, it exists in a set of relations which are essentially inseparable from it. The general patterns of existence may be expressed as

$$(6) \qquad y = f(x_1, x_2, \ldots x_n)$$

where y are physical phenomena, as recognized properties, x_1 are physical objects, or realities, and its essential attributes (in some cases), $x_2, \ldots x_n$ are different relational parameters, but in most cases, only one parameter is the principal one which is determined by our perspectives. So, its simplified form is

$$(7) \qquad y = f(x_1, x_2)$$

As we have known from QM, when the coupling of x_1 and x_2 is not neglectable and reducible, we must admit that relations are realities, phenomena are realities, the quantum objects can be represented by a conceptual reconstruction of all concerned complementary quantum phenomena, by a set of inverse functions like

$$(8) \qquad x_1 \rightarrow \{g(y, x_2)\}$$

According to this view of reality, quantum properties are not intrinsic, but relational. As the results of interaction, their reality depends on the measuring set up, and even relates to the type of measuring setup of the next time. Similarly, quantum states, as Jammer analyzed, are relational. They should be interpreted as the catalogue of potential interactions with all concerned measuring setups, as the probabilistic spectra in these relations, as the relativity requirement with respect to the means of observation; the relationalization, relativization and projectivization of reality are the very features of quantum reality. In fact, in the macro-field, a physical object is also in relations with its surrounding, as a part of a much larger system. Only in such a larger system including instrument and observer, could the object get its phenomenological manifestation and become an object of measurement and theoretical investigation. In a word, reality always gets its peculiar ontological and epistemological positions by its relations with its surroundings.

(2) Relations are real, the reality of relations is precisely in its materiality and its objectivity. We have emphasized the relational character

of reality, but no matter whether those relations are between moving body and reference system in the theory of relativity, or those between micro-object and measuring instrument in QM, they are all realistic relations between material systems. Although reference systems and measuring instruments are chosen, even made by human beings, as soon as they are chosen, or made-up, the results of experiment are definite or probabilistically predictable. Similarly, the relativity we have emphasized is the relativity of object to different material systems like frames of reference, measuring instruments, etc. But in each special case, namely each special material relation, the information provided to all scientists is equivalent, objective; there is nothing related to their individual subjective characteristics such as personality, will etc.

(3) Relation is prior to relata in many cases. "One can give a strong sense to the idea that a relation does not 'reduce' to non-relational properties by saying that a relation does not supervene upon the non-relational properties of its relata" (P. Teller, p. 71). Systematic properties of subsystems in system theory, quantum properties in QM, and some high-energy particles produced by cyclotrons, are all examples of such relata. So contrary to the trend of going deeper and deeper to find the most 'elementary' particles and their intrinsic fundamental properties, we find at least some such properties, and even particles, are related or come from their relations with other material systems. A famous *sutra* of Chinese Buddhism says: Things have no inherent attributes, which emerge from their relations. We think at least we can say some of them emerge from their relations with others.

(4) Relations and attributes are mutually transformable in different perspectives. In the general pattern of existence, when we define the different sorts of relational parameters x_2, we face different phenomenal worlds, and for each sort of x_2, when we give it different values, we get different phenomena. Then we find not only that attributes become relations, but also relations become attributes when x_2 is defined. So, when x_2 is a measuring setup in QM, we face the objective world of quantum properties which may be called the zero-th qualities. When x_2 is the frame of reference, we face the physical world of the primary qualities. In this way, we can also confine x_2 to the state of human senses (in the sense of normal people), and then face the common-sense world of the secondary qualities, even confine x_2 to the common aesthetic (or ethical) experience and then face the aesthetic (or ethical) world of the third qualities, although for the latter we see the narrower and the weaker

nature of their objectivity, and the stronger of their relativity. In so doing, the view of relational realism could embody and reconcile the views of reality at different levels and fields, and manifest a theory of multi-worlds, as a much more fruitful and broader view of reality.

5. RELATIONAL LOGIC

Relational realism came from the reform of the view of physical reality. In order to give it a much rigorous and general expression, we must enter into the logico-linguistic field in order to develop the relational logic which is suitable to relational realism.

Different from the subject-predicate (s-p) logic which was set up by Aristotle on the base of substance-attribute realism, relational logic stresses the feature of centering its attention on the relations between objects. It restricts or replaces the classical judgement formula 's is p' by a formula 'aRb', adds a relational operator in the standard system of propositions to serve as a necessary condition for getting the truth value of a proposition. So corresponding to the existence pattern of relational realism, a sentence is meaningful in the syntax structure of relational logic only in form of

$$y = f(x, r) \text{ or } Pr(p)$$

where r is relational operator. That will lead to a series of important conclusions.

Obviously, s-p logic is suitable to substantial realism which was centered upon entities, but in such a logical frame of monadic predicates combined by the principle of bivalence, substantial realism will go to its opposite – antirealism in many cases. When some p (red) couldn't be attributed to s (the flower) itself (attributes couldn't be attributed), or some contradictions arise like s is both p and not p (the flower is one-meter high and not one-meter high), we have to deprive s of these p, reduce them to human subjective perceptions and go to the anti-realistic conclusions of p. In this way, we had deprived the physical realities of the secondary qualities, we should also deprive them of the primary qualities, and quantum properties because most of them are also relational and could not be attributed to realities themselves; they are also secondary qualities as many philosophers have indicated. In the end, we would be metaphysical realists with a bare, abstract concept of reality but anti-realists concerning most of the physical properties.

Contrary to the traditional logic, by maintaining the description of physical reality by binary (multivariate) predicates, relational logic is compatible with relational realism through their common symbolic expression of binary (multivariate) functions. So, it provides a possibility to overcome metaphysical realism and avoid anti-realism. For example, 'x is one-meter length' in the theory of relativity and 'x is particle' in QM are meaningless according to classical logic, but meaningful in relational logic because relational operator r was introduced as an essential element in its syntax structure. Namely, 'length', 'particle property' are not monadic predicate but relational predicate, are functions of an ordered couple (x, r), each of them is separately a projection of a 4-dimension invariable to a reference system or of a quantum object to a measuring setup. Corresponding to such a change of ontological category, the syntax structure of the concerned judgments is correct when introducing r, i.e., 'in the reference system r' or 'on the setup r'. That is the meaning of the symbolic expression Pr(p), namely "the proposition P is tenable in the relation r". Strictly speaking, the generating of a proposition and the getting of its definite truth value depend on the indication of the value of r.

As we stressed the relational character of relational realism contrasts to classical realism, the essential feature of relational logic is to introduce a relational operator in the standard system of propositions.

The relational operator is similar to the tense operator in tense logic, but it has much broad implications. It can express the relational items like 'to whom', 'when', 'where', 'what situation' and 'how', etc., and it includes the tense, spatial, contextual factors, etc.

By introducing r, combined with the principle of bivalence, we can accept that contradictory statements are true at the same time (both 's is p' and 's is not p' are true for different r, e.g. that 'the flower is red' is true for a normal retina and that 'the flower is not red' is true for an abnormal retina) and thus avoid the tendency of antirealism (p is subjective), the tendency of dogmatism (only 's is p' is true and reflects the real features of thing), and the tendency of so-called dialectic logic (s is both p and not p). From the formula of phenomenal structure, we can see, neglecting the parameter r (or x_2 . . .) is the cause of difficulties in the above three tendencies or views.

By introducing r, a variation of the view of truth is necessary contrary to such a abstract view of truth as the classical correspondence theory, simply regarding a proposition as the seat of the truth; generally, we should regard truth as a property of a function of the triple of a speaker, conditions of speaking and a sentence. In this way, the concept of truth could be used in natural language.

By introducing r, we can go to a much broader view of reality. On the one hand, because as soon as the relations are defined, the sentence has truth value, and truth value is determined by the objective characters of the system $f(x_1, x_2, \ldots)$, irrelevant to personal will and characteristic, so in the first place, relational realism is a sort of realism (p can be attributed to s). But on the other hand, instead of a metaphysical external realism, it is an internal realism. This is so because without the participation of x_2 (e.g., retina) on the ontological sense, according to relational realism, not only the p (red) could not emerge, but also it could be said there is no such a P (e.g., colour). Colour as a property of the system is internal in the system $f(x_1, x_2, \ldots)$ and will change with the $x_2 \ldots$ Because relational logic provides us a new method of regarding all those properties affected by evaluating the subject (such as 'cool', 'warm', 'red', 'yellow', 'good', 'evil', 'beautiful', 'ugly') not as monadic predicates but as relational predicates through introducing new variables relating aspects which are concerns of the judging subject, it provides us with a possibility of describing them and the relations between them objectively too. For example, 'x is good' should be expressed completely as 'for r, x is good' and thereby receive its truth value. Such a general pattern of sentence 'for r, s is p' implies that a sentence of evaluation can become a conditional sentence of a statement, namely a descriptions of facts (of cause in reverse, a so-called statement sentence is always a description of fact in relation r, and it contains the evaluation). In this way, 'the flower is beautiful', 'the flower is red', 'the flower is one-meter high' are all objective statements when confined to the relational operator, with the degree of objectivity of each proposition determined by the r concerned. It is helpful for our understanding the objective significances of aesthetic and ethical judgements and provides us with a logical foundation for constructing a much broader, richer and more colourful view of reality.

6. ON SOME PROBLEMS

6.1. *On the Paradox of the Grue Emerald*

The puzzle of the grue emerald is that the same set of observations, if made prior to t, supports two incompatible generalizations ('All emeralds are green' and 'All emeralds are grue') and leads to two incompatible predictions ('The emerald is green after t' and 'The emerald is blue after t') (Goodman, N.). In the view of relational logic, the puzzle comes from the fact that two sorts of language are combined in describing the same world, that the r which constricts the range in which each language is suitable is neglected. Let us provide a formalized analysis. As we have said, colour is a relational predicate generally (especially in the scientific world) and has an expression f(x, r). In the present situation, r is t, then c = f(x, t). For the commonsense world, colour is a constant and becomes a monadic predicate, c = f(x, t) = f(x) = G(green). But for a special possible world expressed by 'Grue' language,

$$C = f(x, t) = \begin{cases} \text{G when } t < 2000 \text{ A.D.} \\ \text{B when } t > 2000 \text{ A.D.} \end{cases}$$

Therefore, two sorts of language, corresponding to two possible worlds, are confined separately by two values of t, and the combined usage of them will cause contradictions. It might be much popularly accessible and vivid if we replace t with v (velocity), for according to the Doppler effect, colour is a function of velocity, and we will see a differently colourful world-picture in different moving systems, just as the length of a rod is different in different frames and cannot be combined.

Up to now, most resolutions to this paradox held that only one world (the commonsense world) exists. That led inevitable to the choice of different sort of language in describing the world and only gave us such pragmatic criteria as 'habit', 'entrenchment' in the choice of predicates. But in the view of relational logic, each language corresponds to a possible world to which we should not refuse its reality, in principle. Then we can give a much formal rule to allot the predicates and the validity of inductive references.

6.2. *Relational Realism v. Metaphysical Realism*

Obviously, relational realism is a challenge to the metaphysical realism represented by Kripke. In the view of Kripke, the bearer of a name, i.e., the individual and its essential properties, are identical in all possible worlds. Identity will be definite forever as soon as it is defined. The essence of the individual is sole [unique], necessary and absolute. This metaphysical view of identity and essence is the cornerstone of Kripke's metaphysical realism (see Kripke).

But in the view of relational realism, things (as relata) and their essence are defined by special relations. They would be changeable in certain conditions, with the variation of these relations. The objective connections and relations between objects and surroundings include the relation between human practice and object, to form a context of recognition. So the definition of things often reflects human practical action and its end, the 'essential properties' described in the 'definition' could not be absolute properties separated completely from certain relations and conditions, and neither could they be pure ontological, metaphysical necessary properties which are completely irrelevant to a human being's existence and mode of understanding. On the contrary, because of the manifold patterns of a thing connected with its surroundings, and the manifold ways of a human being's practice and recognition, it should be possible to reveal the different 'essential properties' of the thing in different 'definitions'.

In contrast to essentialism, relational realism shows the complexity of classifying and naming of things. There is a relativity not only in the differentiating of thing, varieties, and essential properties, but also, to some extent, in the problem whether they exist or not, related to certain language communities, criteria and modes of understanding. We can give definite answers to questions such as 'what is a object?', 'What essential properties does it have?' only after all these conditions are defined.

In our linguistic practice, the relativity of objects and essential properties is shown by the relativity of references, which means that all objects referred in the variety of language communities are concrete and describable. It is this describability that provides those predicates, which have the same extension and different intension in the given world (often a certain scientific world), some conditions by which to manifest their different intensions in other possible worlds.

Again, different from Kripke's admitting only one world, relational realism holds a position of realistic comprehension of (relationally relevant) possible worlds, i.e., admits of the existence of many possible worlds with very different properties and features. Such a view of multi-worlds not only finds its supports from the advances in physics and cosmology, but also may rationally satisfy modern people's request for a much richer and more colourful view of reality in their practical and spiritual lives.

6.3. *Oriental Thought as Background of Relational Realism*

Relational realism has been based on modern science, mathematics and logic, but it has a fundamental connection with traditional Oriental thought.

Contrary to Western substance thinking and logic, in the Orient, especially in China, a mode of thinking centered on the relativity of things originated even in ancient times. At the origin of Chinese philosophy, the Yi-king (*Book of Changes*) had distinctly manifested such a feature. According to it, every phenomenon could be interpreted by the combinations of a pair of elements, Yin and Yang which are negativity and positivity respectively (symbolized as – and + respectively). All phenomena possess negative and positive aspects and all judgements have negative and positive values. But it should be stressed that both of them are neither substances nor inherent essences of things, but jointly ways for expressing some relations between things. For example, when such a pair of concepts applies to relations between male and female, a man is Yang to his wife (in husband-wife situation) and he is Yin to his mother (in mother-son situation). That means Yin and Yang are not monadic but binary or relational predicates. They are constrained by the relational operators, such a relativity of Yin and Yang could be expressed as following,

$$[y]\ \text{Yin}\ [a, y]\cdot[z]\ \text{Yang}\ [a, z]$$

namely 'thing a is Yin relative to thing Y and Yang relative to thing Z'. It is easy to see that Y and Z are two values of relational operators. Of course, such embryos of ancient relational logic were clarified through the view of modern logic, while in the history of more than two thousand years, and even for many current Chinese scholars, philosophers were unaware of the roles of relational operators, and the relativity of things were expressed as 'being both Yin and Yang'.

Later, relational thinking was drowned entirely by so-called dialectical thinking.

It is an easy passage from the relativity of Yin and Yang to a relativity of 'existence' and 'nothing', and everything. That means we cannot talk about 'existence' or 'nothing' in isolation. They are mutually promoting and restricting, complementary and interexclusive. Reality is always relative to a certain relation and condition. Such a relative view of reality in Chinese philosophy was expressed in Chinese Buddhism as 'impermanence', 'endless genesis', namely everything is transient, there is neither permanent, invariant self-nature, nor an independent, isolated thing-in-itself. Everything depends on other things, generates, and emerges in certain conditions and relations. Quite the contrary, from Parmenides to Bradley, the Western absolute view of reality pursued the ultimate reality which is behind common sense, behind variation, behind phenomena and particulars. That is the reality before God's-eye.

The relational view of reality underlines both the objectivity of reality and the relational limitation of reality, namely the multiplicity of reality. It holds onto grasping reality by different abstract methods, and also holds that phenomena are realities, processes are realities, reality is in the unity of human practice and its objects.

Institute of Philosophy,
Chinese Academy of Social Sciences,
Beijing, China.

BIBLIOGRAPHY

Bohm, D. (1951). *Quantum Theory*, Prentice-Hall.

Bohr, N. (1934). *Atomic Theory and the Description of Nature*, Cambridge University Press.

Bohr, N. (1935). 'Can the Quantum Mechanical Description of Reality be Considered Complete?' *Physical Review* **48**.

Bohr, N. (1937). 'Causality and Complementarily', *Phil. Sci.* **4**, pp. 292–293.

Bohr, N. (1947). 'Newton's Principle and Modern Atomic Mechanics', *Newton Tercentenary Celebrations*, Cambridge University Press, pp. 59–60.

Castaneda, H. N. (1982). 'Leibniz and Plato's *Phaedo* Theory of Relations and Predictions', in M. Hooker (ed.), *Leibniz: Critical and Interpretive Essays*, Manchester University Press.

Folse, H. J. (1988). *The Philosophy of Niels Bohr*, North-Holland.

Goodman, N. (1955). *Fact, Fiction and Forecast*, Harvard University Press.

Harré, R. (1986). *Varieties of Realism*, Basic Blackwell.

Harré, R. (1991). *Lectures in the Sino-British Summer School of Philosophy.*

Heisenberg, W. (1970). *Physics and Philosophy; The Revolution in Modern Science*, Harper & Row.

Holcomb III, H. R. (1986). 'Latency Versus Complementarily: Margenau and Bohr on QM', *Brit. J. Phil. Sci.* **38**, pp. 193–120.

Hu, X. (1988, 1991). In N. Bunnin (ed.), *Best Examination Essays, Sino-British Summer School of Philosophy.*

Jammer, M. (1966). *The Conceptual Development of QM*, McGraw-Hill.

Jammer, M. (1974). *The Philosophy of QM*, John Wiley.

Kripke, S. (1980). *Naming and Necessity*, Basil Blackwell.

Margenau, H. (1950). *The Nature of Physical Reality*, McGraw-Hill.

Sternfeld, R. and Zyskind, H. (1987). *Meaning, Relation and Existence in Plato's Parmenides: The Logic of Relational Realism*, Peter Lang.

Teller, P. (1986). 'Relational Holism and QM', *Brit. J. Phil. Sci.* **37**, pp. 71–81.

MAO SHIYING

A REALISTIC INTERPRETATION OF QUANTUM MECHANICS

1. INTRODUCTION

The contemporary debate between scientific realism and anti-realism prominently presents the question whether or not a realistic interpretation of quantum mechanics (q.m., afterwards) can be given adequately when it is reflected in concrete scientific theories. Scientific realism generally demands that a scientific theory aims to be a true or approximately true description of entities, properties, structures, processes, etc. that exist in an objective world independently of our observation and consciousness. All classical physical theories satisfy the demand. However, the Copenhagen interpretation, as the orthodox interpretation of q. m., is usually regarded as an instrumentalist interpretation with an epistemological meaning alone. It avoids answering the ontological questions such as whether the quantum theory really describes the microworld. Just as Bohr wrote: "The entire formalism is to be considered as a tool for deriving predictions, of definite or statistical character, as regards information obtainable under experimental condition described in classical terms" [1]. It is obvious that this interpretation is in accordance with the anti-realist doctrine of Van Fraassen, which holds that "science aims to give us theories which are empirically adequate, and acceptance of a theory involves a belief only that it is empirically adequate" [2]. The primary reason for the Copenhagen school to take the instrumentalist position instead of the realistic position, I think, lies in the fact that in a quantum measurement the measuring device or the observer can no longer be separable from the observed object, as it could in classical measurement; the measuring device and the object system constitute an inseparable unified whole together, the entire experimental situation is already of an inseparable wholeness or nonseparability. It is according to the concept of wholeness that Bohr refuted the EPR argument against the completeness of q. m. [3]. In fact, the experimental results of Aspect *et al.* about the Bell-type inequalities in recent years have undoubtedly verified that the microreality possesses nonseparability, i.e. a microsystem cannot be divided into two local parts which are independent of each

R. S. Cohen, R. Hilpinen and Qiu Renzong (eds.), Realism and Anti-Realism in the Philosophy of Science, 381–393.

other in space, the principle of separability, that is based on classical measurements and used as a basic principle of the realistic interpretation of classical physics, no longer holds for the quantum theory.

However, the presence of nonseparability does not certainly shows that scientific realism cannot hold any more. The basic demands of scientific realism such as objectivity, causality, and completeness, should be still upheld, or perfected further on the basis of new scientific discoveries, but they should not be abandoned simply, otherwise the scientific enterprise has no way to resist the aggression of subjectivism. In fact, it is due to one-sided depending on the nonseparability, that the Copenhagen interpretation cannot give a realistic interpretation to the phenomenon of the 'wavepacket reduction' of measurements, and only considers it as a sudden change of knowledge of the observer. Thus this viewpoint cannot guarantee objectivity independent of any observers, causality and completeness of the theoretical statements, and leaves a backdoor for the subjectivist interpretations of E. Wigner *et al.* Actually, scientific realism is not certain to be in conflict with nonseparability. As I shall show in the following, it is not impossible to construct a realistic interpretation of q. m. which is compatible with nonseparability. The key to the settlement of the question is to construct a new type of correspondence theory of truth on the basis of quantum measurement. In what follows I shall first give an analysis of the quantum measurement.

2. THE QUANTUM MEASUREMENT IS THE UNITY OF NONSEPARABILITY AND MACROINTERVALITY

Because of the presence of nonseparability of the entire experimental situation, the measuring device (including the observer as a recorder) and the object system constitute an inseparable composite system together, and in the case of a measurement the disturbance which is imposed on the object system by the measuring device, can no longer be neglected or compensated as it could in a classical measurement in principle, and various measurement results, such as prepared quantum states, the phenomena of wavelike or particlelike, and etc., are formed in the measuring interaction between object and device. Therefore, the measuring device already has the ontological function of acting as a macroscopic physical system or macroscopic surroundings to participate in the objective world, and the measurement process should be in principle seen to form a physical process, and the measuring interaction should be considered as an objective physical interaction. E. Wigner holds that in the quantum measurement the consciousness of the observer plays a central role, but this point of view cannot explain how the universe existed and evolved before man was born.

Besides nonseparability, I think, the quantum measurement still possesses another universal character: macrointervality, which is what the macroscopic spatio-temporal conditions of the measuring device constituting a kind of relatively independent region show in the measurement. The conclusion is drawn by contrasting the theoretical description with the empirical facts about a microobject. The nonclassical phenomena appear in quantum measurements, in which the theoretical description cannot in principle correspond to the empirical fact directly and wholly. Firstly, the phenomenon of the wavepacket reduction shows that during a measurement the superposition of states of an object transforms into one term from the superposition, and the other eigenstates in the theoretical description do not appear. Secondly, the theoretical description of the wave-particle dualism of an object cannot correspond to the phenomenon of the mutual exclusion of wavelike property and particlelike property directly and simultaneously. The nonclassical phenomena mentioned above actually show that every observable phenomenon of an object is limited within some relatively independent space-time interval. Due to the space-time region of an observable phenomenon being determined by a macroscopic measuring device, the interval is a macroscopic one. The phenomenon of wavepacket reduction is actually thought to occur when the measuring device possesses a macrointerval which has the same spatial size as some term from the superposition of states, thus resulting in the so-called wavepacket reduction. A detailed analysis about this will be given later. The phenomenon of simultaneously mutual exclusion of wavelike property and particlelike property in measurement shows the two properties have a different macrointerval respectively, and want us to use two kinds of different devices to measure respectively. The relation between the macrointerval in an experiment and the spatio-temporal region of a microobject in the theoretical description, is one between part and whole. Because the objects of classical measurements are macroscopic physical systems, the macrointervality of measuring devices could not show up in classical measurements. It can only be manifest in the measuring interaction between device and microsystem. The macrointervality of measuring devices determines a device in a quantum measurement to possess its nonclassical peculiar epistemological function of selecting and manifesting some physical property of a microsystem which is latent relative to our sense organs, where the function of selection can be classified as two types:

(1) Active selection: determination of wavelike or particlelike properties of a microsystem within some macrointerval. In the measurement we can estimate the measurement result in advance. Due to the complementarity principle, the two selections are mutually exclusive and complementary.

(2) Stochastic selection: determination of some eigenvalue of an observable of a microsystem. Since the eigenstates are generally countless, and the macrointerval of the device only accommodates one eigenstate in a measurement, we cannot estimate the result of each measurement in advance, thus the selection is wholly passive and random.

A selecting act of a measuring device for some physical property is also a manifesting act in the meantime a measuring device as a carrier of physical information, can fetch and manifest some piece of latent physical information of a microsystem in an active or a stochastic selecting way within a macrointerval. Moreover, there exists the potentializing (or concealing) act in the meantime i.e. of the manifesting act, i.e., the other pieces of physical information of the microsystem simultaneously lose the possibility of appearing within the interval. For instance, the wavelike property and the particlelike property of a microsystem are manifested respectively through measuring devices. And the mutation-like measurement actually shows that in a measurement one and only one eigenstate is able to be manifested by a device, and all the other eigenstates are potentialized.

The macrointervality of the measuring device actually constitutes the mechanism of macrointervalized action, which is able to make the information of a microsystem manifest partly and potentialized partly. The quantum measurement is actually a physical process in which the microsystem as an indivisible whole manifests some interval part of itself corresponding to a localized observable phenomenon; meanwhile the rest of the whole is potentialized. The empirical structures coming from empirical data are actually the limited macroscopic localized manifestations of the physical structure of the world in measurements. The macrointervalized action generally results in the uncontrollable loss of a certain amount of energy and negative entropy, so the measuring process, by and large, is an irreversible process, just as a die moving in the air, due to the constraint of a table plane, finally only manifests one of six numbers at rest.

3. A NEW TYPE OF CORRESPONDENCE THEORY OF TRUTH

The very core of scientific realism is the correspondence theory of truth. It claims that there exists a direct one-to-one correspondence relationship between correct scientific theory and reality. Whether or not a scientific theory is true lies in whether the correspondence relation can be tested by experiment. In fact, the correspondence relation is one relating a theoretical model, empirical facts and the objective world, and is constructed on the basis of measurements. The traditional correspondence theory of truth, I think, is actually constructed on the basis of classical measurements subject to Einstein's principle of separability. This is to say, during the measurement the device or the observer can be in principle separable from the macrosystem, and be spatially isolated from it; thus the measurement outcomes basically are purely objective records about the object, and a theoretical model can directly correspond to the objective world. The separability of classical measurements actually reflects the separability of the macroworld. The traditional correspondence theory of truth actually only holds good for classical theories about the macroworld, i.e., whether a classical theoretical model is true lies in whether we can in principle verify the isomorphic relations between the empirical structures constructed from empirical data and all the substructures of the structure of the theoretical model on the basis of classical measurements (Historical tests are of course always limited and incomplete). Because an object of study of classical physics is the same macroscopic physical system as a measuring device, they have the common macroscopic property, so the physical structure of the macroworld that can be represented by the structure of a certain theoretical model, can in principle manifest wholly as various observable phenomena through various experimental means, and be directly confirmed by all observers, i.e. intersubjectively. There are in principle no unobservable quantities and substructures in classical physics, which cannot be forever examined by macroscopic experiments. Historically, the concepts of absolute space and time had no empirical support for a long time, but they were finally falsified only by experiments.

However, as to the quantum theory, due to the nonseparability of the experimental context, we cannot already talk about the observed object under the condition of getting rid of the measuring device or the observer; moreover, the macrointervality of the device makes it impossible for us to obtain the whole information of the microobject, therefore the traditional correspondence theory of truth does not hold for the quantum theory. It cannot show the direct correspondence relations between quantum theory and objective world under the condition of crossing experiment. The failure of the traditional correspondence theory provides us with two options: either give up realism and take up instrumentalism, or reconstruct it. The Copenhagen school takes the first, de Broglie, Bohm *et al.* take the second. I favor the second one, and think that a new type of correspondence theory of truth – the quantum correspondence theory – which is compatible with nonseparability, should be constructed on the basis of the quantum measurement. First I favor D. Dieks' point of view: whether a theoretical model of q.m. is empirically successful, consists in whether a number of substructures of the theoretical structure can be isomorphic to empirical structures constructed from empirical data [4].

But the key to the question lies in how to treat the unobservables of the theory. As indispensable elements of a theoretical model, unobservables come from the free invention of scientists on the basis of experience. Whether unobservables are of reality, is already a focus of the realism-antirealism debate. For realists, it is logical coherence and the unity of theoretical structure itself that ensure the relative reliability of the causal relation between unobservables and observables, and make unobservables be probably of reality. Historically, the success of a lot of predictions of classical physics verifies this point of view. Anti-realists completely deny the relative reliability of abductive inference from observables to unobservables, thus completely deny the possibility of the existence of unobservables. Clearly, the point of view of thinking in terms of absolutes restricts the powers of interpretation and prediction of a theory, and makes it unacceptable to many working scientists. But there is a unresolvable fundamental problem for realists. Since there actually are various theoretical models that possess common empirical substructures corresponding to observable phenomena, e.g., quantum mechanics have various theoretical models, such as the many-worlds interpretation, the hidden-variable theory, etc. that agree with the orthodox interpretation in experience, each contains its own ontological commitment, and possesses some unobservables, then which theoretical model is closer to the real structure of the objective world on earth? Of course, this cannot be judged only according to the criterion of empirical adequacy. It is necessary to synthesize according to other criteria, such as logical consistency, simplicity, usefulness, economy, elegance, etc. meanwhile. Perhaps we cannot obtain a unique correct answer forever. This paper takes the conventional quantum theory as the standard model, and tries to construct its realistic interpretation, because according to various different criteria, the prevalent theoretical model is easiest to accept rationally and universally.

As to the unobservables of q. m., I think, they are unobservable in principle. This is different from the case of the unobservables in classical physics, which can be in principle examined sooner or later by experiments. Firstly, the uncertainty relations show that some pair of canonical conjugates in the theoretical description cannot simultaneously have the two empirical counterparts. That is to say, relatively to a canonical conjugate being observed, another conjugate is uncertain, namely an unobservable. We can call them 'relative unobservables'. Secondly, the phenomenon of the wavepacket reduction shows that during a measurement one and only one outcome is actual, and all the other eigenstates (which are possibly countless) in the theoretical description do not emerge. These disappeared eigenstates can be called 'absolute unobservables'. According to the tremendous success of q. m. in experience, the logical consistency and the unity of the theoretical structure itself, I think, these unobservables have real existence. Because of the macrointervalized action of devices (see Sec. 2), the theoretical observable corresponds to the manifested part of an object in a measurement, therefore the relevant unobservable should be thought to correspond to the potentialized part of the object beyond the macrointerval in the measurement. The unobservable actually is of the non-macroscopic reality (concrete analyses about this will be given afterwards). Whether the magnitude of an unobservable will change after a measurement, needs to be concretely analyzed in terms of the quantum theory. Therefore, investigating whether the quantum theory corresponds to the physical reality, is to investigate not only whether the observables correspond to the manifested part of the world, but also whether the unobservables can represent the potentialized part of the world in terms of relatively reasonable inference. In this way, Einstein's demand of completeness: 'Every element of the physical reality must have a counterpart in the physical theory' [5], is concretely ensured. After passing the investigation mentioned above, we can get a realistic interpretation of q. m.: the mathematical structure of q. m. corresponds to the physical structure of the world, and is of the same inseparable wholeness transcending the macrointerval as the latter. The empirical substructure of the quantum theory corresponds to the manifested localized part, and the relevant unobservable part of the theory corresponds to the potentialized part of the physical structure of the world in measurement. According to the point of view, as I will show in the following, the uncertainty relations and the mutation-like measurement are given a realistic interpretation respectively.

4. THE UNCERTAINTY RELATIONS REFLECT THE DUAL STRUCTURES OF SOME MICROSCOPIC PHYSICAL QUANTITIES

Heisenberg's uncertainty relations are usually regarded as mathematical expressions of Bohr's complementarity principle. They show that the canonical conjugate physical quantities such as momentum and space, energy and time, due to the constraint of Planck's constant, can no more be measured simultaneously and accurately as they could in classical measurements. An uncertainty relation as a substructure of the quantum theory, actually reflects an internal relation between two canonical conjugates, in which two sides are independent of each other and mutually restrictive, but no longer absolutely independent of each other as in classical physics. I call the relation the 'dual structure'. It is a kind of physical structure independent of observation by man. The conjugates in the dual structure are relative unobservable, and can be measured respectively in experiments. The measurement outcome about a dual structure is actually only the concrete manifestation of some canonical conjugate within a macrointerval, meanwhile another conjugate is potentialized, i.e. loses the possibility of accurately manifesting itself in the macrointerval. The dual structure makes the space-time coordination and the claim of causality (generally referring to the law of conservation of momentum and energy) be mutually exclusive and complementary, but not freely combined as in classical physics [6]. The dual structure also shows that the wavelike property (represented by position, time or frequency, etc.) and the particlelike property (represented by momentum, energy, etc.) are mutually exclusive and complementary. Therefore, we are no longer able to construct an intuitive picture of the dual structure, and can only construct an abstract composite picture about it – a 'picture of abstraction' about the microworld – in complementary measuring and describing ways. Bohr's complementarity principle should be regarded as not only a description of some observable phenomenon, but also a description of a dual structure with an ontological status.

5. THE REALISTIC INTERPRETATION OF THE MUTATION-LIKE QUANTUM MEASUREMENT

The key to the construction of a realistic interpretation of q. m. lies in the treatment of the mutation-like measurement. The conventional treatment about this is to consider the measurement as a process of the 'wavepacket reduction', in which the microsystem transforms from a superposition of states to a certain term from the superposition. Suppose the initial state of the object system is $|\psi\rangle = \Sigma\, C_n|\psi_n\rangle$, where $|\psi_n\rangle$ is an eigenvector from the Hilbert space spanned by the complete orthogonal set of eigenvectors $\{|\psi_n\rangle\}$ of some observable to be measured. Suppose a position of the pointer of the measuring device is represented by a statevector $|A_n\rangle$ from the space spanned by the complete orthogonal set of vectors $\{|A_n\rangle\}$, and the zero position of the pointer before the measurement is labeled $|A_o\rangle$. Before the measurement there is no interaction between the two systems, two systems possess together an initial total statevector $|\psi\rangle\,|A_o\rangle$. After the measurement begins, two systems affect each other, the initial statevector at last transforms into the final total statevector $|\psi_m\rangle\,|A_m\rangle$, where $|\psi_m\rangle$ refers to the final state of the object system. The whole process of measurement can be represented as:

$$(1)\qquad \sum_n C_n|\psi_n\rangle\,|A_o\rangle \rightarrow |\psi_m\rangle\,|A_m\rangle$$

not

$$(2)\qquad \sum_n C_n|\psi_n\rangle\,|A_o\rangle \rightarrow \sum_n C_n|\psi_n\rangle\,|A_n\rangle.$$

According to the linearity of the evolution of a statevector with time, the measurement process should be (2). However, the actual final outcome of the measurement is not the superposition of the right-side of (2), but a single term of its component terms. Where do the other eigenstates go? This constitutes the largest obstacle for a realistic interpretation of q. m.. Historically, von Neumann generalized the phenomenon as the famous 'projection postulate' [6], which supposes that during the measurement the initial statevector is 'projected' onto one eigenvector of the Hilbert space. It becomes one of the basic postulates of the conventional quantum theory. However, it is after all an additional postulate coming from outside the quantum theory, and is unable to give the measurement a genuine realistic interpretation in terms of the intervally valid quantum theory itself. As is well-known, the postulate results in the violation of the causal continuous evolution of the statevector governed by a Schrödinger equation, breaks down the symmetry of time evolution, makes the theoretical description fail to be in a one-to-one correspondence with the empirical fact rationally, and leaves a back door for the subjective interpretations of E. Wigner *et al.*

According to scientific realism, the quantum theory should be able to really reflect such a physical process as the measurement. Only a theory can determine what we are able to observe. Because the theoretical description does not show that the statevector reduction will occur, if we still hold that the quantum theory is correct, then we should not hold that the statevector really 'reduces'. Moreover, there is no time scale for automatic reduction of the statevector, which is compatible with all observations [7]. Besides, the "statevector reduction" is said to be relative to the apparatus. Since the device and the object already constitute an inseparable composite system, we should analyze from the perspective of the framework of the composite system. In this way, the eigenstates "disappeared in measurement" cannot be logically considered to have disappeared. They should unchangedly exist beyond the macrointerval of the device and inside the composite system. The so-called statevector reduction, I think, actually shows that there is the mechanism of macrointervalized action made by the measuring device. This is to say, the measuring device constitutes the macrointervalized action to the statevector of the microsystem, which makes a certain eigenstate manifest as an actual macrostate, i.e., leaves a permanent macroscopic localized record in a measurement instrument that can be confirmed by all observers. Meanwhile, it destroys the coherence of physical states, and makes the other eigenstates be potentialized beyond the macrointerval. For instance, in the Stern-Gerlach experiment about a spin 1/2 atom, the total spin about the z-axis has $\pm 1/2$ two values corresponding respectively to two opposite oriented statevectors. A mark emerging in the upper or lower side of the detecting screen along the z-axis, reflects a manifesting act of the device to some eigenstate of spin of the atom, meanwhile another eigenstate of the spin is potentialized beyond the macroscopic screen. Suppose the macrointervalized action is represented by the operator I, which is equivalent to the projection operator in the mathematical status.

Therefore, the measurement process can be represented as follows:

$$\text{(3)} \qquad I|\psi\rangle \rightarrow \sum_{n\neq m} C_n|\psi_n\rangle_p + C_m|\psi_m\rangle_a$$

where 'p' refers to potentialized states relative to the measuring device, and 'a' represents the manifested actual state alone. It is obvious that the operator I can definitely illustrate the overall measurement process. In contrast to the projection operator, it has really the realistic meaning.

When a microsystem manifests one state of its superposition of states through a measuring device, the measuring device itself manifests one state, i.e. the pointer points to some position corresponding to the eigenvalue of the manifested state of the object. Since the device and the object already constitute an inseparable whole, the other states representing the other positions of the pointer should be considered to be potentialized out of the macrointerval represented by the actual position of the pointer meanwhile, they correspond to the eigenvalues of the potentialized states of the object respectively. Therefore, the overall measurement process can be seen as a process in which the microsystem and the measuring device constitute a composite system, and two sets of statevectors are finally unified as a set of composite statevectors governed by a Schrödinger equation. The measurement process can still be represented as follows:

$$\text{(4)} \qquad \sum_n C_n|\psi_n\rangle|A_o\rangle \rightarrow \sum_{n\neq m} C_n|\psi_n\rangle_p|A_n\rangle_p + C_m|\psi_m\rangle_a|A_m\rangle_a.$$

The right side of (4) is the state of the composite system (supposing it is represented by Ψ), which has been the superposition of states consisting of a manifested state and potentialized states simultaneously, and in which the macrointervality of the measuring device has been reflected. In this way, the theoretical description corresponds wholly to the measurement process, only corresponds to the empirical fact partly.

According to (4), the probability of the actual eigenvalue is:

$$\text{(5)} \qquad |C_m|^2 = |\langle\psi_m A_m|\Psi\rangle|^2/|\langle\Psi|\Psi\rangle|^2.$$

From this expression it is thus clear that the probability of the actual eigenvalue actually means the squared modulus of the relative proportion of the manifested macrostate of the composite system, in the superposition of all states as a whole with the weight (Ψ). It is already verified by the statistics of a number of experimental results in experience.

In this way, the quantum measurement is given a realistic interpretation: the quantum measurement is a process of transformation of a physical structure, in which the structure of the Hilbert space of a microsystem and the spatial structure of the adequate coordinates of a device constitute the structure of a composite Hilbert space through the mutual coupling where the macrointervalized action is included, and there is no time scale. Relative to the measuring device, the measurement is an irreversible physical process in which the spatial structure of the microsystem manifests itself as the limited macroscopic localized empirical structure, and which is of the statistical law about the measurement outcome determined by a statevector subject to a Schrödinger type of equation. The quantum measurement is just like a process in which a two-dimensional living thing measures a die moving in the air with a table-like device with the function of manifesting-number. When the die falls down on the 'table plane', the device measures one number, but the other numbers of the die as the component elements of the spatial structure correspond to the other coordinates of the device potentializedly. The probability of emergence of some number is one sixth. This can be derived by the distributive relations of the numbers of the die in the spatial structure, and be verified by a lot of observational results. The quantum measurement actually shows the gap between the limited marcoscopic measuring power of man and the abstract power of cognition of man about the limitless microworld in terms of the quantum theory. The interpretation restores the symmetry of time evolution, and ensures that the measurement is still governed by a Schrödinger type of equation after going through the macrointervalized action.

In summary, this paper proposes a consistent realistic interpretation of q. m. on the condition of the conventional formalism of q. m.. In contrast to the many-worlds interpretation of Everett [8], which supposes that all eigenstates correspond to countless different worlds, along with their measuring devices and observers respectively, the present interpretation holds that there is only one actual world, one device and one observer, but permits the presence of countless relatively potentialized property worlds corresponding to the potentialized eigenstates in the measurement respectively. Because the countless eigenstates of the object belong to some macroscopic observable operator, which corresponds to a macroscopic observable in classical physics, these eigenstates should correspond to one apparatus and one observer together, so I think, the postulate of the many-worlds interpretation is not adequately reasonable. Bohm's hidden-variable theory [9] agrees with the orthodox interpretation in experience, but its additional concepts and structure of determinism, due to satisfying the demand of nonseparability, maybe are forever unable to be tested by localized experiments, and appear like a surplus burden in practical application. The present interpretation keeps the conventional formalism of q. m. invariant, and of course agrees with the Copenhagen interpretation empirically. Moreover, it satisfies the demand of simplicity, and is logically consistent. Not only does it keep the epistemological meaning of the formalism of q. m., but it endows the formalism with an ontological meaning. In this sense, so to speak, the present interpretation is a realistic 'Copenhagen' interpretation.

Institute of Technology & Society,
Northeastern University,
Shenyang,
China.

REFERENCE

1. Bohr, N. (1948). 'On the Notion of Causality and Complementarity', *Dialectics* **2**, pp. 312–319.
2. Fraassen, B. C. van (1980). *The Scientific Image*, Oxford University Press, Oxford, p. 12.
3. Bohr, N. (1935). *Phys. Rev.* **48**, pp. 695–702.
4. Dieks, D. (1989). 'Quantum Mechanics Without the Projection Postulate and Its Realistic Interpretation', *Found. Phys.* **19**, p. 1402, p. 1416.
5. Einstein, A., Podolsky, B. and Rosen, N. (1935). *Phys. Rev.* **47**, pp. 777–780.
6. Jammer, Max (1986). *The Conceptual Development of Quantum Mechanics*, McGraw-Hill Book Company, p. 351, p. 371.
7. Squires, Euan (1986). *The Mystery of the Quantum World*, Adam Hilger Ltd, Bristol and Boston, p. 53.
8. DeWitt and Graham (1973). *The Many-Worlds Interpretation of Quantum Mechanics*, Princeton University Press, Princeton.
9. Bohm, D. (1952). 'A Suggested Interpretation of the Quantum Theory in Terms of 'Hidden' Variables', *Phys. Rev.* **85**, pp. 166–179, 180–193.

MORTON L. SCHAGRIN

RUMFORD'S EXPERIMENTS ON THE MATERIALITY OF LIGHT[1]

More Heat Than Light

Rumford's experiments on the heat generated during the boring of cannons, from which he argued for the "inexhaustible" creation of heat by friction, were cited and used by James Prescott Joule many years later to calculate roughly a mechanical equivalent of heat.[2] I shall suggest in what follows that the fact that this work was so cited and used by a *later* work that itself provided the reference point for *even later* research is a key factor in the growth of science.

In contrast to this use of Rumford's earlier work on heat, his researches on the materiality of light[3] made little impression on his contemporaries, and are far less known today than his work on heat. We recall that Rumford's interests always turned to practical improvements in living. His work on heat was not purely theoretical: he designed fireplaces with improved efficiencies; he devised a coffee percolator; he explored ways to improve cooked field rations for soldiers. In his studies regarding light he also devised improved lighting arrangements; the use of frosted glass in lamps rather than clear glass is due to Rumford. To assist him in measuring the amount of light produced by different sources, Rumford constructed a sensitive photometer based on the earlier models employed by Pierre Bouguer (1729) and Johann Heinrich Lambert (1760).[4]

As were the earlier ones, Rumford's photometer was based on two simple principles:

(1) The intensity of light diminishes as the square of the distance travelled.

(2) If two sources illuminating a single body each cast a shadow of the same observable darkness, then the intensities of the light from each source ar the same.

The instrument (Fig. 1) had a white screen upon which the shadows fell, an opaque object for casting a show,[5] and two tracks leading away from the screen at an acute angle to each other bisected by a perpendicular to the screen. These tracks were calibrated by "degrees" marked off in proportion to the square of the distance. One track was twelve

R. S. Cohen, R. Hilpinen and Qiu Renzong (eds.), Realism and Anti-Realism in the Philosophy of Science, 395–403.

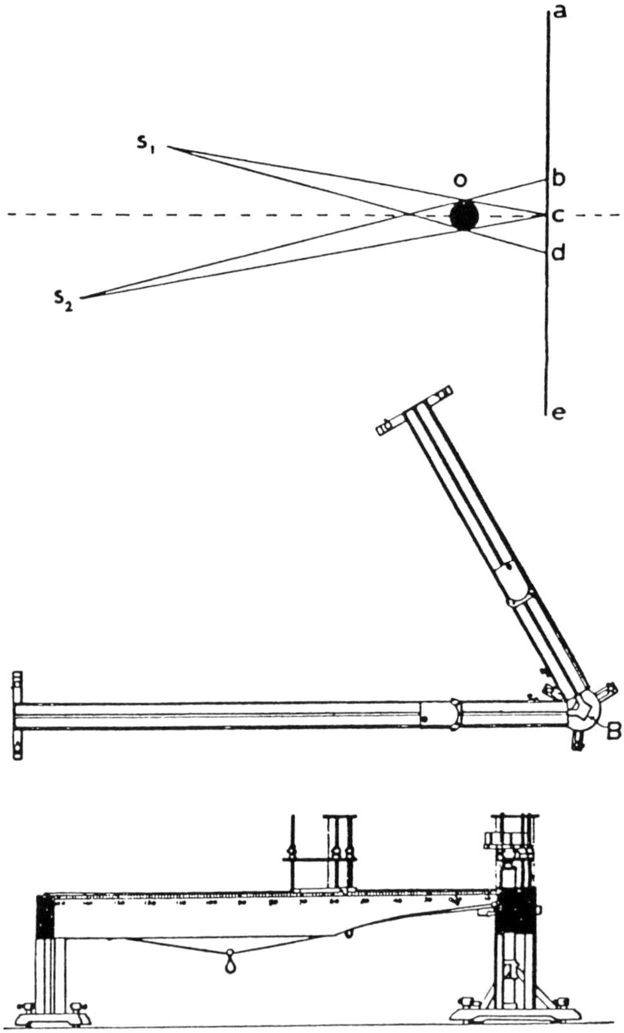

Rumford's photometer

By permission from: W. J. Sparrow, *Count Rumford of Woburn, Mass.* (1964)

Fig. 1.

feet and the other was twenty feet long. The procedure employed a standard source for comparison, say a candle, placed on one arm at, say, 10°. Another source was then moved up and down the second arm until the shadows were equally dense. For instance, if the second source were to cast an equally dense shadow at the position marked 400°, then the test light would be forty times as intense as the standard candle.

The experiment proceeded by using a standard wax candle at 100° on one arm of the photometer. Next an Argand lamp, cleaned and trimmed, was placed successively at 100° intervals on the other arm. The wick was adjusted so that the shadows were equally deep. Thus the lamp shone with intensities of 100° to 900° in 100° steps. The amount of oil consumed in thirty minutes at each station was carefully measured. (The lamp was weighed before and after the thirty minute period, with the difference in weights recorded.) After some arcane juggling of figures in order to get a basis for comparison, Rumford arrived at the following table, showing how much intensity of light can be attributed to the consumption of 100 "units" of oil per hour at the different intensities of flame:

Intensity of the lamp	Light furnished per hour per 100 parts of oil
100°	48°
200	74
300	98
400	112
500	121
600	138
600	149
800	155
900	160

The first row tells us that if the Argand lamp were to burn for one hour at 100° of brilliance, then 100 "units" of oil would furnish 48° of the brilliance of the light. And in the last row, if the lamp were to burn for one hour at a brilliance of 900°, then 100 units of oil would contribute 160° of the brilliance.

Rumford remarks:

> On comparing the results of these nine experiments, it appears that the quantities of light furnished were very far from being in a constant ratio to the quantities of oil consumed, as they would doubtless have been, were light one of the chemical products of combustion,

In the terminology of Bogen and Woodward,[6] Rumford is claiming there is no *phenomenon* where one should be expected: the *data* should show a relation between the light given off the oil consumed. The data in this case are presented in the table above. Yet, *pace* Bogen and Woodward, those are not Rumford's *observations*. What I shall call Rumford's *personal experiences* were such things as the intensity of shadows, the pointer of a balance coinciding with the null mark on a scale, the position of a carriage on a track, etc. His *observation reports* used such notions as the brilliance of an Argand lamp, the weight of oil consumed, etc. These observation reports were further refined to produce the table of data, which according to Rumford does not reveal the expected phenomenon: the constant ratio of light produced and oil consumed.

It is difficult for me to understand how anyone who has conducted an actual experiment relevant to a genuine hypothesis could have accepted such positivistic accounts of scientific data in terms of observation reports such as "Red here now", with the additional claim that theories are confirmed or refuted by such subjective experiences.

But what about the key experiment described above? Is there *no* relation between the amount of oil consumed[7] at given lamp intensities and the amount of light released? And if there is not, what was Rumford entitled to conclude?

First of all, if one assumes that the amount of light produced per unit of oil consumed depends on the brightness adjustment of the wick, that is, the "chemical" reaction is more efficient with more intense flames, and that the amount of light given off asymptotically approaches a maximum, we might conclude that the amount of light produced depends on the brilliance of the flame as a logarithmic growth curve with a maximum.[8] It does not seem implausible that the reaction releasing light becomes more complete as the flame burns more intensely. A plot of such an equation with actual data noted is as follows (Fig. 2).

Yet suppose my *ad hoc* hypothesis is not confirmed by further observations at less intense or more intense flames? Must we then embrace Rumford's conclusion? Not at all! Because the experiment deals only with *visible* light. Recall that W. Herschel had discovered infra-red light in 1800 and Ritter had found ultra-violet rays in 1801. Since more (invisible) light might be present in his experiments, Rumford cannot (logically) refute a theory of light as a material ingredient of chemical compounds by using observations on visible light alone.

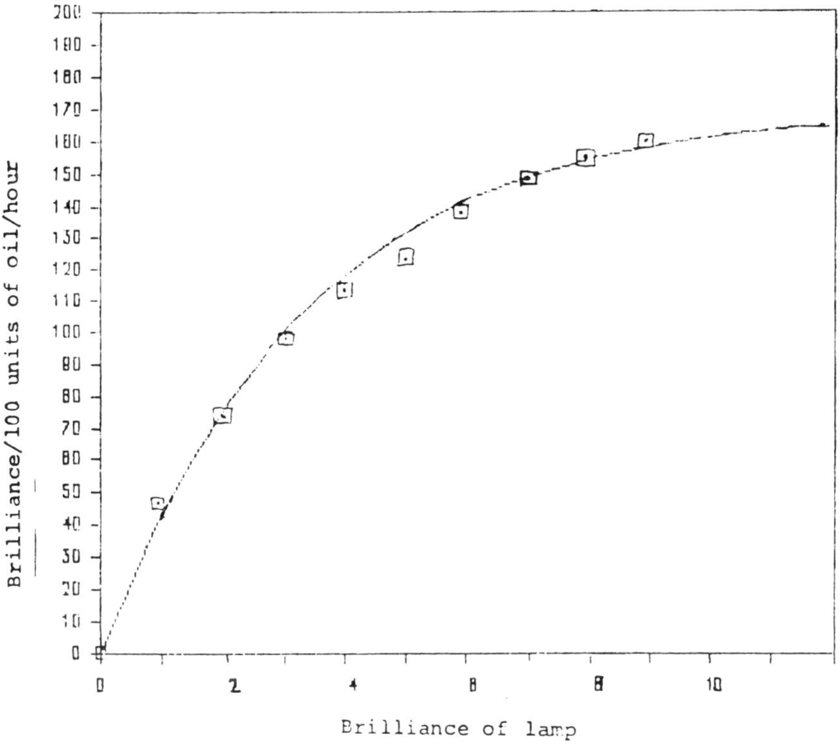

$$y = M(1 - e^{-kx})$$
$$y = 170(1 - e^{-0.003x})$$

Fig. 2.

Regardless of the reasons for its neglect, the fact remains that the paper was neglected, and so remains in the archives of scientific literature. But Rumford's experimental results never became a part of the corpus of science (scientific knowledge, scientific truths). The contrast with the cannon-boring paper is striking. There Rumford subscribed to a(n untenable) vibratory theory of heat (not random motion of particles). There Rumford's theory was incapable of handling radiant heat phenomena. Yet Rumford's work there is enshrined in the corpus of science, even though his thoughts were quite at odds with the finally accepted theory of heat. Here Rumford's experiments on light are conceptually more compatible with the emerging theories of light. Here Rumford makes substantial practical suggestions. Yet unfortunately no one cites Rumford's work on light, and thus it never becomes a part of science.

PART II

A practitioner of science may cite an earlier work in order to give credence to his own current efforts. Numerous sociologists beginning with Merton himself have remarked on this feature of citations.

I call attention to a reverse effect: the *cited* work gains in scientific stature because someone can build upon (an interpretation of) it:

> One function of reference is therefore to act as a device which establishes the authority on which the author's argument is founded. At the same time, in supporting his argument by referring to a particular knowledge claim, the author implies that he accepts the validity of the claim; in short, a citation used to justify an argument suggests that the author *recognizes the cited claim as a contribution to knowledge.*[9] (emphasis added)

But more than this

> A public expression of approval is given to a paper only when it is cited by some further paper, which uses its findings as evidence to support a new knowledge claim.[10]

Yet, so far, we have no closure. There is a cited paper and a citing paper that approves of the first. I suggest that what is needed is a *third* paper that claims to build on the second citing paper. That is, a research finding becomes an item of scientific knowledge only if it is cited by a report that is itself cited in turn by another report. Of course, mere citation is insufficient: it must be favorable citation, and the new work must *use* the cited information in a constructive manner.

My "double citation" proposal as a necessary condition for acceptance into science is similar to a recursive definition except that there is no base case and it is forward looking rather than backward based. I coin the new term *procursive* to characterize the necessary condition:

> A published item x is in the corpus of science only if x is favorably cited by publication y and y is favorably cited by publication z.

An important consequence of the proposed procursive condition should be noted. Suppose the status of x is secured by y, and the status of y is secured by z, but z is explicitly refuted or simply not cited by future work in science. The whole sequence of items is threatened with "unzipping". That is, y's status is now threatened and *a fortiori* x's. But it usually will be the case that some items in the sequence will be connected to other items in other sequences. Such multiply-connected items will remain in the corpus of science. Other supposed results that cannot find new connections, that do not, that is, provide a basis for new work, are indeed rejected from science. One is reminded of spurious phenomena such as N-rays, polywater, or Bohm plasma-diffusion. A small flurry of mutually supporting experimental results began to penetrate science, but eventually all of this work was extirpated because it led nowhere; it couldn't be built upon.

Sometimes when a line of investigation becomes halted through refutation or neglect, and when all the items of a sequence consequently are threatened, a new idea or hypothesis may lead to a reconstruction of the linkages. A new sequence of results can emerge, but with some items missing. The "lost" items may have to wait a while before they can be used again in some viable sequence of results. Moreover, items may be permanently lost; not all items find their way into a future reconstruction. And while they are so lost they are not in the corpus of science. They are not taught to novices; they are not favorably mentioned in current research. Notice that these reports have not been refuted in any strong sense. They just are no longer important or interesting or accepted. These remarks may help us to understand the phenomenon, emphasized by Kuhn and others, of non-cumulative paradigm shifts.

The above remarks also show why the argument for Realism based on the so-called "inference to the best explanation" for why science is successful is simply the Argument from Design in modern clothing. Again D. Hull makes the point succinctly:

> All scientists are trying to understand the world in which they live, but only a small percentage succeed. Science as a process looks so directional because of the retrospective bias that we bring to it. It can be made to *look* globally maximising by careful editing. Most research fails or leads nowhere. Of the research that finds its way into print, most has no appreciable effect. Once we weed out all the failures and false steps, science looks very efficient, goal-directed, and globally maximising.[11]

Rejecting the design argument for the existence of God does not make one an atheist. Likewise, rejecting the inference from the fact that science enables us successfully to control and predict features of our world to the conclusion that the uncognized world *is* as we represent it to be does not force one to be an anti-realist. The fact that our currently accepted representations are successful is due to the fact that we do not accept representations unless they are successful, and can be built upon – the survival of the fittest. Yet, just as the presence of natural evil suggests to some the atheistic conclusion, so too do the past failures of previously successful representations (e.g., caloric theory, phlogiston, aether) suggest to some that the elements of our currently successful representations are not necessarily similar to elements of an uncognized world.[12]

Department of Philosophy,
SUNY at Fredonia, NY,
U.S.A.

NOTES

[1] An earlier, and more complete, version of this paper will appear in *Synthése* **99**(1) (1994) 111–121.

[2] *Phil. Trans.* **140** (1850). Joule *mentions* Rumford in his earlier papers, but this is the first time he *uses* Rumford's reported observations.

[3] 'On the Light Manifested in Combustion', *Phil. Trans.* (1812)

[4] A brief description of these photometers can be found in A. Wolf, *A History of Science. Technology, & Philosophy in the 18th Century*, Vol. 1 (Harper Torchbook, 1961, New York) pp. 167–170. According to W. J. Sparrow, *Count Rumford of Woburn, Mass.* (Thomas Y. Crowell, 1964, New York), Rumford developed his instrument in ignorance of the earlier work. Since the principles upon which all of them were constructed are so simple, Rumford may have learned about the earlier models through conversations. The point is not important here and will not be pursued.

[5] Rumford actually had two opaque objects so that the shadows from the two sources would be next to each other on the screen. This variation increased the sensitivity of the judgment of equal darkness, but did not affect the point of the photometer.

[6] James Bogen and James Woodward, 'Saving the Phenomena', *The Philosophical Review* XCVII **3** (July 1988) 303–352. Jim Woodward, 'Data and Phenomena', *Synthése* **79** (1989) 393–472.

[7] It is not possible to estimate the errors of measurement in these experiments, since Rumford did not mention the weights of the Argand lamps nor how much oil was used. Because he determined the oil consumed by subtracting the weight of the lamp after burning for a half hour from the weight of the full lamp at the beginning, it is clear that the error in the difference of two large numbers can be quite substantial.

[8] My thanks to S. Brown, J. McKenna, and D. Jelski, colleagues at SUNY at Fredonia, for helping me with a little curve fitting experiment.

[9] G. Nigel Gilbert, 'The Transformation of Research Findings into Scientific Knowledge', *Soc. Stud. of Science* **6** (1976) 281–306, see especially p. 287. This extremely important paper came to my attention after the first draft of the current essay, well after my earlier attempts in 1987 to persuade the Staff Seminar at Fredonia that my insight had merit.

[10] *Ibid.*, p. 294.

[11] D. L. Hull, *Science as Process* (University of Chicago Press, 1988, Chicago) p. 474.

[12] I must confess that I am equally at a loss as to how to establish (prove) that God does not exist as I am to prove that the uncognized world is *not* as we currently represent it to be. And why would I want to prove either result?

RICHARD H. SCHLAGEL

RESOLVING THE REALIST-ANTIREALIST DILEMMA

The realist-antirealist dilemma consists in deciding whether the results of current research in the physical sciences (such as genetics, neurophysiology, and particle physics) justify belief in the *approximate* truth of the respective theories and in the *conditional* or *contextual* reality of the physical structures designated by the theories, or whether the fallible and transitional nature of scientific explanations preclude believing in anything other than the *observable* results of experimental inquiry. For realists, accepting the confirmed predictions deduced from theories while denying reality to the insensible, underlying mechanisms and interactions from which the predictions are derived makes scientific inquiry miraculous or magical.

Antirealists, on the other hand, following a long tradition from Hipparchus and Osiander to Mach and van Fraassen according to which astronomical systems and physical theories were derided as mathematical fictions, argue that anomalies in quantum mechanics (such as the wave-particle duality and the fuzzy status of the electron), impasses in the general theory of relativity (pertaining to the continuity or discontinuity of space-time), as well as the continuous revision of theories, preclude theoretical concepts from designating any actually existing physical entities, states, or processes. As Bas van Fraassen states:

> It is the intensional correlate of the conceptual framework through which I perceive and conceive the world. But our conceptual framework changes, hence the intensional correlate of our conceptual framework changes – but the real world is the same world.[1]

From this he concludes that "to accept a theory is . . . to believe that it is empirically adequate – that what the theory says *about what is observable* . . . is true".[2] Because theories are ultimately verified by observable data observation constitutes the limits of what can be known, conforming to the ancient dictum of "saving the phenomena" while refusing to believe in the reality of the explanatory mechanisms. Antirealist philosophers like van Fraassen are the Osianders of contemporary physics.[3]

Arthur Fine, who describes himself as a "nonrealist", states that realists

R. S. Cohen, R. Hilpinen and Qiu Renzong (eds.), Realism and Anti-Realism in the Philosophy of Science, 405–417.

mistakenly make "the question-begging move from explanatory efficacy to the truth of the explanatory hypothesis",[4] and erroneously "suppose that . . . confirmation is a mark of an approximately correct ontology and approximately true laws".[5] He even denies that scientific inquiry offers any evidence of "a [progressive] change where we learn more accurately about the *same things*".[6] In my opinion these antirealist assertions are so implausible given the meticulous competitive confirmations in physics (for example, the joint confirmation of the J/psi particle by Burton Richter's group at SLACK and Ting's group at Brookhaven), the frequent theoretical convergences among the sciences (as in molecular biology), and the remarkable technological inventions and applications (such as the radio telescope, tunnelling microscope, and current use of computers in simulating physical experiments), that we simply can reject the antirealist horn of the dilemma.[7]

It requires an implausible credulity to believe that all the remarkable consequences of current experimental inquiry, purported to be based on discovered entities and properties, are actually derived from fictitious theories and nonexistent entities: e.g., the magnetic acceleration of electrons and protons in particle accelerators, genetic engineering and the genome project based on the atomic-molecular structure of the DNA-RNA, high speed transit utilizing electromagnetic fields, and the role of anions in superconductivity. Rather than imposing arbitrary *a priori* limitations on the possibility or actuality of scientific discoveries, what is required is a conception of science that preserves and accounts for its achievements while also acknowledging the idealized, approximate, and fallible nature of scientific theories.

From the beginning of modern science with Kepler's discovery of the three planetary laws, Galileo's experimental confirmation of the correct law of free fall, and Newton's derivation of the universal law of gravitation from Kepler's third law, it was apparent to the investigators that their laws applied to idealized states and approximate conditions where the extraneous factors had been abstracted and eliminated. This does not mean, however, that these laws are false but that they are true within the limiting conditions specified. In fact, as science progresses we discover why previous laws only held approximately, as Newton's law of gravitation demonstrated with respect to Kepler's astronomical laws and Galileo's gravitational law, and as Einstein's general relativity theory showed with respect to the anomalous shift in the perihelion of Mercury, unexplainable on Newton's law.

While it may be dramatic to declare "how the laws of physics lie",[8] false laws are difficult to reconcile with the precision of scientific predictions (as with the spacecraft Voyager that flew billions of miles over a period of eight years and arrived within seconds of the predicted time), or with the construction of extremely complex apparatus (as in the design of the magnets required to accelerate the electrons or protons in the previously funded but now cancelled superconductor-supercollider), or the exact prediction of particles with specific properties (as the omega minus). As Kepler stated: " 'It is a most absurd fiction . . . that the phenomena of nature can be explained by false causes' ".[9] While not all the properties and interactions of every model used in science are considered real, as Cartwright affirms, this does not mean that the laws of science on which confirmed theories are based are false.

In addition to idealized states and approximate conditions, scientific inquiry utilizes abstract concepts such as mass, force, instantaneous velocity, and inertia, as well as isotopic spin, symmetry, charm and color, and quarks. In so far as these concepts are precisely defined in terms of experimentally detectable properties and magnitudes – and that is what mainly differentiates twentieth century science from past science – they have exact meanings and designations within the theoretical framework despite their abstractness and remoteness from ordinary experience.

Furthermore, because natural processes are understood as functional relations among constants and variables as expressed in mathematical equations ($F = MA$, $E = MC^2$, $e = h\nu$, etc.), it is the formalism that conveys our most precise representation of the underlying physical structures and interactions producing phenomena. What anchors this formalism in the real world is the *experimental* determination of the magnitudes of the theoretical structures and their properties, with the discovery of constants (like the invariant velocity of light and Planck's constant of action) attesting to some encounter with an inherent feature of physical reality. It is hardly credible that these constants, any more than Mendeleev's periodic chart, are merely phenomenal uniformities with no underlying physical significance or causes.

The reason experimental inquiry is necessary is that ordinary occurrences are not self-existent or self-explanatory, but require explication in terms of deeper causal structures and processes or inherent mechanisms. Moreover, because sensory observation discloses the world relative to those sense modalities, the use of optical and radio telescopes, spectroscopy, refractive, electron, and tunnelling microscopes, cloud or bubble chambers, particle accelerators and supercolliders, etc., have been invented to augment or supplement our sensory knowledge. Just as the qualitative macroworld appears as it does due to the nature of the human organism, as Kant maintained, so the detected states and properties of the microworld essential for scientific theorizing are manifested experimentally, as Bohr claimed.

Like all knowledge, our access to and understanding of deeper domains of physical reality depend upon the type, power, and precision of the instruments used in probing nature as interpreted by the conceptual framework and mathematical formalism.[10] To claim that only the world as displayed within the limitations of human sensory observation is worthy of belief is to succumb to a naive anthropomorphism. Even the distinction between what is a "theoretical construct" as opposed to "empirical phenomena" constantly shifts with improvements in technology: e.g., the prediction of Neptune based on Newton's laws to account for the irregular motion of Uranus was later telescopically confirmed; the molecular structure of DNA-RNA can now be observed with the tunnelling microscope; and even the lattice structure of atoms comprising the surface of metals can be 'seen' using the latest microscopic technology aided by computer enhancement. The later *observable* confirmation due to improved technology of what at one time were purely theoretical entities should reinforce the belief that theories can represent actual, if insensible, structures.

So rather than concluding that none of our theories has any verisimilitude or referential validity beyond what is optically observable and tactually detectable, it should be recognized that physical reality itself has induced scientists to develop improved investigative techniques and forced them to assimilate the striking, unexpected results obtained – typified in the remark of the American physicist I. Rabi, "Who ordered this?", when the muon was correctly identified.[11]

This brief justification of a realist interpretation of science is based on actual scientific achievements. However, in the realist-antirealist debate two theoretical arguments opposed to realism are often introduced which, if shown to be mistaken, can remove much of the opposition to realism. The first, related to Quine's thesis of "the underdetermination of theories by empirical evidence",[12] claims that because the empirical support for theoretical explanations is almost always inconclusive it is *possible* that alternative theories *could* account for the same phenomena. So if it is *possible* that any present theory *could* be replaced by another theory that would serve just as well, why should we believe that any one theory is true? This is a variation of the argument Pope Urban VIII stated to Galileo which he repeated in *The Two Chief World Systems*;[13] namely, that however persuasive the empirical evidence might be it should be remembered that God in his omniscience and omnipotence could have caused things to occur by many means inconceivable to man beyond those they appear to have. The argument has a certain heuristic value (as in Galileo's case where he was mistaken about the cause of the tides), in that it acknowledges the fallibility of scientific inquiry and cautions against premature acceptance of a theory, but does it warrant the conclusion that we never are justified in believing that any theory is true (within certain domains and relative to certain approximations)? – for example, the kinetic-molecular interpretation of gases, electrical conduction, the explanation of chemical reactions in terms of molecular structures and bonds, the chemical-electrical interpretation of synaptic discharges and neurotransmitters, etc.

What weight should be given to an argument based on a theoretical possibility as compared to actual historical precedents? For if we consider modern scientific controversies, invariably there were two or three competing theories, not an indefinite number of possible contenders. While during the proto stage of scientific development (as in Presocratic philosophy) a wide diversity of explanations was possible, as scientific inquiry advanced with its methodology becoming more selective and its theories more refined, the possible interpretations became more restricted. Consider the classic example of the Copernican revolution where there were just three competing theories, the Aristotelian-Ptolemaic-geocentric view, the Copernican-Keplerian-heliocentric system, and the Tychonic compromise. Would anyone claim that the subsequent evidence provided by Galileo's new telescopic observations and Newton's universal law of gravitation (finally "resolving the problem of the planets") did not justify astronomers in believing that the heliocentric theory of the planetary system was true? Should Galileo, Kepler, and Newton have remained skeptical as to the truth of heliocentrism because another explanation theoretically was possible?

Or consider another classic example, the "hypotheses" offered to explain light in terms of particles or corpuscles and pulses or waves? Based on the evidence of the sharp outline of shadows and the refraction of light in denser mediums (which he attributed to attraction of the corpuscles), Newton concluded that the corpuscular hypothesis was correct, even though he was aware that other phenomena such as color films and diffraction seemed to support the pulsation or wave hypothesis of Hooke and Huygens. Newton's prestige, along with the success of atomism in chemical explanations, was such that the corpuscular hypothesis prevailed until the early nineteenth century when experiments by Young and Fresnel demonstrated that diffraction patterns supported the wave theory of light, and Foucault proved that light is retarded in denser mediums contrary to Newton's corpuscular prediction. As a result, the wave theory was accepted until Einstein's explanation of the photoelectric effect reintroduced the conception of discrete quanta of light as photons, which was confirmed by the scattering experiments of Arthur Compton. Now light is believed to have both particle and wave characteristics, as detected under different experimental conditions.

While this brief history suggests that it would have been imprudent earlier to have decided conclusively in terms of either particles or waves, it also shows that the experimental evidence constrained the possibilities of explanation. This was true of other controversies, such as the fluid or particle interpretation of electrical conduction, the caloric versus the kinetic theory of heat, Stahl's phlogiston versus Lavoisier's oxygen explanation of combustion, Darwin's natural selection conception of evolution versus Lamarck's inherited characteristics, the Golgi reticulum versus the Cajal neuronal theory, or Linus Pauling's protein versus Watson's and Crick's nucleic acid interpretation of genes, etc. In each case the *evidence*, along with the theoretical *constraints* based on accepted theories, decided the issue, despite the *possibility* of other unknown explanations!

At the frontiers of science (as today in quantum mechanics) there is considerable uncertainty as to what is the correct interpretation, but this was always true. When reading the works of Galileo and Newton[14] one is struck by the fact that they were very doubtful as to whether sufficient evidence ever would be forthcoming to decide whether the universe was spherically finite or indefinitely extended (Galileo) or to explain the nature of gravity and of attractive and repulsive forces (Newton). While it is true that some of these problems are still with us, many others have been settled. Thus the mere *theoretical possibility* of alternative theories should not be an impediment to believing a theory once the evidence is compelling, as it is in many instances. It surely is an unwarranted inference from the fact that it is always *possible* that another theory *could be found* to explain the same phenomena to the conclusion that *no scientific theory*, regardless of the evidence, *can be considered true.*

Newton himself was aware of the deficiency of this argument as indicated in the following quotation, although the point of his argument was to oppose the introduction of merely hypothetical explanations.

> If any one offers conjectures about the truth of things from the mere possibility of hypotheses, I do not see how any thing certain can be determined in any science; for it is always possible to contrive hypotheses, one after another. . . . Wherefore I judged that one should abstain from considering hypotheses as from a fallacious argument, and that the force of their opposition must be removed, that one may arrive at a maturer and more general explanation.[15]

The second objection pertains to theory change, reference, and incommensurability. Ever since Kuhn's influential monograph on paradigm shifts in science,[16] and Feyerabend's "anarchistic" thesis that "anything goes"[17] in scientific inquiry, it has been argued that scientific revolutions are incommensurable because the concepts in the opposing theories refer to different entities. Because the competing theories describe different universes they cannot be compared, nor can one be the result of a transformation of the other. Revision and development of concepts occurs within paradigms, but paradigms themselves are so self-contained and disparate that they cannot be compared, only rejected and replaced. While one can find some instances illustrating this, as when the corpuscular-kinetic, mechanistic universe was developed by Boyle and Newton to replace the outmoded substance-attribute, organismic cosmology of Aristotle, or when the theory of oxygen replaced that of phlogiston, it is extremely rare and does not characterize most scientific developments, even revolutionary ones. This notion of scientific revolutions as involving rationally incomparable theories is based, I believe, on a mistaken interpretation of conceptual frameworks and how they are revised.

A world-view is contained in a theoretical framework consisting of a system of basic concepts and a network of various semantic implications (plus the syntactic structure) based on the meanings of the concepts as stated by the sentences composing the framework. A *scientific* world-view consists of the conceptual system derived from the accepted methodology. These conceptual-linguistic frameworks comprise the background reference system used in communicating or exchanging views, as Whorf and Quine pointed out. Without any commonality in the background framework no meaningful exchange is possible. Moreover, it is these entailment connections among the meanings of scientific concepts and laws that make possible scientific reasoning, inference, and prediction.

Intellectual development, whether personal or historical, consists of a reorganization of the conceptual framework by revising the meaning of older concepts and introducing new ones, which in turn modifies the implicative or entailment connections composing the strands in the network, and by adjustments in the weights given to the evidential significance of the statements. This is why theory change, even if revolutionary, is never wholesale. Some of the concepts and/or their meanings are modified while new ones are added disrupting the implicative relations. Whereas in the order framework the inferential links tended to be consistent or coherent, modifications introduce incongruities and anomalies that require reinterpretation of the meanings as well as the rejection of older, and the creation of newer, concepts. But if *all* the underlying concepts and their entailment relations were changed, the system would be completely unintelligible. As Paul Thagard states: "Thus even revolutionary conceptual change occurs against a background of concepts that have relative stability".[18]

This is why ordinary language plays such a central role in any discourse – it provides the broadest and most familiar background by means of which new concepts can be explicated and defined. Bohr made the same claim for classical physics, asserting that discourse in quantum mechanics presupposes the description of laboratory equipment and experimental phenomena in the language of Newtonian physics. There must be some underlying conceptual-symbolic system that remains relatively stable and intelligible if one is to talk sensibly of paradoxes and anomalies.

This can be illustrated best in terms of actual historical developments, such as the Copernican Revolution, the favorite example of Kuhn and Feyerabend. Certainly there was considerable ambiguity, misunderstanding, and irrationality underlying the opposing arguments in the early seventeenth century. But even granted this, the transformation of the traditional geocentric universe into the heliocentric system was a gradual process, as Kuhn himself argued in his earlier book.[19] Copernicus left the traditional universe essentially as it was, arguing that a simpler more interconnected explanation of planetary motion could be obtained, and artificial computing devices such as the equant eliminated, if one described the orbital revolution of the planets with the sun rather than the earth at the center – as Aristarchus had suggested nearly two millennia earlier. This interchange of the sun and the earth had its jarring conceptual implications affecting the meanings of 'terrestrial' and 'celestial bodies', since the stationary earth would revolve among the celestial bodies and the rising and setting sun would be stationary and centrally located in the terrestrial world. In addition, terrestrial objects would no longer fall to the center of the universe but to the center of the earth.

The motion of the earth introduced further problems: if the earth had an orbital revolution why was no stellar parallax observed and why are not clouds and birds left behind as the earth revolves? If the earth rotates on its axis diurnally causing the apparent rising and setting of the sun, why are not objects thrown from the earth's surface, why do objects propelled straight upward return to the same location, and why do cannon balls fired in opposite directions with the same force land an equal distance away (since the earth should have rotated further from one and towards the other during their flights)? These objections are indicative of disrupted semantic or entailment relations based on the changed meaning of the concepts induced by the heliocentric conception, yet the general framework of the older system remained intact: e.g., the assumption of a finite spherical universe with a center and periphery bounded by the fixed stars, the inferior and superior planets revolving around the center with a circular uniform motion, etc. Most of the difficulties had to do with reconciling the earth to its new position as a revolving, rotating planet.

It was Kepler who, in order to describe precisely the orbital revolution of the planets based on a sun-centered system and Tycho's astronomical observations of Mars, challenged the ancient simplifying assumptions that the planets as celestial bodies must have perfect circular and uniform motions. By discarding these assumptions he was able for the first time in history to formulate precise laws depicting and predicting planetary motions along with their relative orbital distances from the sun. Then Galileo introduced new astronomical data derived from his telescopic observations that undermined the ancient distinction between the celestial and terrestrial worlds: the observation of mountains and craters on the moon resembling those on the earth, spots revolving on the surface of the sun suggesting clouds, satellites revolving around Jupiter resembling the earth and its satellite the moon, and the evidence of the phases of Venus supporting the predictions based on the heliocentric system and refuting those derived from the geocentric view.

In addition, Galileo skilfully devised arguments rebutting the objections opposing the heliocentric system based on terrestrial implications and the Aristotelian distinction between circular and rectilinear motion as definitive of celestial and terrestrial bodies. These dialectical arguments demonstrate irrefutably that for Galileo the Aristotelian-geocentric and Copernican-heliocentric frameworks were not incommensurable because the arguments presuppose a thorough understanding of both systems (not surprising since he taught the Aristotelian system at the University of Pisa before committing himself to heliocentrism).

So gradually the theoretical framework containing the classical geocentric cosmology became unravelled and a revised network of concepts woven into a new astronomical system. But this was a gradual process of change that took nearly a century from the publication of Copernicus's *De Revolutionibus* to the eventual confirmation and acceptance of the heliocentric system. Yet even this radical transformation involved some continuing stable designations regardless of the shifts in meaning: e.g., despite the change in significance of 'celestial bodies' the term 'planets' still designated six of the seven traditional bodies; regardless of the altered position of the sun and the earth the terms referred to the same entities; although elliptical, the planets still revolved in orbits, etc. Attributing the inability to change frameworks to being incapable of shifting conceptually from one paradigm to another because they are incommensurable overlooks the ancillary role of additional beliefs. Even when one clearly understood the nature of and rationale for the two systems there were all kinds of reasons for resisting the change beyond the astronomical evidence: the displacement of man from the center of the universe, the challenge to Catholic doctrine and authority, the refutation of scriptural passages claiming the sun moved, the threat to university professors teaching the Aristotelian system, and the disruption of common sense and religious convictions.

I think a similar analysis would show that Einstein's relativity theory, despite the radical change in meaning of the basic concepts, was not incommensurable with Newtonian science if that is construed as not being able to compare rationally the two systems or understand the reasons for the introduction of Einstein's theory, especially in connection with the null results of the Michelson-Morley experiments. Einstein did not describe his relativity theory as incommensurable with Newton's cosmological system, but as reducing to it in the limiting case where the velocity of light is infinite or instantaneous, as it is for all practical purposes on the earth. Similarly, Bohr described quantum mechanics as reducing to classical mechanics when the magnitude of interactions is large relative to Planck's constant.

The main flaw in the interpretation of theory change as incommensurable, rather than transitional, is that it does not recognize that the intensional or connotative meanings of words can change while their denotations or designations remain the same. This misunderstanding underlies Fine's assertion quoted earlier that the history of science offers no evidence of "a change where we learn more accurately about the same things". But can it sensibly be maintained that when we talk about the rings of Venus we are not referring to the same planet as did Copernicus, that when we describe the heart as a pump producing the circulation of the blood we are not designating the same organ meant by Aristotle or Galen, that when we refer to neurons in the brain we are not denoting the same organ as did Descartes, or that the reference to atoms is not continuous with that of Leucippus and Democritus, Epicurus, Lucretius, Dalton, Rutherford, and Bohr?

If learning more about the inner structures and properties of atoms, cells, neurons, and planets precluded referring to the same entities, then this would have to be true of our increasing knowledge of ordinary objects as well. The term 'whale' which once meant a fish could not be used to designate the same aquatic creature when it was learned that whales are mammals. We could not claim bread consists of carbohydrates, vitamins, and minerals, that the composition of water is H_2O, or that genes are actually chromosomes which in turn are DNA-RNA structures because the terms 'bread', 'water' and 'gene' would have different referents from those of their ingredients. But denotation can remain fixed while the intensional cluster of properties change. When the earth became a planet, it was still the earth that was designated.

Unless this is recognized we *would* have to admit that our knowledge of the physical universe had not advanced beyond that of Aristotle and Newton, or that our conception of the atom had not progressed beyond Democritus or Dalton, or that our understanding of how the nervous system functions is no better than Galen and Descartes because each investigator would have referred to a different physical phenomenon. And if "explanatory efficacy is not a sign of the partial truth of explanatory hypotheses" and "predictive confirmation not a mark of an approximately correct ontology and true laws", as Fine claims, then why are not concepts of crystalline spheres, the five Platonic solids, phlogiston or genes just as effective in scientific explanations as gravity, oxygen, W and Z particles, and the DNA-RNA? As Sheldon Glashow claims regarding the Grand Unified Theory (GUTs), " 'I'm convinced the theory is sound . . . the essential framework of GUTs simply explains too many things for it to be fundamentally wrong' "[20] – the basic claim also of realists. Since it is experimental inquiries as interpreted by these theories that account for the achievements of science, an adequate conception of science must include an explanation of how this occurs. The basic choice, then, is between accepting an interpretation of science (as antirealists or nonrealists do) that precludes actual scientific discoveries, approximately true explanations, and continuous progress, or accepting the obvious fact of the latter and formulating an interpretation of scientific inquiry that accounts for it. After all, what is actual must be possible.

Department of Philosophy,
The George Washington University,
U.S.A.

NOTES

[1] Fraassen, Bas van (1980). *The Scientific Image*, Clarendon Press, Oxford, p. 81.
[2] *Ibid.*, p. 18.
[3] Cf. Schlagel, Richard H. (1988). ' "Critical Study": "Experimental Realism: A Critique of Bas van Fraassen's "Constructive Empiricism" ', in *Review of Metaphysics* **41**, June 1988, pp. 789–814.
[4] Fine, Arthur (1986). *The Shaky Game: Einstein Realism and the Quantum Reality*, The University of Chicago Press, Chicago, p. 120.
[5] *Ibid.*, p. 117.
[6] *Ibid.*, p. 130. Brackets added.
[7] Cf. Schlagel, Richard H. (1991). ' "Critical Notice": "Fine's 'Shaky Game' (And Why NOA is No Ark for Science)" ', *Philosophy of Science* **58**, June 1991, pp. 307–323.
[8] Cf. Cartwright, Nancy (1983). *How the Laws of Physics Lie*, Oxford University Press, Oxford, p. 152.
[9] Kepler, Johannes (1963). *Astronomia Nova*, Prefatory Matter. Quoted from Koestler, Arthur, *The Sleep Walkers*, Universal Library, New York, p. 169.
[10] Cf. Schlagel, Richard H. (1985). *Contextual Realism: A Meta-physical Framework for Modern Science*, Paragon House, New York, Ch. IX.
[11] Cf. Ne'eman, Yuval & Kirsch, Yoram (1986). *The Particle Hunters*, Cambridge University Press, Cambridge, p. 86.
[12] Quine, W. V. O. (1970). 'On the Reasons for Indeterminacy of Translation', *The Journal of Philosophy* **67**, pp. 178–179.
[13] Galilei, Galileo (1970). *Dialogue Concerning the Two Chief World Systems*, trans. by Stillman Drake, University of California, Press, Berkeley, p. 464.

[14] This is based on my research for Volume II of *From Myth to the Modern Mind, Copernicus through Quantum Mechanics.*

[15] Letter to Oldenburg, Secretary of the Royal Society, in reply to Pardies' objections to Newton's theory of colors. Quoted from Sir Isaac Newton (1962). *Principia*, Vol. 2, "The System of the World", Motte's trans., revised by Cajori, University of California Press, Berkeley, p. 677.

[16] Cf. Kuhn, Thomas S. (1962). *The Structure of Scientific Revolutions*, University of Chicago Press, Chicago.

[17] Cf. Feyerabend, Paul (1978). *Against Method*, Verso, London, p. 23.

[18] Thagard, Paul (1992). *Conceptual Revolutions*, Princeton University Press, Princeton, p. 56.

[19] Cf. Kuhn, Thomas S. (1957). *The Copernican Revolution*, Harvard University Press, Cambridge, pp. 180–184. In my opinion his earlier analysis is more correct than the later one.

[20] Quoted by Crease, Robert P. and Mann, Charles C. 'How the Universe Works', *The Atlantic Monthly*, August 1984, p. 92.

ZHOU JI HONG

THE MATHEMATICIZED PRACTICAL ATTITUDE

Arthur Fine suggested 'The Natural Ontological Attitude', which accept the results of scientific investigation as 'true', on a par with more homely truths. He believed that both realists and antirealists all accept scientific truth at this 'core position'. The natural ontological attitude (acronym NOA) is just the core position itself and all by itself, neither realist nor antirealist but mediates between the two, and it is for commonsense epistemology. Considering that realism adds an outer direction to NOA (that is, the external world and the correspondence relation of approximate truth), and antirealism (typically) adds an inner direction (that is human-oriented reductions of truth, or concepts, or explanation), Arthur Fine tried to conciliate the controversy between realist and antirealist through NOA: 'NOA suggests that the legitimate features of these additions are already contained in the presumed equal status of everyday truths with scientific ones, and in our accepting them both as truths. No other additions are legitimate and none are required.'[1] Thus, the controversy between realist and antirealist seems to be nonsignificant, and the puzzling problems about modern physics and the challenge they give to philosophy all will be covered up by the empirical view of NOA.

In my opinion, NOA is too near a naive empiricism to deal with the problems about modern scientific realism. Although NOA correctly realizes the fundamental defects in realism and antirealism, it does not offer a better means to draw an alternative scheme but recedes to a nonactive position, evidently not enough for our debate. So I put forth a 'Mathematicized Practical Attitude' (acronym MPA) to supplement and modify NOA. In this essay I want to analyse scientific theory in detail and elaborate MPA, which argues that scientific theory actually consist of a mathematicized structure – it is an operational framework – from an inner-view, while scientific explanation as an abductive inference just reflects a subjective perspective. Finally, MPA suggests that, from a point of view of an operating level, science provides a mathematicized operational system for getting knowledge of what the world is like and for controlling nature, and thus antirealism is approximate; on

R. S. Cohen, R. Hilpinen and Qiu Renzong (eds.), Realism and Anti-Realism in the Philosophy of Science, 419–426.

the other hand, from a point of view of an onlooking level, science is the only way through which we can touch the world beyond our senses, and the operational framework which scientific theory provides can reveal an image about the world. This image's rightness is obvious within a related scientific theoretical system and a scientific practical system, and thus a special realism is the very reasonable attitude. According to MPA, realism and antirealism seem to be incommensurable and are complementary at different levels.

I

The Empiricist and Popperian views of science take scientific laws and theories to be statements of the form; for all X, if all X has the property F, then X has the property G. The first step is conceptually to delineate a universe of X's having property F, and then it is to be empirically determined whether such X's also have property G. A number of these X's are thus to be observed and found either to have or not to have this property.[2] But this kind of view, based on a Deductive-Model, was rejected by later philosophers after the criticism of Kuhn and Feyerabend who offer a Gestalt-Model as the substitute. From the point of view of a psychologist and historian, Kuhn used the term 'incommensurable' in order to express what is perhaps the same particular insights concerning the nature of the relation between certain scientific theories.[3] So scientific theories are not taken to be statements corresponding with a truth value, but are instead likened to individual empirical concepts or predicates which are intended to apply only to certain phenomena. Thus theories are conceived of as being intended to apply to certain states of affairs, and to be such that they may be judged to be more or less successful in their application.[4]

The progress of scientific philosophy of Post-Kuhnian and post-Feyerabendian [analysis] shows that restricting attention to syntactical at the expense of semantic problems is a lopsided view, which does not conform to the history of science (particularly the history of physics). Although Kuhn's explanation is on a correct path by which scientific theory is regarded as a correlated and historical unit, it has so strong a psychological tint that it is apt to introduce people to an indefinite and imprecise state.

Considering that these views are not all satisfactory I put forward an alternative view about scientific theory as the basic idea of MPA. In my opinion, science is rather a mathematicized practice than a theory-system which would exhibit a more and more precisely natural picture, or than a subjective paradigm which mixes together many non-scientific factors. This practice is neither merely a magnificent organizing tool nor just corresponds with realism itself, but provides us with an operational framework; my core position depends upon the practical notion rather than on a commonsense epistemology. The difference between MPA and NOA is that the former always considers scientific theory from the angle of view of practice.

MPA holds that the basic components of scientific theory are mathematical deductive systems, dimensional systems and their frames of reference. Many basic concepts of theory are operationally defined. In formulations of laws, these concepts appear as corresponding parameters with certain dimensions. By means of these basic components and parameters which have specific space-time features, scientific theory mathematically defines and expresses the properties and dispositions of analysed objects or phenomena, and further makes it possible for them to go into an experimental operational process in order to investigate and test their structures or relations, and so on. Thus the consistency of intersubjects based on this operational process can offer a stable and reasonable way for scientific communication and development. If we do not rigidly adhere to linguistic analysis of scientific theory, but take MPA's practical viewpoint, we should agree that scientific theory consists of an especially abstract element such as model and hypothesis which cannot simply be reduced or separated into isolated statements.

Before the twentieth century, many scientists and philosophers believed that there were some laws of nature laying hidden in the night, and scientists could discover them. Now, however a law is conceived as the reason which accounts for uniformity in nature, not the mere uniformity or regularity itself.[5] MPA suggests that a law is just a working hypothesis for scientific practice and a temporary achievement of theoretical explanation. A theory might include many elements such as postulates, axioms, principles and theoretical terms, but its core content is mathematical formula and physical (or chemical or other special) parameters and working hypothesis. The theory's axioms, principles, and basic postulates determine its operational structure together with its experimental laws and its core content.

Of course, models embedded in scientific theory would provide an explanation of mathematical systems, but they are not enduring and can only play a role of enlightenment. The really core part of a theory lies in its operational framework which is determined by some basic elements of the theory. For example, as to the mechanical law: F = m, Δv/Δt. While in appearance this law seems to explain the cause of an object's motion and its regularity, the formula itself gives us nothing else than an instrument by which we can measure interrelated parameters, examine relations between objects or motions and predict the result of a given mechanical process. Here all these terms (F, m, v, t) are operable in experiment although they may be attached with unnecessary principles or metaphors. In short, it is an operational framework within which people can experiment and forecast to understand the relations between elements and dispositions of an object.

In the case of classical physics, no postulate contains the term 'measurement' and in whatever expressions the term may appear, it is always eliminable in favor of other theoretical terms. Now in empirical science, a theoretical statement must be confirmed either directly or indirectly by empirical evidence.[6] According to MPA's opinion, a scientist always brings experimental data into a correlative methematicized connection or context, the cluster of laws, or a specific model, and harmonizes them. The viability of theory depends rather on the fact that it indicates the mathematicized angle of view, and the operational program of thus observable phenomena can be treated by mathematical means and experiments, more than that giving a kind of plausibly speculative hypothesis. For scientists, theory is the means which they design to solve those problems left unresolved. For instance, force to Newton was never an obscure qualitative action, as the sympathies and antipathies of Renaissance Naturalism had been. He set it in a precisely mechanical context in which force was measured by the quantity of motion it could generate.[7] Within the operational framework and mathematical system of theory, the parameters, terms and theoretical entities would take a defined denotation which is operational directly or indirectly in experiments. When going into the operational framework and mathematical system, the property and disposition of object can be described, tested and predicted during the mathematicized practice.

II

From the surface, we can only see that scientific theory seems constituted of principles, and explanatory model(s); we cannot see its core content. It is obvious that scientific theory does not directly describe measurement processes and operational structures, but the mathematical system, dimension system and basic frame of reference which theory contains will jointly indicate the operational framework. This in-formed framework also exists in the explanational model, the inductive hypothesis. Even if the theoretical parts, more precisely, the inferential and explanatory parts are denied, the in-formed framework will remain and still be the base of the development of theory, as it is stable in its operational style and experimental data/phenomena. The operational framework makes it possible for scientists to go further to construct mathematical laws, universal principles, and to 'abduct' a working hypothesis and a theoretical explanation.

However, MPA is different from instrumentalism, for it does not regard scientific theory only as a useful means, but argues that theory can exhibit a scientific picture of the world through its operational framework. From scientific theory's axioms, deductive – hypothesis system, and other theoretical components we can infer some universal conclusions. Based on these conclusions people can explain and comprehend matter's structure and motion, phenomena's cause and result. In other words, people can understand what the world is and why it so happens, relying on abductive inference within the operational framework. In short, scientific theory is a cognizing apparatus of mathematicized practice: it constructs mathematical models and an operating program by which it would reflect a picture about nature. So MPA not only treats scientific theory as a kind of means which is a practical medium but also treats it as an ideological mediation between subject and object.

From this opinion of MPA, I maintain that theory is the 'eye' of human cognition, that we can realize matters underlying the natural phenomena only by this 'eye'. This is by means of scientific theory and the human eye in order to reflect the abstractly rational image and the perceptual image which human being can 'see' in the world. The rational image of the world goes far beyond the commonsense image, for it is a kind of abstract structure which often renews and replaces, and it not only comes from abductive inference and conjecture based on scientific laws and theories but also comes from subjective impressions based on the explanations of such theories.

Of course, this scientific image is not as stable as a sensory image which depends on human sensory experience, while also not as trustworthy for it is the result of mental activity as to what originates from body-activity. However, what can we know about this world except by our sense organs? Scientists rely on a large number of tests to confirm their hypotheses and to examine the efficiency of the hypotheses, and thereby the abductive or inductive inferences drawn from those theories would be acceptable. Considering that scientific theory and language always go along within its operational framework, so scientific terms would correspond with specific denotations, and in the terms of practice, the well-confirmed theory would be 'acceptable truth' within the operational framework. Here, the acceptable truth is neither homely truth nor pragmatic criterion; it takes scientific practice as its standard and also relies on its theoretical framework. According to MPA, this kind of truth would be nonsense if beyond the theoretical operational framework, and the framework itself transcends commonsense and the problem about truth. Meanwhile, the acceptable truth does not use an unobservable, transcendental criterion as approximate truth does.

In addition, when some correlated theories and experiments progress, the angle, scale and degree of cognizance of the previous operational framework will change correspondently. MPA argues that scientific revolution is not only a change of mind-model and the conclusion of law; it would accompany change in theoretical statement, particularly the shift of the operational framework – including space-time framework, dimensional system, and operational-model. As MPA sees it, from Newtonian mechanics to Einstein's theory of relativity, what had changed was indicated not only by the shift of theoretical conclusions, but more important, was a change from a three-dimensional operational space-framework to a four-dimensional space-time-framework. In terms of history of science, here the narrowly operational framework would be broken and an obscure hypothesis would be modified, for more elaborate mathematicized practice after a long-term scientific development. Although during a special period the result of mathematicized practice could not be regarded as absolute truth, we can tentatively accept the image under theory's operational framework as reliable.

III

By now, it is apparent that the distinctive feature of MPA, which separates it from NOA and other similar views, is MPA refusal to analyse scientific theory (or terms) from the angle of a linguistic or a semantic view. Rather, MPA recognizes science at the point of practice and offers a new idea about truth.

Inside scientific practice, that is, at the operating level, we cannot regard scientific theory as a literally true story about the world and should realize that it is just constructed by scientists, grounding it on basic mathematicized elements which correspond with certain operations. As a subjective structure, it has no necessary relation with the natural world; realists cannot provide full reasons in favor of themselves, so antirealism is more applicable. On the other hand, outside scientific practice, that is, at an external, onlooking level, scientific theory is the only way with which we can aim to understand the natural world, and it can be treated as 'acceptable truth' when its results are well-confirmed and its mathematical model and operational framework are empirically adequate. MPA does not oppose Tarski's definition of truth, but it believes that, as an anthropocentric practice, science can acquire a kind of acceptable truth. Based on this notion of truth, scientific theory can be true and, then, realism's attitude is unavoidable.

So realism and antirealism are incommensurable and complementary at different levels in MPA. In fact, we can find that realism and antirealism cannot always offer abundant reasons in defence of their opinions, and their strong attitudes have obvious shortcomings which are difficult to accept; in fact NOA had analysed them clearly. Here I suggest MPA to go a step further. Certainly, MPA cannot completely solve the divergences between realism and antirealism, but it helps to provide a unified thought for us to clear up some confusion in the realist and antirealist controversy.

Center for Science and Society,
Beijing University,
China.

NOTES

[1] See Fine, Arthur (1984). 'The Natural Ontological Attitude', in Jarrett Leplin (ed.), *Scientific Realism*, University of California Press, California, pp. 97–101.

[2] Dilworth, Craid (1986). *Scientific Progress*, D. Reidel Publishing Company, Dordrecht, p. 77.

[3] Kuhn, T. S. (1962). *The Structure of Scientific Revolutions*. University of Chicago Press, Chicago, p. 102.

[4] *Ibid.*, p. 147.
[5] Fraassen, Bas van (1989). *Laws and Symmetry*, Oxford, Clarendon Press, p. 22.
[6] Cf. Cohen, R. S. and Wartosky, M. W. (eds) (1984). 'Physical Sciences and History of Physics', *Boston Studies in the Philosophy of Science* **82**, p. 148.
[7] Westfall, Richard S. *The Construction of Modern Science*. Cambridge University Press, London, p. 143.

USKALI MÄKI

SCIENTIFIC REALISM AND SOME PECULIARITIES OF ECONOMICS*

1. INTRODUCTION

I suggest that the issue of realism about science should be contextualized in terms of the peculiarities of particular disciplines and kinds of theories. Instead of any absolute and universal assertions for or against scientific realism we end up with a sort of relativization of realism. This amounts to a defence of concrete and local as against abstract and global philosophy of science. This suggestion is supported by using the case of economics as evidence.

Without pretending that there obtains a single standard formulation of scientific realism, let us nevertheless list a few general statements characteristically attached to this doctrine: Scientific theories depict a realm of entities which (i) lie beyond the boundaries of the ordinary realm of commonsense entities and (ii) exist independently of the human mind. Ideally, theories succeed in (iii) referring to those entities and (iv) describing them correctly, approximately at least. Finally, (v) the facts stated in (iii) and (iv) explain the uncontroversial technological success of science. According to the inference to the best explanation, this is construed as an argument for scientific realism. Let (i) to (v) constitute one standard statement of scientific realism.

It will be shown that the ideas (i) to (v) become problematic if 'science' is replaced by 'economics' and 'scientific theories' by 'economic theories'. It is argued that these ideas, created with the characteristics of physics in mind, are at odds with what has to be required of a scientific realism capable of accommodating economics. Several themes of relevance emerge.

First, and perhaps least harmfully, there is the non-philosophical usage of 'realism' amongst economists. They use the term to designate a set of properties attributable to economic theories and their constituent parts in phrases such as "the realism of the assumption that business firms maximize their profits". Here, 'realism' does not designate a philosophical doctrine. In many cases, this usage of 'realism' is not even

R. S. Cohen, R. Hilpinen and Qiu Renzong (eds.), Realism and Anti-Realism in the Philosophy of Science, 427–447.

conceptually related to realism understood as a family of philosophical doctrines.

Secondly, the fundamental thesis of ontological realism needs a careful adjustment if it is to apply in the case of economics. The most abstract formulation of ontological realism is the thesis that 'X exists'. Different more specific versions can be gained by specifying both 'X' and 'exists'. The concepts of mind-independent existence and mind-external existence, characteristic of many formulations of scientific realism, are not sufficient to accommodate the ontology of economic theories, since these theories typically invoke mental and social entities.

Third, whereas in physical sciences the issue of realism has typically been formulated in terms of the existence of nonobservational postulated theoretical entities (electrons, photons, electromagnetic fields, etc.), economics, for the most part, deals with observational or commonsense entities (business firms, households, goods, money, prices, and folk psychological entities such as aims and beliefs). Economic theories accomplish various rival modifications and rearrangements among such "folk" entities and their attributed properties. Relative to the situation in the physical sciences, this implies two things. First, ontological and referential realism are not or should not be major issues regarding the basic entities of economics. Second, some sort of commonsense realism becomes particularly relevant in economics.

Fourth, while popular arguments for scientific realism in the case of physical sciences involves appeals to technological success and experimental manipulability, it is very problematic to use these arguments in the case of economics. Questions of existence and truth in economics are not typically settled by invoking manipulability or success. In most cases, the existence of economic entities seems to be pre-experimentally unproblematic. Also, economic theories do not seem to be the sort of things of which we could relatively unambiguously say that they have or have not had technological success. No wonder then that judgements based on intuitive plausibility or compatibility with commonsense experience play such a prominent role in economics.

2. REALISM AND REALISTICNESS

The first characteristic of economics to point out is one of terminological usage. Indeed, the common usage of 'realism' itself in economics does not fit with the conceptual substance of scientific realism. On the one hand, this is a trivial point, but on the other, it is significant since only by making it are we able to see the crucial issues clearly.

Economists talk about realism on a daily basis. They argue about "the realism" of this or that theory or model or assumption, such as those of profit maximization, stable preferences and technology, full information on the part of agents, perfect competition, full employment equilibrium, benevolent state. Some economists – such as Herbert Simon – hold the view that the "realism" of such assumptions is a major issue and that more "realism" is better than less. Some others – such as Milton Friedman – think that the "realism" of assumptions does not matter a bit if only they help construe theories or models that provide good predictions or possess other pragmatic virtues.

Philosophers have a long tradition of talking about realism. They discuss the problem of the existence of universals, of natural necessity, of the mental, of subatomic particles, of moral values. They also argue about how language relates to the world, of reference and truth. Those who espouse realism may hold the view that either properties, moral values, the theoretical entities of physics or all of them may really exist; and that language can be used to refer to such realities and to formulate true statements about them. The anti-realists reject such views.

Most of the time, these two discourses, by economists and by philosophers, do not overlap. Indeed, when economists talk about the "realism" of their theories and their constituents, they do not talk about realism in the philosophers' sense at all. Using the term economists characterize various properties of economic representations. But realism as a philosophical doctrine is not a property of economic theories and their constituent parts. Thus, using the term economists must mean something else. Perhaps due to the poverty of the English language, there has not been available a separate expression for this something. My suggestion is that when using the term 'realism' economists are talking about what may be called *realisticness*. Realisticness and unrealisticness are properties attributable to economic representations. Given this stipulation, 'realism' can be reserved for denoting a philosophical doctrine or many such doctrines (Mäki, 1989).[1]

The introduction of the distinction between realism and realisticness has two significant consequences. First, it facilitates the recognition that neither of these categories is unambiguous, that there are a number of relevant kinds of both. Second, it helps us understand that within economics there are widely discussed kinds of realisticness (and unrealisticness) which are not conceptually connected to realism at all (such as testability, comprehensiveness, plausibility, practical relevance). There are others, however, which are so connected, such as referentiality ("referential realisticness") and truth ("veristic realisticness"), suitably understood. This means that, as against a misguided habit, the dispute between those who favor more "realism" in economic theory and those who are against it cannot always be construed as a simple controversy between the realists and the instrumentalists. For instance, the disagreement between Simon and Friedman does not seem to be a standard case of a realist quarrelling with an instrumentalist. In what follows the focus will be on those sorts of realisticness that are conceptually connected to the realism issue.

3. ONTOLOGICAL AND SEMANTIC REALISM

The ontological issue of existence is the place to begin when dealing with the questions of realism. The most general definition of ontological realism goes as follows:

> Def 1. Ontological realism is the thesis that X exists or that Xs exist.

This formulation permits a vast number of alternative specific versions of ontological realism. Here X is a variable which has to be given appropriate values. 'Exist' is also in need of specification. Its presumed extension and intension can be specified separately. I will make a suggestion to this effect in the next section.

The next natural step in our clarification of realism is to face the semantic issue of truth and its prerequisites, reference and representation:

> Def 2. Semantic realism is the thesis that (not any, but many) linguistic expressions refer to entities that exist (referential realism) and represent their features (representational realism) and are either true or false about them partly by virtue of what they are like, i.e., by virtue of the way the world is (veristic realism).

The respective semantic forms of (un)realisticness are the ones that are connected to semantic realism. They are referential, representational and veristic (un)realisticness, attributable to some linguistic expressions. A referentially (un)realistic expression is one which refers (does not refer) to entities that exist. A representationally (un)realistic expression is one which represents (does not represent) features that existing referents possess. A veristically (un)realistic expression is one which is true (false) about its referents partly by virtue of what they are like.

There seems to be an interesting difference between economics and physics regarding their respective focus on the various aspects of the issue of realism. In physics, existence and reference constitute a major issue, whereas in the context of economics, existence and reference would appear to be relatively unproblematic and, where the issue of semantic realisticness is posed at all, the emphasis seems to be on truth. While the existence of quarks and black holes is construed as problematic in the philosophy of physics, the existence of business firms and households, preferences, goods and their prices, appears as more or less unproblematic. The issue in economics is rather geared to, for instance, the representation of the actors as having rational expectations and transitive preferences and taking the single price of a homogenous good as given. In Section 5 below I attempt to give reasons for this difference of emphasis between physics and economics. Before that, we have to address the ambiguity of 'existence'.

4. CONCEPTS OF EXISTENCE

It is customary to define the ontological side of realism as "the thesis that there is a domain of mind-independent existence" (Rescher, 1987: p. 1). It is then natural to suggest that knowledge of this domain is acquired by natural science. The objects of astronomy, physics and chemistry such as galaxies, positrons and molecules, might indeed exist independently of the human mind. A realism qualified for accommodating economics cannot build upon such a concept of existence. Economics clearly deals with objects which do not exist independently of the human mind; it even deals with the contents of the human mind itself. A realism about economics has to employ a different concept of existence.[2]

I defined ontological realism as the thesis that X exists or that Xs exist. The *extension* of the predicate 'exist(s)' consists of the set of objects to which the predicate can be truthfully applied. Partial specifications of the extension of 'exist(s)' are partial specifications of the set of appropriate values of X. The set of candidate members in that extension include sorts of objects such as universals and particulars, essences and appearances, material and mental entities, observables and unobservables, all of which are relevant for studying social and economic theories from a realist point of view. The issue of ontological realism in economics and other social sciences involves the question of whether there are objects of those sorts in the social world.

Partial specifications of the *intension* or meaning of 'exists' are provided in terms of three specifications in the order of descending strength (Mäki, 1990b: p. 294).

Def 3. *Independent existence.* X exists independently iff it exists independently of the human mind. (Only material entities may exist independently.)

This concept of existence leads to a materialist or physicalist realism, since it leads to the exclusion from the realm of real existents things such as mental states and processes (e.g., preferences and expectations) and anything the existence of which is dependent on the mental (e.g., social institutions such as money and the market). Only material entities (such as neurophysiological entities as well as photons, plants, and planets) may exist independently. Rescher's version of realism subscribes to this concept.

Def. 4. *External existence.* X exists externally iff it exists externally to or independently of any individual human mind. (Both material and social entities may exist externally.)

Again, this (more or less Durkheimian) notion excludes mental entities from the set of existing items. Contrary to what is implied by ontological realism employing the concept of independent existence, here social entities gain a membership in the set of real existents. Social entities can variously be characterized in terms of social rules, collective patterns of behavior, unintended consequences of individual action mediated through a social mechanism, or the like. The set of such entities relevant to various streams of economics comprises the market, money, debt, property rights, classes, corporations, etc. Social entities in this sense depend on individual human minds for their existence (in this they are unlike material entities), but exist externally to them (just as material entities do). I here ignore attempts to do without irreducible social entities and thereby without the concept of external existence. Whether or not those attempts are deemed successful does not affect my main argument, which hinges upon the concept of objective existence and the centrality of mental entities in much if economics.

Def. 5. *Objective existence.* X exists objectively relative to a given representation iff it exists unconstituted by that particular representation. (Both material, social, and mental entities may exist objectively.)

In accordance with this specification, something exists objectively only relative to particular representations of that something. Objective existence relative to representation *R* is denied those entities whose existence depends on *R* or is constituted by *R*. Along with material entities (such as the molecular structure of gold) and social entities (e.g., money, the market mechanism), also mental entities (such as preferences, purposes, and expectations) may exist objectively. This is because mental entities represented from a third-person point of view are not constituted by such representations even though they may be constituted by other (first-person) representations. The preferences of a consumer are not brought to existence by being represented by an economic theoretician.

It used to be a popular view in the nineteenth century, from Nassau Senior and J. S. Mill onwards, to regard economic theory as based on two kinds of premises, material and mental, such as the law of diminishing returns and the desire for wealth respectively. The objects of material premises may exist independently, but the referents of the mental premises may exist objectively at most.

This means that a realism about economics and a realism about physics do not have a single shared concept of existence. A realism about economics needs the notion of objective existence and finds that of mind-independent existence too restrictive. A realism about physics is able to combine the two.

5. SCIENTIFIC THEORIES AND FOLK VIEWS

In commonsense experience, we confront our physical surroundings as being composed of hard rocks and green trees and drinkable water and other such visible and tangible macroscopic objects. The science of physics hypothesizes an entirely different realm of objects, composed of configurations of subatomic particles, gravitational and electromagnetic fields, black holes, curved space-time. This theoretical realm of scientific objects radically transcends the boundaries of ordinary commonsense experience. The issue of scientific realism about physics is one of the reality of the scientific realm and the truth of statements about it.

In economics, the situation seems to be drastically different. The constituents of the worlds hypothesized in economic theories appear to be more or less the same as those inhabiting the realm of our ordinary conceptualized experience. No radical gap seems to prevail between the scientific realm and the ordinary realm. Economic theories speak about objects that are confronted in our ordinary experience about economic matters – business firms and households, their aims and expectations, money and goods and their prices, land and labor and capital, wages and profits and taxes. Therefore, the existence of the objects of the scientific realm should not be a major issue in economics. The referential realisticness of the fundamental elements of economic theories is more often than not beyond doubt: since the terms of economic theories seem to refer to entities with which economists and others are familiar on the basis of ordinary experience, the referential status of these terms should not be taken as suspect.

The relevant issue lies elsewhere, namely in how the ordinary objects are represented in economic theories. Even if the referents of ordinary experience and economic theories are pretty much the same, the ways in which they are represented often dramatically diverge. There are two main ways in which the theoretical representations seem to deviate from commonsense representations. I call them *modification* and *rearrangement*.

To see the point more clearly, let me postulate the existence of three simplified and hypothetical levels of thought, characteristic of an imaginarily pure Western culture. First, there are *general folk views* of "man" and society at large. They comprise folk psychology, among other elements.[3] These folk views are widely and flexibly applicable. Second, there is what may be called *folk economics*. It is a modification and application of the general folk views to economic matters. Third, there is *scientific economics*. Scientific economics accomplishes a modification and rearrangement of the ordinary objects of folk economics.[4] We can now imagine a quasi-historical tale which delivers a message of the systematic relationships between the three frameworks.

The general folk views depict human life and social phenomena in terms of men and women, families and nations, castes and classes, desires and aims, hopes and fears, values and norms, beliefs and expectations, deliberation and routine, producing and consuming, working and eating, tools and what you make with them, love and hate, duty and honor, kinship and partnership, gifts and saleables, right and wrong, freedom and subordination, etc. Elements such as these constitute the "ontic furniture" of the general folk views of "man" and society. Folk economics modifies this furniture by eliminating some of the elements and focusing on some others. For instance, gender is excluded in favor of "man", the generalized economic actor. Emotions are excluded in favor of rational deliberation. Making love is excluded in favor of making money. Gifts are excluded in favor of saleable goods. Kinship is excluded in favor of partnership. And so on.

Scientific economics goes further in modifying the elements of the ordinary realm. A fairly trivial aspect of this modification is the *reformulation* of the linguistic expressions used for the elements in the ordinary realm in terms of technical vocabulary (such as characterizing buying decisions in terms of 'indifferent curves' and 'budget constraints'; and expressing the idea that a change in the price of a good has a variable impact on how much more or less people then wish to buy it as 'the price elasticity of demand') and in terms of formal languages (such as replacing 'price' by 'p'; and 'price elasticity of demand' by '$\partial x_i/\partial p_i$'). Modifications of substance include *selection* (such as including case-by-case means-ends deliberation and excluding routines in the picture of economic agents; and focusing on trade between two countries and forgetting about the rest of the world); *abstraction* (such as talking about labor power in general instead of you and me; and about the representative firm instead of particular business firms); *idealization* (such as the assumptions of perfect information; zero transaction costs; perfect substitutability between goods and factors of production; and firms as price takers rather than price makers); *exaggeration* (such as the assumption of maximization; and that of transitive preferences); *projection* (such as the notion of expected permanent income, equal to the actual measured income of an individual minus a transitory part); *aggregation* (such as summing up unemployed people to yield the rate of unemployment); and their various combinations.

There are a number of ways to perform these operations in the construction of economic theories, but none of them accomplishes a major departure from the ontic furniture of the ordinary realm. No new *kinds* of entities or properties are introduced. Only the ordinary entities and properties are modified (see Mäki, 1992a).

Take the example of the concept of business firm in standard neoclassical general equilibrium theory. We may take it to refer to real business firms even though it represents these firms by attributing to them properties that they are not ordinarily believed to possess: they are all alike to relevant dimensions; they are devoid of internal structure, they are functionally uniform; they are cognitively omniscient, they are fully informed about the market conditions; they are impotent regarding price setting, they take prices as given in the market; they maximize their profits by using refined marginalist calculations; they are not engaged in active entrepreneurial rivalry but rather react passively to changes in cost and demand conditions. Some commentators are led by such attributions to the conclusion that this version of neoclassical theory does not refer to real business firms. My point is that folk economics and neoclassical economics have real business firms as their shared referent even though they represent these firms differently.

Or take the folk psychological notion of preference. The preferences of consumers are represented by the axioms of standard neoclassical theory as complete (the consumer has a preference order over any two bundles of goods), reflexive (any bundle is at least as good as itself), and transitive (if the consumer prefers bundle x to bundle y and y to z, then she prefers x to z). It would be a mistake to conclude that if consumers do not have preferences with these characteristics, they do not have preferences at all. There may be other reasons to doubt the reality of preferences (along with the rest of the folk psychological realm), but, say, the intransitivity of preferences should not be one such reason. The axioms of consumer theory may refer to real entities irrespective of how these entities are represented.

I am suggesting that, ontologically speaking, a major part of economics is a modified form of economic folk views. The implication is that the scientific status of modified folk economics cannot be articulated similarly to that of theoretical physics.While in physics the ontic furniture of the theoretical realm represents a radical *departure from* that of the ordinary realm, in economics it is a *modification* of the furniture of the ordinary realm.[5] As regards realism, the implication is that, concerning the individual referents of theories, ontological scientific realism may fit physics, while economics seems to require some sort of *ontological commonsense realism.* At this level, no major abyss is opened between folk economic and scientific economics.

I suggest that there is a more significant difference though. The key to understanding this difference between folk economics and scientific economics is connected not so much to the modifications of the folk objects, but rather to their *arrangement*. This is where entire theories or theory-like systems of ideas enter the picture. I mention such semi-theoretical formations, since it is not clear that folk economics is a theory in any genuine sense (cf. the situation regarding folk psychology). In any case, while the modifications of the folk objects are accomplished by the individual concepts and statements of a theory or a theory-like formation, their arrangement is a matter of a theory as a whole.

To begin with, we can postulate a form of economic thought that can be treated as having some of the typical characteristics of folk views in other fields as well. It is based on the idea that the functioning of national and global economies can be considered analogically to the functioning of household or business management. To introduce another analogy, while folk psychology is akin to Aristotelean psychology and folk physics is similar to Aristolelean physics, folk economics could be considered as akin to Aristotelean *oikonomike*. Recall that *oikonomike* was understood as the study of household management. Folk economics involves essentially the idea that the economy is managed, that is, that economic phenomena are typically intentionally brought about by people in charge: it is the government in power or the big business which is taken as responsible for what takes place in the economy. Mercantilist and cameralist streams of economic thought may be taken to exemplify this belief.

It is only when questioning this belief that we enter the realm of scientific economics as understood since Adam Smith or so. Notions such as "the invisible hand" and the "spontaneous order" or the "character masks" and "the natural law of motion" of the capitalist economy exemplify this questioning. This idea is brought out by Ullman-Margalit (1978, pp. 268–270) who points out that an invisible-hand explanation of social phenomena is not the most "natural" or "plausible" account, because it is "an explanation in terms of intentional design [which] will naturally suggest itself". This may be reformulated by saying that an intentional-design explanation is a kind of commonsense account for social patterns and institutions. Let us formulate the two perspectives as follows ([FE] stands for 'folk economics', [SE] for 'scientific economics') (see Mäki, 1990a: p. 337).

[FE] Economic phenomena and institutions are results of intentional design by human individuals and their groups.

[SE] Economic phenomena and institutions are unintended consequences of human actions mediated by a social mechanism.

We may say that [FE] and [SE] are two principles of social constitution. In any economic doctrine, one of them typically predominates. I am suggesting that in moving from the predominance of [FE] to the predominance of [SE] no mere modification of folk objects but their major *rearrangement* takes place. The ontic furniture of folk economics remains pretty much the same, but the elements have been reorganized. This means that the nature of the causal relation between human action and economic phenomena is depicted differently in the two views. In folk economics, the outcomes are intended results, while in scientific economics, the outcomes are unintended consequences.

There are a number of alternative theoretical specifications of such rearrangement, ranging from Adam Smith's depiction of the market as an arena of relatively harmonious coordination of self-interested economic activities through the invisible hand down to Karl Marx's presumed picture of the downfall of capitalism due to the tendency of the rate of profit to fall, unintended by any single capitalist. Such alternative further arrangements within [SE] involve rival isolations of causally relevant objects and features from those that are supposed to be causally less potent or relevant, using idealizations of various kinds, including the *ceteris paribus* clause. Any given rearrangement presupposes certain modifications concerning the elements. Different rearrangements presuppose different modifications of the elements. For example, if the modifications do not serve to exclude entrepreneurship, ignorance and learning on the part of economic actors, the resulting rearrangement is different from one based on the assumption of perfect information and the exclusion of entrepreneurship (see Mäki, 1992b).

It is now easy to see why the issue of realism in economics is centered on representation and truth, while existence and reference should be regarded as relatively unproblematic. There is a potential existential and referential issue, though, related to the social mechanism posited in [SE]. But the issue is genuine only under non-individualist construals of the social mechanism. For instance, we may then ask whether the metaphor of the invisible hand refers to an existing mechanism which is not a folk entity. Even in this case, however, the issue can be dealt with as that of the truth of the metaphor.

For the most part, the issue is that of the truth (or otherwise beneficence) of the many representations involved in scientific economics. The modifications of the folk objects typically involve statements that are obviously false (such as the assumptions of perfect information or transitive preferences or zero transaction costs). Their falsehood, however, may at least sometimes serve the purpose of pursuing true specifications of rearrangement [SE]. Since there are alternative sets of modifications and rival specific rearrangements, the issue of true representation plays a significant role, at least potentially (see Mäki, 1994). It does so on two fronts, as it were. The modification of individual folk objects involves the issue of semantic commonsense realism, i.e., the issue of whether the representations of the folk objects are usefully true or false. The rearrangement of those objects in the form of entire theories involves the issue of semantic scientific realism, i.e., the issue of whether this or that suggested rearrangement succeeds in truthfully representing the detailed mechanism of social constitution.

6. EXPERIMENTAL MANIPULABILITY AND TECHNOLOGICAL SUCCESS

Let us next consider two arguments which relate the correctness of scientific realism to the presumed technological success of science. The first is the argument from experimental manipulability, found in the work of Ian Hacking and others. The second is the popular no-miracle argument for realism about science. Both arguments are problematic in the context of the physical sciences themselves, but I will ignore this here. I just point out why they are particularly problematic in the case of economics.

Hacking's experimental realism is a realism about the ontology of existence. He argues that we can become convinced about the correctness of realism by referring to certain experimental practices in science. In particular, we can be convinced of the existence of subatomic particles if we are capable of manipulating them. Experimental manipulability of an entity is, for Hacking, an indication of the existence of that entity. The example of electrons summarizes the idea: "We are completely convinced of the reality of electrons when we regularly set out to build – and often enough succeed in building – new kinds of device that use various well-understood causal properties of electrons to interfere in other more hypothetical parts of nature." (Hacking, 1983: p. 265).

Again, economics looks different. It seems obvious that no experimental manipulability by economists qua economists is required for them or others to become convinced of the existence of business firms or prices or price expectations and other folk entities. Economists have other sources for most of their existential beliefs. The argument from experimental manipulability does not seem to be needed for an ontological realism about economics.

Let us then move on to the second idea of realism which builds upon the idea of technological success. A popular argument for scientific realism is the no-miracle argument. This argument takes it as an uncontroversial empirical fact about science that scientific theories are successful in the prediction and control of phenomena. Scientific realism is then suggested as an explanation of this fact. It would be a miracle, the argument goes, if scientific theories were not true about the really existing entities and yet would be technologically successful. Scientific realism is offered as a hypothesis which would make the success of science an explainable fact, a non-miracle.

Let us take a closer look at the argument, by using Richard Boyd's formulation, as paraphrased by Hilary Putnam. Realism as an empirical hypothesis is spelled out by Boyd in terms of two principles: (1) "Terms in a mature science typically refer" (i.e., are referentially realistic). (2) "The laws of a theory belonging to a mature science are typically approximately true" (i.e., veristically realistic). Boyd then suggests that "scientists act as they do because they believe (1) and (2) and that their strategy works because (1) and (2) are true" (Putnam, 1975–76: p. 179). The notion of the "strategy" of scientists "working" obviously means that scientists attain success in prediction and control. Simply put, the non-miracle argument is this:

[RSS] (1) Terms in a mature science are typically referentially realistic.

(2) The laws of a theory belonging to a mature science are typically approximately true (i.e., veristically realistic).

---

(3) Science is successful in the prediction and control of phenomena.

Of course, [RSS] (R for 'realism' and S's for 'success' and 'science') is not a complete argument. For instance, premises concerning the beliefs, aims and behavior of scientists are missing. The important point is being delivered though. Realism, in the form of the two premises, is provided as an explanation of the technological success of science. In accordance to the inference to the best explanation, we are invited to conclude that realism about science is true. Since (3) is true, in all likelihood (1) and (2) are true as well.

Applied to the case of economics, we get the following enthymemic argument [RSE] (E standing for 'economics'):

[RSE] (1′) Terms in economic theories are typically referentially realistic.

(2′) The laws of an economic theory are typically approximately true (i.e., veristically realistic).

(3′) Economics is successful in the prediction and control of phenomena.

There are several problems with [RSE] and using it to argue for realism in economics. First, [RSE] substitutes 'economics' for 'mature science', which is problematic in itself. It is not clear what 'maturity' signifies here, but, quite clearly, economics does not count as a mature science in the same way and degree as physics, notwithstanding its celebrated status as the queen of the social sciences.

Secondly, it is doubtful if the explanandum of [RSE] is there at all, waiting for being explained. In any case, it cannot be maintained that (3′) represents an undoubted fact (cf. Hands, 1991, p. 93). It is no doubt true that much of post-Keynes (i.e. post-1936) economics has *pursued* technological success, often with great confidence. Policy interventions have been designed based on quantified estimations of the effects of alternative policy measures using multiple regression and other econometric techniques. Sometimes these measures have been implemented in economic policy, and sometimes they have appeared to work to some degree of satisfaction. More often than not, economists' policy proposals have not been implemented, or have been implemented only partially. Often enough, the outcomes have not been those intended.

No wonder then that, whatever their reasons are, most people seem to believe that economics has not achieved technological success of anything like the same kind and scale as physics (it is possible to take this as one indication of the immaturity of economics). Among the ways of grounding this belief are the following two. It may be maintained that, as a historical fact, economics does not have a comparatively strong predictive and technological record. In addition, or alternatively, it may be argued that, due to the nature of its subject matter, economics is not capable of attaining technological success which would even distantly approximate that in physics. The tradition of Austrian economics and a version of rational expectations macroeconomics exemplify this argument. At the core of these versions of the argument are the ideas that the subject matter of economics is peculiarly complex and that economics is confronted with the problem of reflexivity.[6]

The mere existence of such arguments indicates that the success of economics is not an uncontroversial fact. The status of the explanandum of [RSE] is suspect. Therefore, it is problematic to propose explaining technological success by appealing to semantic realisticness.

Thirdly, the inference to the inverse direction is problematic, too. This is because technological success does not seem to be needed to establish the semantic realisticness of economic theories (nor is technological failure needed to infer to unrealisticness of theory). Regarding (1′), economists are fairly convinced of the referential realisticness of their basic terms and of the existence of their referents independently of whether the theories containing them are technologically successful. This is because those referents are folk objects, as argued above. Business firms and households, their beliefs and aims, money, goods and their prices, etc. are not theoretical entities in the same way as electrons and quarks are.

This also appears to support the old experiential tradition in economists' thinking about the character of their theories, going back to Nassau Senior and J. S. Mill. In 1888, John Elliot Cairnes compared the situation in physics and economics in this way:

> [T]he laws of gravitation and of motion [. . . ,] the undulatory theory of light, the theory of the molecular constitution of matter, the doctrine of *vis inertiœ* – all alike elude direct observation, and are only known to us through their physical effects. [. . .] The economist starts with a knowledge of ultimate causes. He is already, at the outset of his enterprise, in the position which the physicist only attains after ages of laborious research (Cairnes, 1888 [1875]: pp. 84, 87).

Decades later, after major transformations in the substance of economic theory, Lionel Robbins expressed the same idea: The postulates of economics are not such that "the existence of [their] counterpart in reality admits of extensive dispute once their nature is fully realised" (Robbins, 1935: p. 79).

In Economics, [. . .], the ultimate constituents of our fundamental generalisations are known to us by immediate acquaintance. In the natural sciences they are known only inferentially. There is much less reason to doubt the counterpart in reality of the assumption of individual preferences than that of the assumption of the electron.[7] (*Ibid.*, p. 105).

Thus, the justification of (1′) would not seem to require technological success. Regarding (2′), economists also seem to form their judgements about the veristic realisticness or truth of many of the "law" statements of their theories independently of technological success. It seems to be rather their own everyday experience or direct statistical or other empirical evidence that counts, not the controversial technological success. Economists' beliefs about the truth or falsity of, say, the assumption of profit maximization or the expected utility hypothesis or the "law" of diminishing returns or the Jevonian "law" of indifference (there exists only one price for a good in a market) or the "law" of the tendency of the falling rate of profit, are not primarily based on technological considerations.

There is a natural reason for this. Economists systematically practise what may be called the method of theoretical isolation whereby a system of factors is closed from influences from the rest of the universe, this closure being accomplished by assumption. The theoretical method of isolation involves an imitation of a limited portion of the logical structure of experimental research capable of material isolations (see Mäki, 1992a). Besides the disanalogy between theoretical and material isolation, there is a further limitation to the analogy: the feedback from aposteriori experimental evidence to theory is weak or missing in economics. At least part of the reason for this is that all or most of possible technological success in economics takes place in the conditions of open systems. Therefore, such success hardly provides entirely reliable evidence for or against economic theories which depict relations in theoretically isolated closed systems. This turns the focus on the direct plausibility of the basic statements of theories (see also Hausman, 1992).

In Boyd's argument, it is assumed that scientists believe in (2). In economics, it often appears to be the case that economists do not believe in (2′). Even more, many economists appear to believe in the falsehood of many "law" statements in economics. On the other hand, many economists do believe that at least some such statements in their theories are close to the truth. Robbins provides a radical version of this belief: "We do not need controlled experiments to establish their validity; they are so much the stuff of our everyday experience that they have only to be stated to be recognised as obvious." (Robbins, 1935: p. 79). The important thing is that none of these beliefs is usually based on considerations of technological success.

To conclude, an admission to the technological success argument can be made. I have suggested that assessments of the semantic realisticness of a given economic theory T are not typically based on the perceived technological success of T. However, this does not preclude the possibility that it may be based on pretheoretical, that is, pre-T success. In a sense, folk economics is based on the practical experience, the successes and failures, of economic agents in their everyday business operations, whether as "housewives" or "businessmen". If it is the case that T shares with folk economics most of its referents, then whatever technological success folk economics may have, serves as a practical foundation for the ontological commitments of T. In this way, pretheoretical success might play the role of grounding existential belief within scientific economics. Yet, it is not technological success in a more demanding sense and it is not the success of an economic theory itself that is being used as a criterion of existence.

7. CONCLUSION

The above discussion suggests several interrelated points. Ad economics: Economics is not on a par with standard examples of physical sciences on a number of important dimensions. In other words, the suggested arguments illuminate some of the peculiar characteristics of economics, and remind philosophers of science of the fact that science is not uniform. Scientific realism has served as the framework for bringing this point home.

Ad realism: The paper suggests a relativization of scientific realism by using some of the specific features of economics as supporting evidence. It argues for a relativity of realism to kinds of scientific disciplines and theories. Scientific realism already appears in a number of versions, but this multiplicity is not based on the recognition of the multiplicity of types of disciplines and theories.

Ad economics and realism: Standard versions of scientific realism are tailored to suit the characteristics of physics. If we accept the premise that economics is not the kind of intellectual endeavour that would abhor scientific realism tout court, economists have to reject the standard versions. In order to check whether scientific realism is a viable descriptive philosophy of economics, we need to develop a version which is able to accommodate the peculiarities of economics without diverging from some fundamental minimum tenets of realism. Whether this is possible remains to be seen. Some ingredients for such a solution have been provided above.

There is an alternative conclusion. It is to fix a version of scientific realism and to deny any discipline not satisfying the tenets of that version the status of science. For instance, the incompatibility of the characteristics of economics with tenets (i) to (v) of our standard version could be taken as an indication that economics is not a science. The uniformity of science would then be guaranteed by dropping some disciplines from the set of the sciences. The task would not be to adjust our formulation of scientific realism but rather to develop an entirely different economics. This option is also worth examination.

Academy of Finland.

NOTES

* Earlier versions have been presented at the Annual Conference of the History of Economics Society, Fairfax, 30 May–2 June, 1992, and at the conference on Realism and Anti-Realism in Science, Beijing, 15–17 June, 1992, and to audiences at the World Institute for Development Economics Research of the United Nations University and at the Hong Kong Institute of Economic Science. Thanks to the four audiences for lively discussions. Special thanks for helpful comments go to Siddiq Osmani, Nancy Wulwick, Victor Mok and, in particular, Alex Viskovatoff.

[1] The inaccurate usage of 'realism' (and even 'reality') is not restricted to the economics profession. We can find a physicist talking about "the realism of physical theories" (Franklin, 1984, p. 381) and "the reality of theories" (*ibid.*, p. 389), and a philosopher discussing the "reality of assumptions" in economics (Musgrave, 1981). Yet, the scale of such a usage seems to be of a completely different order in the case of economics.

[2] It turns out that economist Tony Lawson's suggestion for a definition of 'realism' is not quite apt for economics: "there is a material and social world that exists independently of any individual consciousness . . ." (Lawson, 1989, p. 61).

[3] Alexander Rosenberg has emphatically made the point that economics and other social sciences are heavily dependent on folk psychological notions (e.g., Rosenberg, 1981).

[4] This perspective on the matter is not devoid of problems. One of them is the very notion of folk economics. We do not have a very accurate idea of the contents of economic folk views; we are much more enlightened about the contents of folk psychology and

folk physics, for example. Some research has been done in the are of a so-called naive economics, and it should be of some help here. Naive economics, however, is mostly concerned with children's views of economic matters. The problem is that a great number of grown-ups have some education in "scientific" economics and use it in their everyday conceptualization of economic phenomena. Thus, the dividing line between scientific economics and actually held economic folk views would not seem to be entirely sharp. Folk economics has to be taken as a simplification itself. Note also that I am using 'scientific economics' as a descriptive, non-evaluative term.

[5] This suggestion implies that those economists in the 1960's were wrong who suggested that 'firm' and other such basic terms be treated as ("partially interpreted") theoretical terms akin to 'electron' (e.g., Ysander, 1961; Puu, 1967; Fabian, 1967).

[6] The possible arguments are not exhausted by these. Yet another possibility is to argue as Rosenberg does, namely that in the background of the predictive and technological impotency of economics lies its dependence on folk psychological categorial framework: the intentional vocabulary of folk psychology fails to carve the nature into the joints, i.e., into natural kinds or causally homogenous classes. He suggests this as an explanation for the predictive and technological failures of economics which he takes as an uncontroversial fact (Rosenberg, 1981).

[7] Note that if Hacking is right with his argument from the manipulability of electrons, then Robbins's statement would have to be qualified.

REFERENCES

Cairnes, John Eliot (1888 [1975]). *The Character and Logical Method of Political Economy*, Macmillan, London.

Fabian, Robert (1967). 'An Empirical Principle for Deductive Theory in Economics', *Southern Economic Journal* **34**, pp. 53–66.

Franklin, Allan (1984). 'The Epistemology of Experiment', *British Journal for the Philosophy of Science* **35**, pp. 381–390.

Hacking, Ian (1983). *Representing and Intervening*, Cambridge University Press, Cambridge.

Hands, D. Wade (1991). 'Reply to Hamminga and Mäki', in Neil de Marchi and Mark Blaug (eds.), *Appraising Economic Theories. Studies in the Methodology of Research Programs*, Edward Elgar, Aldershot.

Hausman, Daniel (1992). *The Inexact and Separate Science of Economics*, Cambridge University Press, Cambridge.

Lawson, Tony (1989). 'Abstraction, Tendencies and Stylized Facts: A Realist Approach to Economic Analysis', *Cambridge Journal of Economics* **13**, pp. 59–78.

Mäki, Uskali (1989). 'On the Problem of Realism in Economics', *Ricerche Economiche* **43**, pp. 176–198. [Reprinted in *The Philosophy and Methodology of Economics*, edited by Bruce Caldwell. Edward Elgar 1992.]

Mäki, Uskali (1990a). 'Scientific Realism and Austrian Explanation', *Review of Political Economy* **2**, pp. 310–344.

Mäki, Uskali (1990b). 'Mengerian Economics in Realist Perspective', *History of Political Economy*, Annual Supplement to Vol **22**, pp. 289–310.

Mäki, Uskali (1992a). 'On the Method of Isolation in Economics', *Poznan Studies in the Philosophy of the Sciences and the Humanities* **25**, pp. 319–354.

Mäki, Uskali (1992b). 'The Market as an Isolated Causal Process: A Metaphysical Ground for Realism', in Bruce Caldwell and Stephan Boehm (eds.), *Austrian Economics: Tensions and New Developments*, Kluwer, Dordrecht.

Mäki, Uskali (1994). 'Isolation, Idealization and Truth in Economics', *Poznan Studies in the Philosophy of the Sciences and the Humanities* **38**, pp. 147–168.

Musgrave, Alan (1981). '"Unreal Assumption" in Economic Theory: The F-twist Untwisted', *Kyklos* **34**, pp. 377–387.

Putnam, Hilary (1975–76). 'What is "Realism"?' *Proceedings of the Aristotelian Society*, pp. 177–194.

Puu, Tonu (1967). 'Some Reflections on the Relation between Economic Theory and Empirical Reality', *Swedish Journal of Economics*, pp. 85–114.

Rescher, Nicholas (1987). *Scientific Realism*, Reidel, Dordrecht.

Robbins, Lionel (1935). *An Essay on the Nature and Significance of Economics*, Macmillan, London.

Rosenberg, Alexander (1981). *Sociobiology and the Preemption of Social Science*, Blackwell, Oxford.

Ysander, Bengt-Christer (1961). 'Truth and Meaning of Economic Postulates', in Hugo Hegeland (ed.), *Money, Growth and Methodology and Other Essays in Economics*, Lund.

TIMOTHY SHANAHAN

REALISM AND ANTIREALISM IN EVOLUTIONARY BIOLOGY

(1) INTRODUCTION

(a) *The Realism-Antirealism Debate*

Do scientific theories provide us with genuine insight into the causal structure of the world, or do they merely provide useful models for organizing observable phenomena into coherent patterns? Scientific realists maintain that scientific theories do, at least approximately and on some occasions, give us insight into the causal structure of the world, including its unobserved and unobservable parts. Antirealists claim that scientific theories serve a number of useful functions, such as organizing empirical data and facilitating predictions, but revealing the hidden structure of the world is not one of them. At issue here is nothing less than the nature of our scientific understanding of the world.

(b) *The Neglect of Biology*

The realism-antirealism issue has been discussed extensively in the context of the physical sciences, with quantum mechanics in particular receiving a lion's share of the attention. One might reasonably expect that philosophers, in order to gain a wider perspective, would explore this issue in the context of other sciences as well. But to a surprising extent this has not happened. Biology, in particular, has received scant attention from philosophers interested in the realism-antirealism issue. This has had two unfortunate effects. First, arguments with regard to the justification of realist vs. antirealist philosophies of science have generally ignored the contribution that biology might make to these debates. Secondly, ongoing controversies in biology have continued without much attention given to the possibility that they might be clarified by exploring the realist and antirealist assumptions undergirding different biological claims and positions.

R. S. Cohen, R. Hilpinen and Qiu Renzong (eds.), Realism and Anti-Realism in the Philosophy of Science, 449–466.

(c) *Aim of this Paper*

My aim in this paper is to begin to rectify this situation by exploring the realism-antirealism debate as it arises in one part of biological science: evolutionary biology. Even within this restricted domain, my discussion will focus on just one specific problem, that of evaluating competing claims about the operation of natural selection. Recent work on this issue in the philosophical literature suggests very strongly that realist and anti-realist assumptions play a major role in ongoing debates about natural selection. Further analysis of this work is needed to make clear precisely what role such assumptions play, and to determine how such assumptions affect the derivation of substantive conclusions in and about evolutionary biology. Although the scope of my investigation will be restricted to a small skirmish in the larger contexts of the philosophy of science and of evolutionary biology, the conclusions reached bear on our understanding of both domains.

(2) THE UNITS OF SELECTION CONTROVERSY

(a) *The Nature of the Problem*

Darwinian evolutionary biology postulates that much of the change in the living world over the last 600 million years or so has been due to the differential survival and reproduction of differentially adapted life forms – that is, to the process of evolution by natural selection. This much is uncontroversial, at least among professional scientists. Where scientists begin to disagree is on the correct account of the *processes* that constitute evolution by natural selection. One of the most persistent debates in evolutionary biology concerns the question of the level(s) of nature at which natural selection operates. The living world is organized hierarchically, with genes, organelles, cells, organisms, kin-groups, populations, species, and ecosystems (and many other entities as well) constituting an ascending scale of biological inclusiveness. At which of these levels does natural selection operate? Which sorts of biological entities should be thought of as in competition with one another? Which sorts of biological entities exhibit adaptations? Biologists have defended a variety of different – and sometimes mutually incompatible – answers to these questions. To date, it has not been possible to reach a consensus on this fundamental issue in evolutionary biology.

At first glance it might seem a relatively simple matter to resolve this problem. Write a grant proposal, get some funding, do field research, analyze the data, write up the results, publish, and move on to the next biological problem. Unfortunately, matters are rarely so simple. First, empirical research is, by definition, based on observations, and hence on what is observable. But natural selection is not *directly* observable; it is an *inference* made from observations of nature conjoined with theoretical principles. What *is* (at least sometimes) observable is phenotypic variation, competition for resources, likeness of offspring to parent(s), and differential survival and reproduction. That nature "selects" some variations over others because of their superior fitness, and that such selection causes a gradual evolution of traits over time, are claims based on observations. Selection itself, however, is not directly observable (For more on this point, see Shanahan, 1992). Evolution by natural selection is a fact, but not all facts are known by direct (non-inferential) observation of nature.

Secondly, identifying the level(s) at which selection operates in a given situation is even farther removed from direct observation than is the process of selection itself. In a given selection process, there may be any number of kinds of biological entities exhibiting variation. Merely noting this variation does not reveal which of these kinds of entities is responsible for evolutionary change. When a kin-group interacts with some aspect of its environment, for example, so do the organisms that compose it, as well as the genes that constitute part of the make-up of the organisms. In addition, all three kinds of entities are causally connected with one another: the properties of the group are at least partially the result of the properties of the constituent organisms, and the properties of the organisms are largely the result of the properties of the genes they contain. Distinguishing the causal role(s) of each of these kinds of entities does not admit of any straightforward empirical determination. Evidently, theoretical (and philosophical) considerations must play an equally important role in resolving this problem.

(b) *Hierarchical Monism and the Pluralist Gambit*

Recognizing this need, biologists and philosophers have offered a variety of theoretical analyses of natural selection designed to identify the essential properties of units and levels of selection (see, e.g., the papers in Brandon and Burian, 1984 and those listed in Shanahan,

1990a, b). None of these proposals has been entirely successful in resolving the problem. When an intellectual logjam of this kind occurs, one begins to suspect that it is due to some mistaken underlying assumption in the way the problem has been formulated. In this case, despite their differences, all of these proposals share a common underlying presupposition, one that is intrinsic to the formulation of the problem as stated. They assume that for any given evolutionary phenomenon requiring a selectionist explanation, it is possible to identify the units (and levels) of selection that are causally responsible for that phenomenon. That is, they assume that units of selection are *real*, that such units can be identified with one or another of the entities making up the biological hierarchy, and that a satisfactory analysis of a given selection process requires the identification of the unit(s) and level(s) of selection causally responsible for that process.

In an important contribution to this debate, Kim Sterelny and Philip Kitcher (1988) challenge this "realist" assumption, and link their challenge to an explicitly *antirealist* philosophy of science. They christen the basic assumption underlying most analyses of natural selection *Hierarchical Monism*, defined as follows:

> Hierarchical monism claims that, for any selection process, there is a unique level of the hierarchy such that only representations that depict selection as acting at that level are maximally adequate. (Intuitively, representations that see selection as acting at other levels get the causal structure wrong.) (Sterelny and Kitcher, 1988, p. 359).

According to Hierarchical Monism, for any selection process there is at most one analysis that correctly captures the causal structure of that process, and this analysis depicts selection as operating on one and only one level of the biological hierarchy.

In place of Hierarchical Monism, Sterelny and Kitcher propose a form of *Pluralism*, according to which there are many "maximally adequate representations" of the causal structure of any given selection process. They explain the difference between these two basic views as follows:

> Hierarchical monism differs from [pluralism] in an interesting way: whereas the pluralist insists that, for any process, there are many adequate representations . . . the hierarchical monist maintains that for each process there is just one kind of adequate representation, but that processes are diverse in the kinds of representations they demand (Sterelny and Kitcher, 1988, p. 359).

Whereas hierarchical monism recommends a "plurality of processes", they note, pluralism recommends "a plurality of models of the same process" (Sterelny and Kitcher, 1988, p. 359 note 14).

Central to their argument for pluralism is recognition of the apparent arbitrariness of segmenting the causal chain constituting selection processes in one way rather than in another. Consider the following biological example: Some spiders construct webs for capturing prey. They do so because they possess certain genes that are causally connected with web-building behavior. When the spider successfully entraps prey in its web, both the spider *and* the genes that caused the web-building behavior benefit. A consequence of this is that webs of the type that such spiders construct become more common in the world. In this example there is a causal chain extending from genes for spiders with web-building behavior to spiders exhibiting this behavior to spider-webs to the capture of prey to the biological success of web-building spiders and their web-building spider genes. By reflecting on such cases, Sterelny and Kitcher conclude that,

> There is no privileged way to segment the causal chain and isolate the (really) real causal story [about a selection process]. . . . We are left with the general thesis of pluralism: there are alternative, maximally adequate representations of the causal structure of the selection process (Sterelny and Kitcher, 1988, p. 358).

On this view, selection can be represented as a function of competition among spiders who build webs *or* as a function of competition among genes for web-building. There is thus no pressing problem in determining the level of the biological hierarchy at which selection "really" operates. Selection can be equally well represented as operating at any one of several levels of the biological hierarchy. We may pick whichever one suits our needs – methodological considerations take over. Pluralism thus (dis)solves the units of selection problem. "[H]ierarchical monism", on the other hand, will always remain "entangled in spider webs" (Sterelny and Kitcher, 1988, p. 360).

(c) *Pluralistic Antirealism*

An especially interesting feature of Sterelny and Kitcher's proposal is their explicit linkage of pluralism and various antirealist doctrines in the philosophy of science. They first align their view with *instrumentalism*:

> Pluralism of the kind we espouse has affinities with some traditional views in the philosophy of science. Specifically, our approach is instrumentalist, not of course in denying the existence of entities like genes, but in opposing the idea that natural selection is a force that acts on some determinate target, such as the genotype or the phenotype. Monists err, we believe, in claiming that selection processes must be described in a particular way, and their error involves them in positing entities, "targets of selection", that do not exist (Sterelny and Kitcher, 1988, p. 359).

Immediately following this passage, Sterelny and Kitcher align their view with *conventionalism*:

> Another way to understand our pluralism is to connect it with conventionalist approaches to space-time theories. Just as conventionalists have insisted that there are alternative accounts of the phenomena which meet all our methodological desiderata, so too we maintain that selection processes can usually be treated, equally adequately, from more than one point of view (Sterelny and Kitcher, 1988, p. 359).

Although some previous analyses have hinted at a conventionalist solution to the units of selection problem (i.e., Cassidy, 1981), Sterelny and Kitcher provide by far the most explicit argument in support of such an interpretation. But is it right?

(3) A CRITIQUE OF PLURALISTIC ANTIREALISM

(a) *Some Questions*

In this section I want to address a number of questions that naturally arise from the foregoing sketch of Sterelny and Kitcher's "pluralistic antirealism". First, is pluralism of the sort that they defend cogent? Is it true that there are (or can be) many maximally adequate representations of the causal structure of any given selection process? Secondly, is there any interesting connection, as they suggest that there is, between their pluralism and antirealism in the philosophy of science? Does their pluralism entail instrumentalism, for example, or vice versa? Finally, have they shown that hierarchical monism is mistaken? Can this view be reformulated to avoid the criticisms they level against it?

These questions are important, because if pluralism is, as Sterelny and Kitcher argue, superior to monism, and if pluralism entails antirealism, then antirealism gains support as a philosophical interpretation of evolutionary biology – and perhaps gains credibility as a global interpretation of natural science as well. On the other hand, if pluralism – as Sterelny and Kitcher argue for it – is not a cogent position, or if it is not linked to antirealism in any interesting way, then their arguments for pluralism are not sufficient to establish the superiority of an antirealist interpretation of evolutionary biology. Hence both a solution to the units of selection controversy as well as support for specific positions in the philosophy of science are at stake. I will address the questions concerning pluralism and antirealism in this section, and reexamine hierarchical monism in section (4).

(b) *On Pluralism*

Sterelny and Kitcher offer two different versions of pluralism, which, while similar, are not identical. According to the "general thesis of pluralism: there are alternative, maximally adequate representations of the causal structure of the selection process" (Sterelny and Kitcher, 1988, p. 358). But a page later they write that, "the pluralist insists that, for any [selection] process, there are many adequate representations" (Sterelny and Kitcher, 1988, p. 359). The first version is considerably stronger than the second. The first version, in fact, entails the second, but not vice versa. If there are a number of maximally adequate representations of a process, then, all of them being maximal, they are necessarily equally adequate as well. But there might be many equally adequate representations of a process, all of which fail to be maximally adequate. (Analogy: If three students do "maximally well" on an exam, by each earning 100% of the possible points, then all three necessarily did equally well. But those three students could have done "equally well" among themselves on another exam, by all earning precisely 80% of the points, without thereby doing maximally well on that exam.) To capture this important distinction, I will refer to these two versions of pluralism as Strong Pluralism and Weak Pluralism. Is either version cogent?

Concerning Strong Pluralism, what does it mean to say that there are many maximally adequate representations of the causal structure of any [selection] process? A *representation* of a process would presumably be a description which highlights certain features of the process as being of special importance and significance. If the representation is of the *causal structure* of a process, then the description would highlight the entities and their properties, along with their interactions, that are constitutive of that process. A *maximally adequate* representation of the causal structure of a process would then be the best description possible of the entities, properties, and interactions that are constitutive of that process. This much seems fairly clear.

The difficult question, of course, concerns the notion of "adequacy". Adequate *for what*? One possibility is: adequate for explaining the outcome of the given process. A representation would be maximally adequate on this criterion if it provides the best possible explanation of the outcome of the given process, by showing how the various parts of the process together produced the outcome in question. Another possibility would be: adequacy for making successful predictions and organizing empirical phenomena. A representation would be maximally adequate on this second criterion if it provides the best possible instrument for predicting the outcomes of specific (kinds of) processes.

Clearly, a given representation could be maximally adequate when judged by one of these criteria but not maximally adequate when judged by the other. The so-called "Symmetry Thesis" is false: Good explanations are not necessarily logically identical to equally good predictions. (Two models of the solar system may be maximally adequate for predicting the next solar eclipse, but may differ in their adequacy for explaining such an event.)

Which kind of adequacy do Sterelny and Kitcher have in mind? From their various remarks it is clear that their pluralism is committed to the idea that a number of different representations of a single selection process can each be maximally adequate in the sense that each representation attributes "causal efficacy" to properties of entities at different levels of the biological hierarchy. This suggests that they are thinking of explanatory adequacy. So, for example, in the case of the spider-web example, an explanation of that process that attributes causal efficacy to organismic properties may be a maximally adequate representation of the causal structure of the process, while an explanation that attributes causal efficacy to genic properties might also be a maximally adequate representation of the causal structure of that process. In short, maximally adequate representations and/or explanations of selection processes may be framed in terms of the entities at any one of several levels of the biological hierarchy.

This way of understanding Strong Pluralism, however, gives rise to a dilemma. If the two (or more) representations are really *different* in attributing causal efficacy to different kinds of properties, then it is not clear how either can really be maximally adequate, because each representation omits the causal role of the properties emphasized by the other representation – properties that presumably play *some* role in the causal process in question. To the extent that a given representation omits causally relevant factors, it is only partially, and hence not maximally, adequate. (In the spider-web example, genic and organismic properties both contribute to the selection process. Omitting either set of properties results in a submaximally adequate representation.) A maximally adequate representation would be one that includes *all* the causally relevant properties constitutive of the causal structure of the process. On the other hand, if we say that there is really only *one* maximally adequate representation of the causal structure of the process, namely, the one that includes all the causally relevant properties (genic, organismic, etc.), then the view is no longer recognizably *pluralist*. Hence the dilemma: Either it is *false* that the different representations in question are each maximally adequate, or it is *false* that there can be more than one representation that is maximally adequate. Either way, it is not clear that Strong Pluralism is a cogent view.

Weak Pluralism, on the other hand, rather than being mistaken, seems to be in danger of being trivial. The claim of Weak Pluralism is that for any [selection] process, there are many (not necessarily maximally) adequate representations. Obviously the status of this claim hangs on what is meant by "adequate representations". If by "adequate" is meant that a given representation completely captures the causal structure of a process, then it is identical with Strong Pluralism, and faces the same objections. On the other hand, if what is meant is that a given representation allows one to capture *enough* of the causal structure of a process to organize empirical data in a coherent way or to make successful predictions, or to explain some feature of a process, then the plurality of equally adequate representations is simply a function of the uniform incompleteness of each. Like the proverbial blind men examining an elephant, each description can be "equally adequate" just because each represents only a partial understanding of the whole. Equivalence can be achieved in this way, but becomes cheapened as well. Despite their provocative claims, therefore, Sterelny and Kitcher have not shown that their brand of pluralism provides a cogent analysis of natural selection.

(c) *On Antirealism*

Instrumentalism in the philosophy of science is the view that while the theoretical terms of a scientific theory may be useful (or even necessary) for structuring empirical data into coherent patterns and/or for making predictions, we are not on this basis required to believe that such terms refer to anything real. Indispensability does not entail existence. Claims about unobservable theoretical entities are merely rules for making predictions about observable states of affairs. They are to be judged on their adequacy or inadequacy for this purpose; hence such claims are not literally true *or* false. According to this view, scientific theories are to be judged for their adequacy in organizing empirical observational data, not for their ability to reveal the unobservable constituents and structures of the world.

As noted above, Sterelny and Kitcher connect their pluralism with an instrumentalist philosophy of science "in opposing the idea that natural selection is a force that acts on some determinate target, such as the genotype or the phenotype." They add that, "Monists err . . . in claiming that selection processes must be described in a particular way, and their error involves them in positing entities, 'targets of selection', that do not exist" (Sterelny and Kitcher, 1988, p. 359). The inference here seems to be this: Monists claim that there is an exclusively correct way to describe any selection process. An exclusively correct description of any selection process will involve positing "targets of selection". But it is false that there is an exclusively correct way to describe any selection process. Therefore "targets of selection" do not exist.

Put this way, the fallacy becomes obvious. Consider an analogy. "Baseball monists" claim that there is an exclusively correct way to describe any baseball game. An exclusively correct description of any baseball game will involve positing entities such as bats, gloves, baseballs, outfielders, umpires, etc. But it is false that there is an exclusively correct way to describe any baseball game. Therefore entities such as bats, gloves, baseballs, outfielders, umpires, etc., do not exist.

Clearly, from the claim that there is no exclusively correct way to describe a baseball game, it does not follow that bats and so on do not exist, or that claims about such things are neither true nor false. Such things *may* not exist, of course, but their non-existence does not follow from the claim in question. Likewise, from the premise that there is no exclusively correct description of any selection process it does not follow that "targets of selection" do not exist, because the existence of targets of selection does not depend in any obvious way on the truth of monism. One *could* be a monist and deny the existence of "targets of selection". Likewise, one could be a pluralist and affirm their existence. Despite their intriguing claim, therefore, Sterelny and Kitcher have not shown that pluralism (of the sort they espouse) and instrumentalism are connected in some interesting way.

It seems clear that pluralism does not entail instrumentalism. It perhaps goes without saying that instrumentalism does not entail pluralism either. An instrumentalist might hold that there is only *one* maximally adequate representation for a given process, and simply maintain that this representation organizes the empirical data without revealing anything about the unobservable structure of this process. "Monistic instrumentalism" is not self-contradictory.

Sterelny and Kitcher link their pluralism with *antirealism* in the philosophy of science. I would suggest, however, that their view actually presupposes a form of *realism*. Recall that their pluralism is the view that there are many maximally adequate representations of any selection process. That is, there are "a plurality of models of the same process" (Sterelny and Kitcher, 1988, p. 359). But talking of a plurality of models of a *single causal process*, as Sterelny and Kitcher do, implies that there is a unitary physical phenomenon serving as the benchmark for scientific representations. If, as argued earlier, the process of natural selection is, strictly speaking, unobservable, then Sterelny and Kitcher are committed to holding that we can have genuine insights into the reality of this process, despite its unobservability. This is, of course, a central tenet of scientific realism. At the very least, talk of a "maximally adequate representation of the causal structure of a process" departs significantly from the spirit of antirealism, because having such a representation suggests that one *does* have access to the unobservable entities and processes responsible for observable phenomena – otherwise why have confidence that there *is* just *one* process being represented in diverse ways?

Ironically, then, Kitcher and Sterelny's Pluralistic Antirealism seems, upon closer inspection, to be based on an underlying assumption of realism about the process of natural selection. If their pluralism concerned only the equivalences of different models of natural selection to organize empirical data and to facilitate predictions, with no attempt to say anything about the causal structure of selection processes, then their view would be unexceptional. If one only restricts one's demands on what a model should do, then it will *always* be possible to find a plurality of "maximally adequate" models of any phenomenon. But establishing the claim that there are a plurality of maximally adequate representations of a single causal process is a different, and much more difficult, task. Pluralism and an ecumenical spirit with regard to how we model and represent natural selection are all very well, especially in cases where different representations exhibit different methodological virtues. But it is well to remember that although methodological and ontological matters are frequently connected in interesting ways, they are just as often conflated.

(4) HIERARCHICAL MONISM RECONSIDERED

(a) *The Problem with Previous Views*

I have been arguing that the Pluralistic Antirealism of Sterelny and Kitcher faces some serious objections. My analysis has been critical and negative. But by reflecting on the problems with Pluralistic Antirealism *and* Hierarchical Monism we can begin to get as glimpse of the broad outlines of a more adequate analysis.

Sterelny and Kitcher's pluralism is an attempt to provide a principled alternative to Hierarchical Monism. Recall that Hierarchical Monism is the claim that, for any selection process, there is a unique level of the biological hierarchy such that only representations that depict selection as acting *at that level* are maximally adequate. Representations that see selection as acting at other levels get the causal structure wrong. Sterelny and Kitcher rightly attack this claim, because it ignores the fact that a given selection process can be understood as the product of causal interactions at any one of a plurality of different levels.

Ironically, however, Sterelny and Kitcher's pluralism embodies a strong monist assumption. Their view still utilizes the idea that a representation focusing on any *one* level of the biological hierarchy can be maximally adequate. Their view differs from the one they reject only in that while Hierarchical Monists hold that a representation of a selection process at a single level will be *exclusively* correct, Sterelny and Kitcher believe that such a representation will be just one of *several* maximally adequate possibilities. But both views hold that an analysis of a selection process at only *one level* may be maximally adequate.

What undermines both formulations is the fact that each selection process is a multi-level causal process. That is, both views fail to accommodate the fact that any selection process will involve causal chains *connecting* entities on a plurality of different levels. Consequently, a representation (be it "monist" or "pluralist") that focusses on one level only will fail to be maximally adequate. A maximally adequate representation of a selection process should depict selection as the product of forces operating on and between a plurality of levels, with no single level standing out as *uniquely* important. This is precisely the kind of view that I believe is correct and wish to defend below.

(b) *Exclusive and Inclusive Hierarchical Monism*

Hierarchical Monism of the sort that Sterelny and Kitcher criticize might better be called *Exclusive Hierarchical Monism*. This view claims that the causal structure of a given selection process can be captured by focusing on just one level of the biological hierarchy. The view I wish to propose might be called *Inclusive Hierarchical Monism*. This view claims that a maximally adequate representation of the causal structure of a selection process will have to take into account a number of levels of the biological hierarchy. Whereas the former view excludes all but one level, the latter view includes many levels in one analysis.

To repeat this important distinction: According to Exclusive Hierarchical Monism, for any selection process there is at most one analysis that correctly captures the causal structure of that process, and this analysis depicts selection as operating on one and only one level of the biological hierarchy. According to Inclusive Hierarchical Monism, for any selection process there is at most one analysis that correctly captures the causal structure of that process, and this analysis depicts selection as operating on a plurality of different levels of the biological hierarchy.

To illustrate Inclusive Hierarchical Monism, consider the spider-web example again. Spider genes are causally connected with spiders and web-building behavior, and thus indirectly to webs and to the capture of prey, resulting in the differential propagation of both the genes and the spiders. Selection in this case is operating on both the genes and the spiders, *and* on the entire gene-spider-web causal process. It is the entire process linking genes and spiders and webs that results in the differential biological success of genes, spiders, and even web-phenotypes. *Selection may be best thought of as operating on the entire causal chain.* A maximally adequate representation of the causal structure of this selection process will include not only a description of how each of the entities involved in the process fares relative to other entities of the same kind (i.e., at the same level), but also how the entire causal chain contributes to the differential representation of genes, spiders, and webs in successive generations. Inclusive Hierarchical Monism *includes* the various multi-leveled entities constituting a selection process in a single representation of the causal structure of that process.

Perhaps an analogy will make this view clearer. Consider the auto industry. The Ford Motor Company builds and sells vehicles such as the Escort, Taurus, and Probe. The sale of these vehicles is the end product of a long causal process, involving engineers who *design* these vehicles, auto workers who use design instructions to *construct* these vehicles, and auto salesmen who *display* these vehicles in the hope of persuading consumers to purchase them. Information about customers's buying habits is then used to improve (or at least change) the design of future models. Through this process the design features (e.g., anti-lock brakes) of very successful models are more likely to show up in future models than are the design features of less successful models.

Within the Ford Motor Company, therefore, there is a kind of selection operating among the vehicles that they build, resulting in more Ford vehicles with features attractive within the current consumer environment. Designers are competing with one another to come up with more attractive features, workers are competing with one another to produce high-quality vehicles, and salesmen are competing with one another to outsell one another. There is selection going on at all levels of the causal process, "driving" it forward.

If we step back and consider the entire auto industry, however, we realize that the Ford Motor Company is just one part of a much larger competitive arena. The Toyota Motor Company produces vehicles such as the Corolla, Camry, and Cressida – models that consumers may choose to purchase rather than the Ford models mentioned above. Toyota, too, has its team of engineers, workers, and salesmen. Within the global auto industry, there is a competition between Ford and Toyota, as well as a competition between other U.S. and Japanese auto makers. Because the final product of each manufacturer is the end result of a long causal chain involving design, production, and marketing, the relative success of rival auto companies can be seen as dependent on the entire causal chain resulting in their products. That is, the entire Ford causal chain is in competition with the entire Toyota causal chain. It is Taurus vs. Cressida *and* Ford vs. Toyota.

Since American and Japanese auto makers use somewhat different processes in the production of their vehicles (e.g., "top-down" management vs. "module-team" organization), it is more than just autos that are in competition – it is also *methods* of producing autos. American auto designers are in competition with Japanese auto designers, and American auto workers are in competition with Japanese auto workers, and so on, but entire companies (and methods and processes) are also in competition with one another. Selection operates within all levels of the "auto hierarchy" as well as between the different causal processes constituting each hierarchy. Understanding "the auto industry" requires understanding each of these processes and how they interact to produce the vehicles television commercials incessantly encourage us to purchase.

Returning now to biological examples, Inclusive Hierarchical Monism holds that selection can best be viewed as operating among entire causal chains. Genes are designers, organisms through development realize these genetic plans, and phenotypes (including behavior) test these designs against rivals. Success depends on each stage in the causal process – the best design in the world will not be successful if it fails to be translated into reality. Selection operates among different spider gene-organism-web causal processes, resulting in the differential representation of these processes in subsequent generations. A maximally adequate representation of this process will *include* the causal role of all the relevant levels, and show how they are causally interconnected.

(c) *Causal Realism*

Inclusive Hierarchical Monism, as described above, presupposes that for any selection process, there may be several equally adequate representations, but there will be at most a single *maximally* adequate representation – that representation that correctly captures the multi-level causal structure of the process. Equally adequate representations are by their very nature only partial descriptions of the causal structure of the process. There may, without inconsistency, be any number of such representations. But a maximally adequate representation, in contrast, will be singular. One of the aims of inquiry on this view is to assemble the various partial representations into something more closely approximating a maximally adequate representation. It need not be assumed, however, that such a maximal representation will ever be achieved.

Such a view is explicitly realist because it assumes that we can have genuine insights into the causal structure of the world or, at least, that there *is* such a causal structure that we can continue to gain information about. Hence one can be a pluralist with regard to the currently available representations of natural selection, but a realist with regard to the causal process itself. That is, one could be a methodological pluralist but a metaphysical monist. It is not that there *is* no causal structure to be discovered, or that claims about the causal structure of a process are neither true nor false (as an instrumentalist would hold), but rather that we often cannot be sure just what the correct causal structure is. This is not a rejection of realism, but rather just an honest assessment of our (current) epistemic limitations. (For a view not unlike the one expressed here, see Cartwright, 1981.)

Note that Sterelny and Kitcher's argument against (Exclusive) Hierarchical Monism does not affect Inclusive Hierarchical Monism. It might be (and indeed is) false that there can be a uniquely correct representation of the causal structure of a selection process that focuses on just one level of the biological hierarchy, but true that there can be a uniquely correct representation of the causal structure of a selection process that includes a number of levels of the biological hierarchy. Sterelny and Kitcher have not shown that Hierarchical Monism is false, but only that one, rather implausible version of it, is mistaken. When Hierarchical Monism is reformulated along the lines suggested above, it not only escapes their criticism but also provides a plausible philosophical analysis of the process of evolution by natural selection.

(5) CONCLUSIONS

According to scientific realists, scientific theories do, at least approximately and on some occasions, give us insight into the causal structure of the world. But even when it is not known *which* representation of the causal structure of a part of the world is correct, realists hold that a correct (i.e., maximally adequate) representation is at least *possible*, that there is a truth of the matter and that, with any luck, we will discover what it is. Such a position does not preclude one from utilizing any number of alternative models of a process in the pursuit of scientific goods: empirical adequacy, predictive success, explanatory scope, etc. But recognizing the (perhaps temporary) empirical equivalence of different models does not entail that there is no underlying reality to learn more about. As Ernan McMullin has noted,

> Scientists have never thought themselves disqualified from pursuing one of a number of physical models that, for the moment, appear empirically equivalent. As metaphors, these models may give rise to quite different lines of inquiry, leading eventually to their empirical separation. Or it may be that one of the alternative models appears undesirable on other grounds than immediate empirical adequacy. If prolonged efforts to separate the models empirically are unsuccessful, or if it comes to be shown that the models are in principle empirically equivalent, scientists will, of course, turn to other matters. But this is not a rejection of realism. It is, rather, an admission that no decision can be made in this case as to what the theory, on a realist reading, commits us to (McMullin, 1984, p. 11).

We cannot, just by casually observing nature, determine the correct causal structure of its processes. Empirical adequacy is a necessary but not sufficient condition for discovering the truth. The difficult task for all scientists is to state clearly the relationship between their theoretical models and the causal structure of nature (as currently understood), and to avoid the temptation to identify the causal structure of nature with the structure of their models just because the models "fit" a certain selected subset of the empirical phenomena and are computationally convenient. Doing so is difficult for an additional reason: It requires that we admit (openly) that our models of natural phenomena are just that – hopeful human constructs – which do not *necessarily* capture the causal structure of nature/reality, no matter how pretty our models may be or how useful we may find them. Theories are nets that we cast to capture the world. But as in fishing, so in science, there is no guarantee that even our best efforts will bring home the catch we are after.

Department of Philosophy,
Loyola Marymount University,
Los Angeles, U.S.A.

BIBLIOGRAPHY

Brandon, Robert and Burian, Richard (eds.) (1984). *Genes, Organisms, and Populations: Controversies over the Units of Selection*, MIT Press, Cambridge, MA.

Cartwright, Nancy (1981). 'The Reality of Causes in a World of Instrumental Laws', in P. Asquith and R. Giere (eds.), *PSA 1980*, Philosophy of Science Association, East Lansing, MI., pp. 38–48.

Cassidy, John (1981). 'Ambiguities and Pragmatic Factors in the Units of Selection Controversy', *Philosophy of Science* **48**, pp. 95–111.

McMullin, Ernan (1984). 'A Case for Scientific Realism', in J. Leplin (ed.), *Scientific Realism*, University of California Press, Berkeley, pp. 8–40.

Shanahan, Timothy (1990a). 'Group Selection and the Evolution of Myxomatosis', *Evolutionary Theory* **9**, pp. 239–254.

Shanahan, Timothy (1990b). 'Evolution, Phenotypic Selection, and the Units of Selection', *Philosophy of Science* **57**, pp. 172–187.

Shanahan, Timothy (1992). 'Selection, Drift, and the Aims of Evolutionary Theory', in Paul Griffiths (ed.), *Trees of Life: Essays in Philosophy of Biology*, Kluwer Academic Publishers, Netherlands, pp. 133–161.

Sterelny, Kim and Philip Kitcher (1988). 'The Return of the Gene', *Journal of Philosophy* **85**, pp. 339–361.

PAUL C. L. TANG AND RALPH W. BROWN III

ANTI-REALISM AND THE COMPLEMENTARITY MODEL OF MIND-BRAIN

I

We introduce the first steps towards advancing a new, anti-realistic, Complementarity Model of the mind-brain. We begin by discussing a very current debate within the philosophy of cognitive science, viz., the debate between Jerry Fodor and Paul and Patricia Churchland concerning the appropriate model of cognition, i.e., the computational model versus the neural net model. We suggest that this debate leads us back to the age-old question of the ontological status of mind and body. At this point we introduce another contemporary representative theory of mind-body, viz., that of Karl Popper. We then argue for the underdetermination and incommensurability of all theories of the mind/body and then present the first steps toward our instrumentalist, complementarity theory. We then identify the virtues of our model within philosophical psychology and suggest that our model may also be capable of helping to resolve the Fodor-Churchland debate.

II

One of the foremost advocates of the Computational Theory of cognitive phenomena is Jerry Fodor, while two of the strongest advocates of the Neural Net/Connectionist theory of cognitive phenomena are Paul and Patricia Churchland. According to Fodor, psychology is concerned with the formal structure of symbols in the mind and of the manner in which these symbols are manipulated. These symbols are mental representations, a so-called *language of thought*, and they play the role of propositions in discourse involving propositional attitudes; moreover, these symbols are held to represent the phenomena about which a subject is thinking. In addition, the mind possesses a set of rules – corresponding to rules of inference – that determine which operations are performed on these representations. Fodor argues that this classical architecture can be used to explain rational behavior, concept learning, and perception. He also argues that this theory can explain, e.g., logical peculiarities

R. S. Cohen, R. Hilpinen and Qiu Renzong (eds.), Realism and Anti-Realism in the Philosophy of Science, 467–479.

of discourse about mental states such as how the mind can understand such relationships as, e.g., "John loves Mary" and "Mary loves John". The ease by which human beings can make these systematic transformations is exactly representable in symbolic terms.

Advocates of connectionism (sometimes referred to as PDP or neural net theories) claim that cognitive phenomena are not computational in the manner of a von Neumann computer. Connectionists, such as the Churchlands, advocate exploring systems known as neural nets. Such systems consist of nodes, each of which possesses a determinate degree of activation at any time, and is connected to many other nodes to which it sends excitatory or inhibitory stimuli. Given an initial pattern of activation, the excitations and inhibitions passing through the system will alter the activation states of the nodes until a stable pattern is reached. Moreover, the strengths of excitatory and inhibitory connections can be altered as a result of local activity within the system. Such systems can learn to respond in new ways and settle into different states at later times. In *Matter and Consciousness*, Paul Churchland provides a simple example of a submarine's neural net computer system training up to learn to recognize the difference between a rock and a mine using sonic echos, hidden units, automated learning by the back propagation of error and the generalized delta rule.[1] Neural nets have been used to model certain cognitive functions; in certain tasks, such as pattern recognition, these systems appear to be much more human-like than those of rule-processing machines. A recent success, NETalk, by Terry Sejnowski, is a project to get a neural net based machine to learn to speak simple, grammatically correct sentences after the researcher has inputted a 1,000-word transcription of an average child's conversation. However, systematic transformations of relationships, guaranteed by classical architecture, are not a natural property of neural nets.

The debate between Fodor and the Churchlands has been vigorous indeed. But the underlying problem, we suggest, is not one of which model – computationist versus neural net – captures the process of cognition in a *realist* sense, i.e., in a sense in which a scientific theory or model provides a literally true (or approximately true) account of the structure of the universe, or of nature, or, as in our present case, of human cognition. Indeed, the problem is not just that of realism versus anti-realism, but one that concerns even more deeply the way in which these two notions of realism and anti-realism pertain to an ontological question, viz., the classical problem of mind/body dualism, so acutely posed by

Descartes. This issue must be addressed first, we suggest, before we can evaluate the Fodorian and Churchlandian cognitive models.

III

On a realist basis, the problem of whether there exists a mind over and above the brain, or whether there exists only a brain, has given rise to a great variety of dualist and monist positions, such as substance dualism, emergentism, epiphenominalism, elemental dualism, interactionist property dualism, popular dualism, functionalism, reductive materialism, eliminative materialism, and so on. These "isms" represent broad classes of philosophical theories, and they indicate the exceedingly complex and subtle nature of the ontological problem. The arguments for these various positions are well known and we shall not rehearse them here.

Nevertheless, which general philosophical position is the correct one? Dualism or monism? This question is important because the former philosophical theory (dualism) grounds – at least from the point of view of *language* – the Fodorian position and the latter philosophical theory (materialism) grounds the Churchlandian position. Granted, Fodor's position is somewhat complex, for if he is a functionalist, then he is a functionalist with very strong Cartesian overtones. This latter characterization stems from his bold talk of the innate language of thought and in his holding that mental states really exist, that they can interact with one another, that it is possible to study them, and that they are nonreductive.

At this point, going beyond just the debate between Fodor and the Churchlands, and in anticipation of the general, alternative complementarity model to be advanced later, we can identify unequivocally contemporary dualists in the persons of Karl Popper and the neurophysiologist, John Eccles, although they would prefer to call their position "interactionism". Popper focuses on aspects of mental activity that could not easily claim to be accomplished solely by physical bodies, e.g., mathematics; in this respect he echoes the philosophy of Descartes. For Popper, abstract objects, such as mathematical objects, scientific theories and works of literature assume a type of autonomy, a "life of their own", and also constitute a distinct realm, which he calls "World 3". This World is to be distinguished from World 1, the realm of physical objects and also from World 2, the world of mental activity, by the fact that World

3 is governed by normative principles, such as the laws and principles of logic. Popper's argument that mental activities are distinct from physical objects follows from the need for an intermediary that can apply information from World 3 to World 1. Popper believes and tries to argue that no purely physical system can grasp the contents of World 3; thus, there must be mental activities that can grasp the objects of World 3 and then causally interact with objects and events of World 1. For Popper, only interactionism can properly explain how World 3 principles regulate World 1 events and objects. For a contemporary defense of Cartesian substance dualism, the reader can look to the work of Richard Swinburne, a distinguished philosopher of religion who has also published much in the philosophy of science.[2]

How shall we answer, then, the ontological question? We shall argue from both an empirically based point of view as well as a conceptually based one. Let us begin with the empirically based view, and, specifically, with a clinical example, which is representative of the vast clinical and experimental literature on mental illness. It is well known that the illness known as *clinical depression* is associated with a decrease in the concentration of neurotransmitters such as norepinepherine and acetylcholine. It is also well known that certain tricyclic drugs, such as lithium, imipramine, amitriptiline, chlorpromazine and phenelzine can restore the level of concentration of these neurotransmitters within 24 to 48 hours, although the antidepressant effects often do not become apparent for one to three weeks (See Dennis L. Murphy *et al.*, 1975). Furthermore, according to A. J. Prange (1975), *all* antidepressants exhibit this lag time.

However, drug therapy, or pharmacotherapy is, apparently, not the sole means of successful treatment, It is also well known that so-call "talk therapy", or psychotherapy, can be solely and successfully employed. According to a study by Ernest E. Beckham and William R. Leber (1985), psychotherapy, in numerous clinical trials, alone yielded superior results in three general types of depression; while in three other types, medication or pharmacotherapy yielded superior results; while in six other types, in which either psychotherapy or pharmacotherapy were used, there was no difference in the success rates. As the investigators write: "The above results suggest that there is currently no evidence to support strict adherence to either a psychotherapeutic or pharmacological model of treating outpatient depression. Each type of treatment has demonstrated its ability to alleviate symptoms of depression" (Beckham and Leber, p. 328). From the point of view of a realist ontology, what do these results indicate?

Before going on to answer this question, we must examine two further concepts, viz., underdetermination and incommensurability (See Quine, 1961, 1975). The first thesis claims that, for any given set of empirical data, or observations, or evidential sentences, an infinite number of theories or models, some inconsistent with each other, can imply (i.e., explain) that set of data, or set of observations, or, better, set of evidential sentences.[3] Moreover, we note that these two theories are incommensurable in Kuhn's meaning-variance sense. For example, within the dualist/psychotherapy model, the term 'cause' is used differently from the way in which this term is used in the materialist/pharmacotherapy model, where 'cause' refers not to some immaterial event, but to a material, physically based event. Following Wittgenstein's theory of meaning as use, we can say that the term 'cause' *means* differently in the two different philosophical theories and in the two different, corresponding psychological models. On the basis of this all-important concept of cause, we conclude that the two models, and the two philosophical positions of dualism and materialism, are in-commensurable.

We now return to the main issue at hand, and explain how underdetermination and incommensurability apply. Referring back to the Beckham and Leber study on depression, we see that, in the case of clinical depression, psychotherapy can effect a successful cure, and this model implies that certain cognitive states, certain mental belief states of the patient are causally responsible for the recovery. This technique of psychotherapy can well be used by, say, a *cognitive* psychologist. However, drug therapy can also be used, and, on this model, chemical action on the brain is ultimately responsible for effecting a cure. This technique of drug therapy can well be used by, say, a *physiological* psychologist. So we see that the two philosophical theories, the dualist theory that underlies the psychotherapy model and the monistic, materialistic theory that underlies the drug therapy model are *underdetermined* by the set of evidential statements, viz., the patient's recovery from depression.

IV

We are at an impasse. From a realist point of view, which philosophical theory is the correct one? The answer is that we may have to accept the possibility that we may *never* answer this question, or that we may never know. Perhaps, again following Wittgenstein, we ought to dissolve the problem. But how? The clue, we suggest, comes from examining the nature of quantum theory. In the famous double-slit experiment, a beam of photons passes through a screen with only one slit open and produces on a metal plate located behind the screen a pattern that can only be interpreted as light rays being fundamentally composed of particles. However, when both slits are open, the photons produce an interference pattern on the metal plate, and such a pattern could occur only if each photon had actually gone through both slits, like a wave. The lesson to be learned from this Complementarity Principle is that light has a dualism, two fundamental characteristics that are incompatible with each other. Moreover, from the viewpoint of scientific realism, one model represents one 'reality'; however, we now have two models (wave model and particle model) that represent 'two realities', and this result refutes again the thesis of scientific realism. The most we can say is that, at times, light is fundamentally particle-like; at other times, it is fundamentally wave-like.

The Complementarity Principle in Quantum Theory gives rise to a philosophical problem, for, if we claim, on the assumption of scientific realism, that our models, or theories, given us *a* literally true description of the universe, conceived as a "single reality", then we have a paradox. For now light, in its fundamental character, is both wave and particle. But this cannot be so.

Within the received Copenhagen Interpretation of quantum theory, this paradox is obviated by quantum physicists treating the theory as simply a tool or instrument for explanation and prediction, noting that the theory *does* very successfully explain and predict phenomena in the microworld. Copenhagenists do not claim that the theory provides a literally true account of these microphenomena. On this anti-realistic, instrumentalist view, Copenhagenists do not find it problematical to say that light sometimes behaves *as if* it were composed of particles and at other times *as if* it were composed of waves. There is no claim that the quantum theory provides a literally true account of the microworld. A key concept is the notion of the 'as if', which forms the basis of Kant's discussion of regulative concepts and regulative principles that help us to order our experiences in the world. Kant's concept of the 'as if' ultimately gives rise to the philosophical position known as 'pragmatism'.

V

We now return to the problem of mind/brain. Given the argument for the underdetermination and incommensurability of dualism and materialism as philosophical theories, and the empirical support provided by the psychotherapy and drug therapy models that are used to treat mental illness (e.g., depression), we suggest that the most fruitful approach to the ontological problem in the context of modern psychology and cognitive science is to follow the lead of the Copenhagen quantum physicists and adopt an anti-realistic, instrumentalist stance towards the mind/body problem. When psychotherapy is used, the psychologist is acting *as if* there exists a mind over and above the brain, and that this entity, immaterial thought it may be, has so-called states, such as belief, that are causally efficacious in effecting a successful outcome, viz., the absence of depression. From the point of view of our anti-realistic Complementarity Model of Mind-Brain that underlies this model of therapy and of psychological explanation, the therapist is acting *as if* the theory or position of dualism is true. However, if psychotherapy is unsuccessful, the psychologist may prescribe drugs, with or without additional psychotherapy. In extreme cases, the psychologist and physician might prescribe electroshock therapy. In these cases, from the point of view of the Complementarity Model of Mind-Brain, the therapist is acting as if the theory or position of materialism is true. What is important to each are the pragmatic results, which can also later be discussed in terms of explanation and prediction.

And now we return to the two competing classes of models of cognition. The Fodorian model accounts well for certain important aspects of cognition, such as systematicity, while the neural net model accounts well for perceptual knowledge. But a realist position gives rise to the dispute as to what 'really' is there, overlooking the important point that Nancy Cartwright makes, viz., that *all* theories abstract from reality and hence are false. This leads us into our anti-realist, pragmatic, instrumentalist Complementarity Model of Mind-Brain. On a realist basis, Fodor and the Churchlands are basically making assumptions, perhaps even value judgments, about which ontological position to take, and then developing their respective, incompatible theories of cognition from that point, even though each model, on its own, is successful in its own domain. The Complementarity Model that we begin to advance here accounts for the success of *both* the Fodorian and Churchlandian models of cognition and places a premium on pragmatics rather than on ontology.

We claim that our use of complementarity in quantum physics is to be taken as an important springboard, not as a perfect analogy with the Complementarity Model of mind-brain, which we have begun to develop in this paper. And we also note that Bohr himself held that some sort of complementarity is to be found in psychology. In a little-known but philosophically insightful paper entitled 'The Quantum of Action and the Description of Nature' (1929), which appears in his collection of essays, *Atomic Theory and the Description of Nature* (1961), Bohr writes about Heisenberg and says that

> . . . we must, in general, be prepared to accept the fact that a complete elucidation of one and the same object may require diverse points of view which defy a unique description . . . The necessity of taking recourse to a complementary, or reciprocal, mode of description is perhaps most familiar to us from psychological problems (Bohr, p. 96).

Later, in the same article, Bohr writes

> It might still be permitted here briefly to refer to the relation which exists between the regularities in the domain of psychology and the problems of causality of physical phenomena. When considering the contrast between the feeling of free will, which governs the psychic life, and the apparently uninterrupted causal chain of the accompanying physiological processes, the thought has, indeed, not eluded philosophers that we may be concerned here with an unvisualizable relation of complementarity. Thus, the opinion has often been expressed that a detailed investigation of the processes of the brain, which, although not practicable, is, nevertheless, thinkable, would reveal a causal chain that formed a unique representation of emotional mental experience (Bohr, p. 100).

And, finally, he writes:

> According to the above-mentioned view on the relation between the processes in the brain and the psychical experiences, we must, therefore, be prepared to accept the fact that an attempt to observe the former will bring about an essential alteration in the awareness of volition. Although, in the present case, we can be concerned only with more or less fitting analogies, yet we can hardly escape the conviction that in the facts which are revealed to us by the quantum theory and lie outside the domain of our ordinary forms of perception we have acquired a means of elucidating general philosophical problems (Bohr, 101).

These passages are very revealing, for they show that, at his most philosophical moments, Bohr believed that complementarity or reciprocity is to be found in psychology and elsewhere.[4]

VI

We began this paper by reviewing the debate between Fodor and the Churchlands on the correct model of cognition. This review led us to a more general position, that of the need to examine more closely the ontological foundations of each position – eliminative materialism, in the case of the Churchlands, and functionalism, with strong Cartesian overtones, in the case of Fodor. We then extended this talk of ontology to include an exposition of Popper's position, viz., a dualist, interactionist view of mind-brain.

Using these three views as representative of the two general positions of dualism and monism, we then argued for the underdetermination and incommensurability of all theories of mind-brain. Our arguments relied on empirical, clinical support as well as conceptual arguments. As a way out of the underdetermination impasse, we investigated complementarity in physics and used this principle as a springboard (not as a strict analogy) to develop our Complementarity Model of mind-brain, which is instrumentalist and anti-realist. We also cited some passages from Bohr, one of the founders of quantum theory, to the effect that he expected to see complementarity in other explanations of the world, including psychology.

We close by indicating the virtues of our model. The Complementarity Model of mind-brain may be able to effect a reconciliation between the Fodorian and the Churchlandian positions. There is no need to make any claim as to what is 'really there', but only to develop models that succeed, on the one hand, in predicting and explaining perceptual knowledge (the Churchlands) or predicting and explaining higher cognition on the other hand (Fodor). On the Complementarity Model, scientists can act *as if* there is a mind, that there is causal interaction between mind and brain, that mind is irreducible to brain, and so on, for the sake of explanation and prediction. Also under our model, scientists can act *as if* there is only brain, which can function in the manner of neural nets (the churchlands). In a strategy somewhat reminiscent of Kant, the Complementarity Model may allow for a synthesis of what appears to be two opposing, irreconcilable points of view.

Indeed, this Kantian synthesis approach reaches even further to the Kantian 'as if' approach, which grounds other areas, such as psychology, psychiatry, and law. We have discussed the first two cases in our section on explaining depression. What is often overlooked is that the law functions as if each individual has mental states such as volitions and beliefs and is thus responsible for his actions. However, the well known insanity defense presupposes a materialist ontology, viz., we act as if chemical, anatomical, and physiological abnormalities causally result in the agent's *not* having free will, and thus not having responsibility for the committed act. On a realist basis, this type of switching poses problems of consistency that are obviated by *not* claiming either that mind and brain both exist or that brain only exists. In other words, the Complementarity Model may not get as 'what's really there', although it does have the virtue of allowing a consistent model to undergird, and comport with, the way we do act when we perform our functions as psychologists, therapists, jurists, lawyers, and so on.

Finally, for those who find that the new model advanced here detracts from the 'mystery' of the mind-brain problem, we would simply say that indeed it does. If the Complementarity Model dissolves (à la the later Wittgenstein) the classical mind-body problem, then so be it. However, our interpretation of any kind of Wittgensteinian dissolution of metaphysical problems involves also the notion of *reposing* of the problem. And that reposing of the problem allows philosophers to join with neuroscientists, psychologists, linguists, biologists, physicists, and so on, to develop *new* questions within a newly emerged, interdisciplinary field. That is precisely what cognitive science is now doing, and the new problems it raises are probably ones that philosophers, we think, can contribute to solving in a significant way. The main import of the Complementarity Model of mind-brain may not be so much the dissolving of a classical question as much as it is the indicator of new ways of conceiving old problems.

Department of Philosophy,
California State University, Long Beach,
Long Beach, U.S.A.

NOTES

1 A network trains up by successive repetitions of a problem with smaller and smaller errors produced with each iteration. The most commonly used function for the error is: $E = 1/2\ \Sigma(y_i - d_i)^2$, where d_i is the desired output of unit i, and y_i is its actual output, where y_i is the sigmoid function $1/(1 + e^{-x})$. To minimize the error we take the derivative of the error with respect to w_{ij}, which is the weight between the units i and j. Thus we obtain

$$\partial E/\partial w_{ij} = y_i y_j(1 - y_j)\beta_j$$

where $\beta_j = (y_j - d_j)$ for output units and $\beta_j = \Sigma_k w_{jk} y_k(1 - y_k)\beta_k$ for hidden units, where k represents the number of units in the next layer that unit j is connected to.

The error can be calculated for the links going into the output units. For hidden units, however, the derivative depends on the values calculated at all levels that come after it; thus β must be back-propagated through the network to calculate the derivatives. Based on these equations, the algorithm is (1) choose a step-size δ (used to update the weights; (2) train the network such that for each sample pattern (a) do a forward pass through the net, producing an output pattern; (b) for all forward outputs, calculate $\beta_j = (y_j - d_j)$; (c) for all other units, from last layer to first, calculate β using the calculation from the layer after it: $\beta_j - \Sigma_k\ w_{jk} y_k(1 - y_k)\beta_k$; (d) and for all weights in the network, change the weight by $\Delta w_{ij} = -\delta y_i y_j(1 - y_j)\beta_j$.

2 See, for example, some of Swinburne's books: *The Coherence of Theism* (1977; *The Existence of God* (1979); *Personal Identity* (with Sidney Shoemaker, 1984); and *The Evolution of the Soul* (1986). All these books are published by Oxford University Press, except *Personal Identity*, which is published by Blackwell's. For another, modern defense of dualism, see J. B. Pratt, *Matter and Spirit* (New York: Macmillan, 1922/1957).

3 The underdetermination thesis can be proved simply as follows: If Theory T implies a set of evidential statements E, then T in conjunction with any sentence not included in T also implies E; but then, since (T & S) $\rightarrow$ E, and, since T & S $\rightarrow$ T, then (T & S) $\rightarrow$ E. Now replace S with S′ and iterate the argument; then replace S′ with S″ and iterate again, and so on to infinity. We conclude that an infinite number of different theories can imply E.

4 The authors are grateful to Professor Henry Folse for informing us – in a brief discussion after the earlier version of this paper was read at the Beijing International Conference for Philosophy of Science 1992 – of this article by Bohr.

BIBLIOGRAPHY

Allman, William F. (1989). *Apprentices of Wonder: Inside the Neural Network Revolution*, Bantam, New York.

Anderson, James A. (1983). 'Cognitive and Psychological Computation with Neural Networks', in *IEEE Transactions on Systems, Man, and Cybernetics*, Vol. SMC-13, No. 5 (September/October 1983).

Bechtel, William (1988). *Philosophy of Mind: Overview for Cognitive Science*, Lawrence Erlbaum, Hillsdale, N.J.

Bechtel, William (1988). *Philosophy of Science: Overview for Cognitive Science*, Lawrence Erlbaum, Hillsdale, N.J.

Block, Ned (1980). *Readings in Philosophy and Psychology*, Harvard University Press, Cambridge, Mass.

Bohr, Niels (1961). 'The Quantum of Action and the Description of Nature', in *Atomic Theory and the Description of Nature*, Cambridge University Press, Cambridge.

Cartwright, Nancy (1988). *How the Laws of Physics Lie*, Clarendon Press, Oxford.

Churchland, Patricia Smith (1988). *Neurophilosophy*, MIT Press, Cambridge, Mass.

Churchland, Patricia Smith (1980). 'A Perspective on Mind-Brain Research', *Journal of Philosophy* **77**(4), pp. 185–207.

Churchland, Patricia Smith (1982). 'Mind-Brain Reduction: New Perspectives from the Philosophy of Science', *Neuroscience* **7**(5), pp. 1041–1047.

Churchland, Paul M. (1988). *Matter and Consciousness*, MIT Press, Cambridge, Mass.

Churchland, Paul M. (1980). 'Critical Notice: Joseph Margolis: *Persons and Minds: The Prospects of Nonreductive Materialism*', *Dialogue* **19**, pp. 461–469.

Churchland, Paul M. (1981). 'Eliminative Materialism and the Propositional Attitudes', *Journal of Philosophy* **78**, pp. 67–90.

Churchland, Paul M. (1985). 'Reduction, Qualia and the Direct Introspection of Brain States', *Journal of Philosophy* **82**, pp. 8–28.

Churchland, Paul M. and Patricia Smith Churchland (1990). 'Could a Machine Talk?' *Scientific American* **262**, pp. 32–37.

Cummins, Robert (1983). *The Nature of Psychological Explanation*, MIT Press, Cambridge, Mass.

Dennett, Daniel (1981). *Brainstorms*, MIT Press, Cambridge, Mass.

Fodor, Jerry A. *et al.* (1988). 'Connectionism and Cognitive Architecture', *Cognition*, pp. 236–258.

Fodor, Jerry A. *et al.* (1975). *The Language of Thought*, MIT Press, Cambridge, Mass.

Fodor, Jerry A. (1981). *Representations: Philosophical Essays on the Foundations of Cognitive Science*, MIT Press, Cambridge, Mass.

Fodor Jerry A. (1968). *Psychological Explanation*, Random House, New York.

Fodor, Jerry A. (1988). 'Connectionism and Cognitive Architecture: A Critical Analysis. Cognitive Science Memorandum', COGMEM 29, 1987, *Cognition*.

Gardener, Howard (1985). *The Mind's New Science: A History of the Cognitive Revolution*, Basic Books, New York.

Garson, James W. (1990). 'Cognition Without Classical Architecture', Unpublished paper delivered at the APA Eastern Division meeting, Boston, Mass. December 1990.

Grossberg, S. (1980). 'How Does the Brain Build a Cognitive Code?' *Psychological Review* **87**, pp. 1–51.

Haugeland, John. (1980). *Mind Design: Philosophy, Psychology. Artificial Intelligence*, MIT Press, Cambridge, Mass.

Hopfield, J. J. (1982). 'Neural Networks and Physical Systems With Emergent Collective Computational Abilities', in *Proc. Natl. Acad. Sci.* **79**, pp. 2554–2558.

Hopfield, J. J. (1986). 'Computing With Neural Circuits: A Model', *Science* **23**, August, pp. 625–633.

Jones, W. T. and Hoskins, Josiah (1987). *Back-Propagation: A Generalized Delta-Learning Rule*, BYTE, October, pp. 155–162.

Kuhn, Thomas, S. (1962). *The Structure of Scientific Revolutions*, University of Chicago Press, Chicago.

Luplin, J. (1984). *Synapses, Circuits, and the Beginnings of Memory*, MIT Press, Cambridge, Mass.

McClelland, J. L. and Rumelhart, D. E. (1986). *Parallel Distributed Processing: Explorations in the Microstructure of Cognition Vol. 2: Psychological and Models*, Bradford Books/MIT Press, Cambridge, Mass.

Murphy, Dennis L. and Costa, Jonathan L. (1975). 'Utilization of Cellular Studies of Neurotransmitter-related Enzymes and Transport Processes in Man for the Investigation of Biological Factors in Behavioral Disorder', in Joseph Mendels (ed.), *The Psychobiology of Depression*, Spectrum Publication, New York, pp. 27–46.

Pinker, Steven *et al.*, 'On Language and Connectionism: Analysis of a Parallel Distributed Processing Model of Language Acquisition', Occasional Paper #33. Request for reprints: Pinker, Dept. of Brain and Cognitive Sciences, MIT Press, Cambridge, Mass., 02139.

Popper, Karl R. and Eccles, J. (1977). *The Self and Its Brain*, Springer-Verlag, New York.

Prange, Arthur, J., Jr., M.D. (1975). 'Pharmacotherapy of Depression', in Frederick F. Flach, M.D. and Susanne C. Draghi, M.D. (eds.), *The Nature and Treatment of Depression*, John Wiley, New York.

Quine, W. V. (1975). 'On Empirically Equivalent Systems of the World', *Erkenntnis* **9**, pp. 313–328.

INDEX OF NAMES

Boston Studies in the Philosophy of Science

Editor: Robert S. Cohen, *Boston University*

1. M.W. Wartofsky (ed.): *Proceedings of the Boston Colloquium for the Philosophy of Science, 1961/1962*. [Synthese Library 6] 1963 ISBN 90-277-0021-4
2. R.S. Cohen and M.W. Wartofsky (eds.): *Proceedings of the Boston Colloquium for the Philosophy of Science, 1962/1964*. In Honor of P. Frank. [Synthese Library 10] 1965 ISBN 90-277-9004-0
3. R.S. Cohen and M.W. Wartofsky (eds.): *Proceedings of the Boston Colloquium for the Philosophy of Science, 1964/1966*. In Memory of Norwood Russell Hanson. [Synthese Library 14] 1967 ISBN 90-277-0013-3
4. R.S. Cohen and M.W. Wartofsky (eds.): *Proceedings of the Boston Colloquium for the Philosophy of Science, 1966/1968*. [Synthese Library 18] 1969 ISBN 90-277-0014-1
5. R.S. Cohen and M.W. Wartofsky (eds.): *Proceedings of the Boston Colloquium for the Philosophy of Science, 1966/1968*. [Synthese Library 19] 1969 ISBN 90-277-0015-X
6. R.S. Cohen and R.J. Seeger (eds.): *Ernst Mach, Physicist and Philosopher*. [Synthese Library 27] 1970 ISBN 90-277-0016-8
7. M. Čapek: *Bergson and Modern Physics*. A Reinterpretation and Re-evaluation. [Synthese Library 37] 1971 ISBN 90-277-0186-5
8. R.C. Buck and R.S. Cohen (eds.): *PSA 1970*. Proceedings of the 2nd Biennial Meeting of the Philosophy and Science Association (Boston, Fall 1970). In Memory of Rudolf Carnap. [Synthese Library 39] 1971 ISBN 90-277-0187-3; Pb 90-277-0309-4
9. A.A. Zinov'ev: *Foundations of the Logical Theory of Scientific Knowledge (Complex Logic)*. Translated from Russian. Revised and enlarged English Edition, with an Appendix by G.A. Smirnov, E.A. Sidorenko, A.M. Fedina and L.A. Bobrova. [Synthese Library 46] 1973 ISBN 90-277-0193-8; Pb 90-277-0324-8
10. L. Tondl: *Scientific Procedures*. A Contribution Concerning the Methodological Problems of Scientific Concepts and Scientific Explanation.Translated from Czech. [Synthese Library 47] 1973 ISBN 90-277-0147-4; Pb 90-277-0323-X
11. R.J. Seeger and R.S. Cohen (eds.): *Philosophical Foundations of Science*. Proceedings of Section L, 1969, American Association for the Advancement of Science. [Synthese Library 58] 1974 ISBN 90-277-0390-6; Pb 90-277-0376-0
12. A. Grünbaum: *Philosophical Problems of Space and Times*. 2nd enlarged ed. [Synthese Library 55] 1973 ISBN 90-277-0357-4; Pb 90-277-0358-2
13. R.S. Cohen and M.W. Wartofsky (eds.): *Logical and Epistemological Studies in Contemporary Physics*. Proceedings of the Boston Colloquium for the Philosophy of Science, 1969/72, Part I. [Synthese Library 59] 1974 ISBN 90-277-0391-4; Pb 90-277-0377-9
14. R.S. Cohen and M.W. Wartofsky (eds.): *Methodological and Historical Essays in the Natural and Social Sciences*. Proceedings of the Boston Colloquium for the Philosophy of Science, 1969/72, Part II. [Synthese Library 60] 1974 ISBN 90-277-0392-2; Pb 90-277-0378-7
15. R.S. Cohen, J.J. Stachel and M.W. Wartofsky (eds.): *For Dirk Struik*. Scientific, Historical and Political Essays in Honor of Dirk J. Struik. [Synthese Library 61] 1974 ISBN 90-277-0393-0; Pb 90-277-0379-5
16. N. Geschwind: *Selected Papers on Language and the Brains*. [Synthese Library 68] 1974 ISBN 90-277-0262-4; Pb 90-277-0263-2
17. B.G. Kuznetsov: *Reason and Being*. Translated from Russian. Edited by C.R. Fawcett and R.S. Cohen. 1987 ISBN 90-277-2181-5

18. P. Mittelstaedt: *Philosophical Problems of Modern Physics.* Translated from the revised 4th German edition by W. Riemer and edited by R.S. Cohen. [Synthese Library 95] 1976
ISBN 90-277-0285-3; Pb 90-277-0506-2
19. H. Mehlberg: *Time, Causality, and the Quantum Theory.* Studies in the Philosophy of Science. Vol. I: *Essay on the Causal Theory of Time.* Vol. II: *Time in a Quantized Universe.* Translated from French. Edited by R.S. Cohen. 1980
Vol. I: ISBN 90-277-0721-9; Pb 90-277-1074-0
Vol. II: ISBN 90-277-1075-9; Pb 90-277-1076-7
20. K.F. Schaffner and R.S. Cohen (eds.): *PSA 1972.* Proceedings of the 3rd Biennial Meeting of the Philosophy of Science Association (Lansing, Michigan, Fall 1972). [Synthese Library 64] 1974 ISBN 90-277-0408-2; Pb 90-277-0409-0
21. R.S. Cohen and J.J. Stachel (eds.): *Selected Papers of Léon Rosenfeld.* [Synthese Library 100] 1979 ISBN 90-277-0651-4; Pb 90-277-0652-2
22. M. Čapek (ed.): *The Concepts of Space and Time.* Their Structure and Their Development. [Synthese Library 74] 1976 ISBN 90-277-0355-8; Pb 90-277-0375-2
23. M. Grene: *The Understanding of Nature.* Essays in the Philosophy of Biology. [Synthese Library 66] 1974 ISBN 90-277-0462-7; Pb 90-277-0463-5
24. D. Ihde: *Technics and Praxis.* A Philosophy of Technology. [Synthese Library 130] 1979
ISBN 90-277-0953-X; Pb 90-277-0954-8
25. J. Hintikka and U. Remes: *The Method of Analysis.* Its Geometrical Origin and Its General Significance. [Synthese Library 75] 1974 ISBN 90-277-0532-1; Pb 90-277-0543-7
26. J.E. Murdoch and E.D. Sylla (eds.): *The Cultural Context of Medieval Learning.* Proceedings of the First International Colloquium on Philosophy, Science, and Theology in the Middle Ages, 1973. [Synthese Library 76] 1975 ISBN 90-277-0560-7; Pb 90-277-0587-9
27. M. Grene and E. Mendelsohn (eds.): *Topics in the Philosophy of Biology.* [Synthese Library 84] 1976 ISBN 90-277-0595-X; Pb 90-277-0596-8
28. J. Agassi: *Science in Flux.* [Synthese Library 80] 1975
ISBN 90-277-0584-4; Pb 90-277-0612-3
29. J.J. Wiatr (ed.): *Polish Essays in the Methodology of the Social Sciences.* [Synthese Library 131] 1979 ISBN 90-277-0723-5; Pb 90-277-0956-4
30. P. Janich: *Protophysics of Time.* Constructive Foundation and History of Time Measurement. Translated from German. 1985 ISBN 90-277-0724-3
31. R.S. Cohen and M.W. Wartofsky (eds.): *Language, Logic, and Method.* 1983
ISBN 90-277-0725-1
32. R.S. Cohen, C.A. Hooker, A.C. Michalos and J.W. van Evra (eds.): *PSA 1974.* Proceedings of the 4th Biennial Meeting of the Philosophy of Science Association. [Synthese Library 101] 1976 ISBN 90-277-0647-6; Pb 90-277-0648-4
33. G. Holton and W.A. Blanpied (eds.): *Science and Its Public.* The Changing Relationship. [Synthese Library 96] 1976 ISBN 90-277-0657-3; Pb 90-277-0658-1
34. M.D. Grmek, R.S. Cohen and G. Cimino (eds.): *On Scientific Discovery.* The 1977 Erice Lectures. 1981 ISBN 90-277-1122-4; Pb 90-277-1123-2
35. S. Amsterdamski: *Between Experience and Metaphysics.* Philosophical Problems of the Evolution of Science. Translated from Polish. [Synthese Library 77] 1975
ISBN 90-277-0568-2; Pb 90-277-0580-1
36. M. Marković and G. Petrović (eds.): *Praxis.* Yugoslav Essays in the Philosophy and Methodology of the Social Sciences. [Synthese Library 134] 1979
ISBN 90-277-0727-8; Pb 90-277-0968-8

37. H. von Helmholtz: *Epistemological Writings.* The Paul Hertz / Moritz Schlick Centenary Edition of 1921. Translated from German by M.F. Lowe. Edited with an Introduction and Bibliography by R.S. Cohen and Y. Elkana. [Synthese Library 79] 1977 ISBN 90-277-0290-X; Pb 90-277-0582-8
38. R.M. Martin: *Pragmatics, Truth and Language.* 1979 ISBN 90-277-0992-0; Pb 90-277-0993-9
39. R.S. Cohen, P.K. Feyerabend and M.W. Wartofsky (eds.): *Essays in Memory of Imre Lakatos.* [Synthese Library 99] 1976 ISBN 90-277-0654-9; Pb 90-277-0655-7
40. Not published.
41. Not published.
42. H.R. Maturana and F.J. Varela: *Autopoiesis and Cognition.* The Realization of the Living. With a Preface to 'Autopoiesis' by S. Beer. 1980 ISBN 90-277-1015-5; Pb 90-277-1016-3
43. A. Kasher (ed.): *Language in Focus: Foundations, Methods and Systems.* Essays in Memory of Yehoshua Bar-Hillel. [Synthese Library 89] 1976 ISBN 90-277-0644-1; Pb 90-277-0645-X
44. T.D. Thao: *Investigations into the Origin of Language and Consciousness.* 1984 ISBN 90-277-0827-4
45. Not published.
46. P.L. Kapitza: *Experiment, Theory, Practice.* Articles and Addresses. Edited by R.S. Cohen. 1980 ISBN 90-277-1061-9; Pb 90-277-1062-7
47. M.L. Dalla Chiara (ed.): *Italian Studies in the Philosophy of Science.* 1981 ISBN 90-277-0735-9; Pb 90-277-1073-2
48. M.W. Wartofsky: *Models.* Representation and the Scientific Understanding. [Synthese Library 129] 1979 ISBN 90-277-0736-7; Pb 90-277-0947-5
49. T.D. Thao: *Phenomenology and Dialectical Materialism.* Edited by R.S. Cohen. 1986 ISBN 90-277-0737-5
50. Y. Fried and J. Agassi: *Paranoia.* A Study in Diagnosis. [Synthese Library 102] 1976 ISBN 90-277-0704-9; Pb 90-277-0705-7
51. K.H. Wolff: *Surrender and Cath.* Experience and Inquiry Today. [Synthese Library 105] 1976 ISBN 90-277-0758-8; Pb 90-277-0765-0
52. K. Kosík: *Dialectics of the Concrete.* A Study on Problems of Man and World. 1976 ISBN 90-277-0761-8; Pb 90-277-0764-2
53. N. Goodman: *The Structure of Appearance.* [Synthese Library 107] 1977 ISBN 90-277-0773-1; Pb 90-277-0774-X
54. H.A. Simon: *Models of Discovery* and Other Topics in the Methods of Science. [Synthese Library 114] 1977 ISBN 90-277-0812-6; Pb 90-277-0858-4
55. M. Lazerowitz: *The Language of Philosophy.* Freud and Wittgenstein. [Synthese Library 117] 1977 ISBN 90-277-0826-6; Pb 90-277-0862-2
56. T. Nickles (ed.): *Scientific Discovery, Logic, and Rationality.* 1980 ISBN 90-277-1069-4; Pb 90-277-1070-8
57. J. Margolis: *Persons and Mind.* The Prospects of Nonreductive Materialism. [Synthese Library 121] 1978 ISBN 90-277-0854-1; Pb 90-277-0863-0
58. G. Radnitzky and G. Andersson (eds.): *Progress and Rationality in Science.* [Synthese Library 125] 1978 ISBN 90-277-0921-1; Pb 90-277-0922-X
59. G. Radnitzky and G. Andersson (eds.): *The Structure and Development of Science.* [Synthese Library 136] 1979 ISBN 90-277-0994-7; Pb 90-277-0995-5

60. T. Nickles (ed.): *Scientific Discovery.* Case Studies. 1980
ISBN 90-277-1092-9; Pb 90-277-1093-7
61. M.A. Finocchiaro: *Galileo and the Art of Reasoning.* Rhetorical Foundation of Logic and Scientific Method. 1980 ISBN 90-277-1094-5; Pb 90-277-1095-3
62. W.A. Wallace: *Prelude to Galileo.* Essays on Medieval and 16th-Century Sources of Galileo's Thought. 1981 ISBN 90-277-1215-8; Pb 90-277-1216-6
63. F. Rapp: *Analytical Philosophy of Technology.* Translated from German. 1981
ISBN 90-277-1221-2; Pb 90-277-1222-0
64. R.S. Cohen and M.W. Wartofsky (eds.): *Hegel and the Sciences.* 1984
ISBN 90-277-0726-X
65. J. Agassi: *Science and Society.* Studies in the Sociology of Science. 1981
ISBN 90-277-1244-1; Pb 90-277-1245-X
66. L. Tondl: *Problems of Semantics.* A Contribution to the Analysis of the Language of Science. Translated from Czech. 1981 ISBN 90-277-0148-2; Pb 90-277-0316-7
67. J. Agassi and R.S. Cohen (eds.): *Scientific Philosophy Today.* Essays in Honor of Mario Bunge. 1982 ISBN 90-277-1262-X; Pb 90-277-1263-8
68. W. Krajewski (ed.): *Polish Essays in the Philosophy of the Natural Sciences.* Translated from Polish and edited by R.S. Cohen and C.R. Fawcett. 1982
ISBN 90-277-1286-7; Pb 90-277-1287-5
69. J.H. Fetzer: *Scientific Knowledge.* Causation, Explanation and Corroboration. 1981
ISBN 90-277-1335-9; Pb 90-277-1336-7
70. S. Grossberg: *Studies of Mind and Brain.* Neural Principles of Learning, Perception, Development, Cognition, and Motor Control. 1982
ISBN 90-277-1359-6; Pb 90-277-1360-X
71. R.S. Cohen and M.W. Wartofsky (eds.): *Epistemology, Methodology, and the Social Sciences.* 1983. ISBN 90-277-1454-1
72. K. Berka: *Measurement.* Its Concepts, Theories and Problems. Translated from Czech. 1983
ISBN 90-277-1416-9
73. G.L. Pandit: *The Structure and Growth of Scientific Knowledge.* A Study in the Methodology of Epistemic Appraisal. 1983 ISBN 90-277-1434-7
74. A.A. Zinov'ev: *Logical Physics.* Translated from Russian. Edited by R.S. Cohen. 1983 [*see also* Volume 9] ISBN 90-277-0734-0
75. G-G. Granger: *Formal Thought and the Sciences of Man.* Translated from French. With and Introduction by A. Rosenberg. 1983 ISBN 90-277-1524-6
76. R.S. Cohen and L. Laudan (eds.): *Physics, Philosophy and Psychoanalysis.* Essays in Honor of Adolf Grünbaum. 1983 ISBN 90-277-1533-5
77. G. Böhme, W. van den Daele, R. Hohlfeld, W. Krohn and W. Schäfer: *Finalization in Science.* The Social Orientation of Scientific Progress. Translated from German. Edited by W. Schäfer. 1983 ISBN 90-277-1549-1
78. D. Shapere: *Reason and the Search for Knowledge.* Investigations in the Philosophy of Science. 1984 ISBN 90-277-1551-3; Pb 90-277-1641-2
79. G. Andersson (ed.): *Rationality in Science and Politics.* Translated from German. 1984
ISBN 90-277-1575-0; Pb 90-277-1953-5
80. P.T. Durbin and F. Rapp (eds.): *Philosophy and Technology.* [*Also* Philosophy and Technology Series, Vol. 1] 1983 ISBN 90-277-1576-9
81. M. Marković: *Dialectical Theory of Meaning.* Translated from Serbo-Croat. 1984
ISBN 90-277-1596-3

82. R.S. Cohen and M.W. Wartofsky (eds.): *Physical Sciences and History of Physics*. 1984. ISBN 90-277-1615-3
83. É. Meyerson: *The Relativistic Deduction*. Epistemological Implications of the Theory of Relativity. Translated from French. With a Review by Albert Einstein and an Introduction by Milić Čapek. 1985 ISBN 90-277-1699-4
84. R.S. Cohen and M.W. Wartofsky (eds.): *Methodology, Metaphysics and the History of Science*. In Memory of Benjamin Nelson. 1984 ISBN 90-277-1711-7
85. G. Tamás: *The Logic of Categories*. Translated from Hungarian. Edited by R.S. Cohen. 1986 ISBN 90-277-1742-7
86. S.L. de C. Fernandes: *Foundations of Objective Knowledge*. The Relations of Popper's Theory of Knowledge to That of Kant. 1985 ISBN 90-277-1809-1
87. R.S. Cohen and T. Schnelle (eds.): *Cognition and Fact*. Materials on Ludwik Fleck. 1986 ISBN 90-277-1902-0
88. G. Freudenthal: *Atom and Individual in the Age of Newton*. On the Genesis of the Mechanistic World View. Translated from German. 1986 ISBN 90-277-1905-5
89. A. Donagan, A.N. Perovich Jr and M.V. Wedin (eds.): *Human Nature and Natural Knowledge*. Essays presented to Marjorie Grene on the Occasion of Her 75th Birthday. 1986 ISBN 90-277-1974-8
90. C. Mitcham and A. Hunning (eds.): *Philosophy and Technology II*. Information Technology and Computers in Theory and Practice. [*Also* Philosophy and Technology Series, Vol. 2] 1986 ISBN 90-277-1975-6
91. M. Grene and D. Nails (eds.): *Spinoza and the Sciences*. 1986 ISBN 90-277-1976-4
92. S.P. Turner: *The Search for a Methodology of Social Science*. Durkheim, Weber, and the 19th-Century Problem of Cause, Probability, and Action. 1986. ISBN 90-277-2067-3
93. I.C. Jarvie: *Thinking about Society*. Theory and Practice. 1986 ISBN 90-277-2068-1
94. E. Ullmann-Margalit (ed.): *The Kaleidoscope of Science*. The Israel Colloquium: Studies in History, Philosophy, and Sociology of Science, Vol. 1. 1986 ISBN 90-277-2158-0; Pb 90-277-2159-9
95. E. Ullmann-Margalit (ed.): *The Prism of Science*. The Israel Colloquium: Studies in History, Philosophy, and Sociology of Science, Vol. 2. 1986 ISBN 90-277-2160-2; Pb 90-277-2161-0
96. G. Márkus: *Language and Production*. A Critique of the Paradigms. Translated from French. 1986 ISBN 90-277-2169-6
97. F. Amrine, F.J. Zucker and H. Wheeler (eds.): *Goethe and the Sciences: A Reappraisal*. 1987 ISBN 90-277-2265-X; Pb 90-277-2400-8
98. J.C. Pitt and M. Pera (eds.): *Rational Changes in Science*. Essays on Scientific Reasoning. Translated from Italian. 1987 ISBN 90-277-2417-2
99. O. Costa de Beauregard: *Time, the Physical Magnitude*. 1987 ISBN 90-277-2444-X
100. A. Shimony and D. Nails (eds.): *Naturalistic Epistemology*. A Symposium of Two Decades. 1987 ISBN 90-277-2337-0
101. N. Rotenstreich: *Time and Meaning in History*. 1987 ISBN 90-277-2467-9
102. D.B. Zilberman: *The Birth of Meaning in Hindu Thought*. Edited by R.S. Cohen. 1988 ISBN 90-277-2497-0
103. T.F. Glick (ed.): *The Comparative Reception of Relativity*. 1987 ISBN 90-277-2498-9
104. Z. Harris, M. Gottfried, T. Ryckman, P. Mattick Jr, A. Daladier, T.N. Harris and S. Harris: *The Form of Information in Science*. Analysis of an Immunology Sublanguage. With a Preface by Hilary Putnam. 1989 ISBN 90-277-2516-0

Boston Studies in the Philosophy of Science

105. F. Burwick (ed.): *Approaches to Organic Form.* Permutations in Science and Culture. 1987 ISBN 90-277-2541-1
106. M. Almási: *The Philosophy of Appearances.* Translated from Hungarian. 1989 ISBN 90-277-2150-5
107. S. Hook, W.L. O'Neill and R. O'Toole (eds.): *Philosophy, History and Social Action.* Essays in Honor of Lewis Feuer. With an Autobiographical Essay by L. Feuer. 1988 ISBN 90-277-2644-2
108. I. Hronszky, M. Fehér and B. Dajka: *Scientific Knowledge Socialized.* Selected Proceedings of the 5th Joint International Conference on the History and Philosophy of Science organized by the IUHPS (Veszprém, Hungary, 1984). 1988 ISBN 90-277-2284-6
109. P. Tillers and E.D. Green (eds.): *Probability and Inference in the Law of Evidence.* The Uses and Limits of Bayesianism. 1988 ISBN 90-277-2689-2
110. E. Ullmann-Margalit (ed.): *Science in Reflection.* The Israel Colloquium: Studies in History, Philosophy, and Sociology of Science, Vol. 3. 1988 ISBN 90-277-2712-0; Pb 90-277-2713-9
111. K. Gavroglu, Y. Goudaroulis and P. Nicolacopoulos (eds.): *Imre Lakatos and Theories of Scientific Change.* 1989 ISBN 90-277-2766-X
112. B. Glassner and J.D. Moreno (eds.): *The Qualitative-Quantitative Distinction in the Social Sciences.* 1989 ISBN 90-277-2829-1
113. K. Arens: *Structures of Knowing.* Psychologies of the 19th Century. 1989 ISBN 0-7923-0009-2
114. A. Janik: *Style, Politics and the Future of Philosophy.* 1989 ISBN 0-7923-0056-4
115. F. Amrine (ed.): *Literature and Science as Modes of Expression.* With an Introduction by S. Weininger. 1989 ISBN 0-7923-0133-1
116. J.R. Brown and J. Mittelstrass (eds.): *An Intimate Relation.* Studies in the History and Philosophy of Science. Presented to Robert E. Butts on His 60th Birthday. 1989 ISBN 0-7923-0169-2
117. F. D'Agostino and I.C. Jarvie (eds.): *Freedom and Rationality.* Essays in Honor of John Watkins. 1989 ISBN 0-7923-0264-8
118. D. Zolo: *Reflexive Epistemology.* The Philosophical Legacy of Otto Neurath. 1989 ISBN 0-7923-0320-2
119. M. Kearn, B.S. Philips and R.S. Cohen (eds.): *Georg Simmel and Contemporary Sociology.* 1989 ISBN 0-7923-0407-1
120. T.H. Levere and W.R. Shea (eds.): *Nature, Experiment and the Science.* Essays on Galileo and the Nature of Science. In Honour of Stillman Drake. 1989 ISBN 0-7923-0420-9
121. P. Nicolacopoulos (ed.): *Greek Studies in the Philosophy and History of Science.* 1990 ISBN 0-7923-0717-8
122. R. Cooke and D. Costantini (eds.): *Statistics in Science.* The Foundations of Statistical Methods in Biology, Physics and Economics. 1990 ISBN 0-7923-0797-6
123. P. Duhem: *The Origins of Statics.* Translated from French by G.F. Leneaux, V.N. Vagliente and G.H. Wagner. With an Introduction by S.L. Jaki. 1991 ISBN 0-7923-0898-0
124. H. Kamerlingh Onnes: *Through Measurement to Knowledge.* The Selected Papers, 1853-1926. Edited and with an Introduction by K. Gavroglu and Y. Goudaroulis. 1991 ISBN 0-7923-0825-5
125. M. Čapek: *The New Aspects of Time: Its Continuity and Novelties.* Selected Papers in the Philosophy of Science. 1991 ISBN 0-7923-0911-1

126. S. Unguru (ed.): *Physics, Cosmology and Astronomy, 1300-1700.* Tension and Accommodation. 1991 ISBN 0-7923-1022-5
127. Z. Bechler: *Newton's Physics on the Conceptual Structure of the Scientific Revolution.* 1991 ISBN 0-7923-1054-3
128. É. Meyerson: *Explanation in the Sciences.* Translated from French by M-A. Siple and D.A. Siple. 1991 ISBN 0-7923-1129-9
129. A.I. Tauber (ed.): *Organism and the Origins of Self.* 1991 ISBN 0-7923-1185-X
130. F.J. Varela and J-P. Dupuy (eds.): *Understanding Origins.* Contemporary Views on the Origin of Life, Mind and Society. 1992 ISBN 0-7923-1251-1
131. G.L. Pandit: *Methodological Variance.* Essays in Epistemological Ontology and the Methodology of Science. 1991 ISBN 0-7923-1263-5
132. G. Munévar (ed.): *Beyond Reason.* Essays on the Philosophy of Paul Feyerabend. 1991 ISBN 0-7923-1272-4
133. T.E. Uebel (ed.): *Rediscovering the Forgotten Vienna Circle.* Austrian Studies on Otto Neurath and the Vienna Circle. Partly translated from German. 1991 ISBN 0-7923-1276-7
134. W.R. Woodward and R.S. Cohen (eds.): *World Views and Scientific Discipline Formation.* Science Studies in the [former] German Democratic Republic. Partly translated from German by W.R. Woodward. 1991 ISBN 0-7923-1286-4
135. P. Zambelli: *The Speculum Astronomiae and Its Enigma.* Astrology, Theology and Science in Albertus Magnus and His Contemporaries. 1992 ISBN 0-7923-1380-1
136. P. Petitjean, C. Jami and A.M. Moulin (eds.): *Science and Empires.* Historical Studies about Scientific Development and European Expansion. ISBN 0-7923-1518-9
137. W.A. Wallace: *Galileo's Logic of Discovery and Proof.* The Background, Content, and Use of His Appropriated Treatises on Aristotle's *Posterior Analytics.* 1992 ISBN 0-7923-1577-4
138. W.A. Wallace: *Galileo's Logical Treatises.* A Translation, with Notes and Commentary, of His Appropriated Latin Questions on Aristotle's *Posterior Analytics.* 1992 ISBN 0-7923-1578-2
Set (137 + 138) ISBN 0-7923-1579-0
139. M.J. Nye, J.L. Richards and R.H. Stuewer (eds.): *The Invention of Physical Science.* Intersections of Mathematics, Theology and Natural Philosophy since the Seventeenth Century. Essays in Honor of Erwin N. Hiebert. 1992 ISBN 0-7923-1753-X
140. G. Corsi, M.L. dalla Chiara and G.C. Ghirardi (eds.): *Bridging the Gap: Philosophy, Mathematics and Physics.* Lectures on the Foundations of Science. 1992 ISBN 0-7923-1761-0
141. C.-H. Lin and D. Fu (eds.): *Philosophy and Conceptual History of Science in Taiwan.* 1992 ISBN 0-7923-1766-1
142. S. Sarkar (ed.): *The Founders of Evolutionary Genetics.* A Centenary Reappraisal. 1992 ISBN 0-7923-1777-7
143. J. Blackmore (ed.): *Ernst Mach – A Deeper Look.* Documents and New Perspectives. 1992 ISBN 0-7923-1853-6
144. P. Kroes and M. Bakker (eds.): *Technological Development and Science in the Industrial Age.* New Perspectives on the Science–Technology Relationship. 1992 ISBN 0-7923-1898-6
145. S. Amsterdamski: *Between History and Method.* Disputes about the Rationality of Science. 1992 ISBN 0-7923-1941-9
146. E. Ullmann-Margalit (ed.): *The Scientific Enterprise.* The Bar-Hillel Colloquium: Studies in History, Philosophy, and Sociology of Science, Volume 4. 1992 ISBN 0-7923-1992-3

Boston Studies in the Philosophy of Science

147. L. Embree (ed.): *Metaarchaeology.* Reflections by Archaeologists and Philosophers. 1992 ISBN 0-7923-2023-9
148. S. French and H. Kamminga (eds.): *Correspondence, Invariance and Heuristics.* Essays in Honour of Heinz Post. 1993 ISBN 0-7923-2085-9
149. M. Bunzl: *The Context of Explanation.* 1993 ISBN 0-7923-2153-7
150. I.B. Cohen (ed.): *The Natural Sciences and the Social Sciences.* Some Critical and Historical Perspectives. 1994 ISBN 0-7923-2223-1
151. K. Gavroglu, Y. Christianidis and E. Nicolaidis (eds.): *Trends in the Historiography of Science.* 1994 ISBN 0-7923-2255-X
152. S. Poggi and M. Bossi (eds.): *Romanticism in Science.* Science in Europe, 1790–1840. 1994 ISBN 0-7923-2336-X
153. J. Faye and H.J. Folse (eds.): *Niels Bohr and Contemporary Philosophy.* 1994 ISBN 0-7923-2378-5
154. C.C. Gould and R.S. Cohen (eds.): *Artifacts, Representations, and Social Practice.* Essays for Marx W. Wartofsky. 1994 ISBN 0-7923-2481-1
155. R.E. Butts: *Historical Pragmatics.* Philosophical Essays. 1993 ISBN 0-7923-2498-6
156. R. Rashed: *The Development of Arabic Mathematics: Between Arithmetic and Algebra.* Translated from French by A.F.W. Armstrong. 1994 ISBN 0-7923-2565-6
157. I. Szumilewicz-Lachman (ed.): *Zygmunt Zawirski: His Life and Work.* With Selected Writings on Time, Logic and the Methodology of Science. Translations by Feliks Lachman. Ed. by R.S. Cohen, with the assistance of B. Bergo. 1994 ISBN 0-7923-2566-4
158. S.N. Haq: *Names, Natures and Things.* The Alchemist Jābir ibn Ḥayyān and His *Kitāb al-Aḥjār* (Book of Stones). 1994 ISBN 0-7923-2587-7
159. P. Plaass: *Kant's Theory of Natural Science.* Translation, Analytic Introduction and Commentary by Alfred E. and Maria G. Miller. 1994 ISBN 0-7923-2750-0
160. J. Misiek (ed.): *The Problem of Rationality in Science and its Philosophy.* On Popper vs. Polanyi. The Polish Conferences 1988–89. 1995 ISBN 0-7923-2925-2
161. I.C. Jarvie and N. Laor (eds.): *Critical Rationalism, Metaphysics and Science.* Essays for Joseph Agassi, Volume I. 1995 ISBN 0-7923-2960-0
162. I.C. Jarvie and N. Laor (eds.): *Critical Rationalism, the Social Sciences and the Humanities.* Essays for Joseph Agassi, Volume II. 1995 ISBN 0-7923-2961-9
Set (161–162) ISBN 0-7923-2962-7
163. K. Gavroglu, J. Stachel and M.W. Wartofsky (eds.): *Physics, Philosophy, and the Scientific Community.* Essays in the Philosophy and History of the Natural Sciences and Mathematics. In Honor of Robert S. Cohen. 1995 ISBN 0-7923-2988-0
164. K. Gavroglu, J. Stachel and M.W. Wartofsky (eds.): *Science, Politics and Social Practice.* Essays on Marxism and Science, Philosophy of Culture and the Social Sciences. In Honor of Robert S. Cohen. 1995 ISBN 0-7923-2989-9
165. K. Gavroglu, J. Stachel and M.W. Wartofsky (eds.): *Science, Mind and Art.* Essays on Science and the Humanistic Understanding in Art, Epistemology, Religion and Ethics. Essays in Honor of Robert S. Cohen. 1995 ISBN 0-7923-2990-2
Set (163–165) ISBN 0-7923-2991-0
166. K.H. Wolff: *Transformation in the Writing.* A Case of Surrender-and-Catch. 1995 ISBN 0-7923-3178-8
167. A.J. Kox and D.M. Siegel (eds.): *No Truth Except in the Details.* Essays in Honor of Martin J. Klein. 1995 ISBN 0-7923-3195-8

Boston Studies in the Philosophy of Science

168. J. Blackmore: *Ludwig Boltzmann, His Later Life and Philosophy, 1900–1906.* Book One: A Documentary History. 1995 ISBN 0-7923-3231-8
169. R.S. Cohen, R. Hilpinen and R. Qiu (eds.): *Realism and Anti-Realism in the Philosophy of Science.* Beijing International Conference, 1992. 1996 ISBN 0-7923-3233-4
170. I. Kuçuradi and R.S. Cohen (eds.): *The Concept of Knowledge.* The Ankara Seminar. 1995 ISBN 0-7923-3241-5
171. M.A. Grodin (ed.): *Meta Medical Ethics*: The Philosophical Foundations of Bioethics. 1995 ISBN 0-7923-3344-6
172. S. Ramirez and R.S. Cohen (eds.): *Mexican Studies in the History and Philosophy of Science.* 1995 ISBN 0-7923-3462-0
173. C. Dilworth: *The Metaphysics of Science.* An Account of Modern Science in Terms of Principles, Laws and Theories. 1995 ISBN 0-7923-3693-3
174. J. Blackmore: *Ludwig Boltzmann, His Later Life and Philosophy, 1900–1906* Book Two: The Philosopher. 1995 ISBN 0-7923-3464-7
175. P. Damerow: *Abstraction and Representation.* Essays on the Cultural Evolution of Thinking. 1996 ISBN 0-7923-3816-2
176. G. Tarozzi (ed.): *Karl Popper, Philosopher of Science.* (in prep.)
177. M. Marion and R.S. Cohen (eds.): *Québec Studies in the Philosophy of Science.* Part I: Logic, Mathematics, Physics and History of Science. Essays in Honor of Hugues Leblanc. 1995 ISBN 0-7923-3559-7
178. M. Marion and R.S. Cohen (eds.): *Québec Studies in the Philosophy of Science.* Part II: Biology, Psychology, Cognitive Science and Economics. Essays in Honor of Hugues Leblanc. 1996 ISBN 0-7923-3560-0
    Set (177–178) ISBN 0-7923-3561-9
179. Fan Dainian and R.S. Cohen (eds.): *Chinese Studies in the History and Philosophy of Science and Technology.* 1996 ISBN 0-7923-3463-9
180. P. Forman and J.M. Sánchez-Ron (eds.): *National Military Establishments and the Advancement of Science and Technology.* Studies in 20th Century History. 1996 ISBN 0-7923-3541-4
181. E.J. Post: *Quantum Reprogramming.* Ensembles and Single Systems: A Two-Tier Approach to Quantum Mechanics. 1995 ISBN 0-7923-3565-1
182. A.I. Tauber (ed.): *The Elusive Synthesis: Aesthetics and Science.* 1996 ISBN 0-7923-3904-5
183. S. Sarkar (ed.): *The Philosophy and History of Molecular Biology: New Perspectives.* 1996 ISBN 0-7923-3947-9
184. J.T. Cushing, A. Fine and S. Goldstein (eds.): *Bohmian Mechanics and Quantum Theory: An Appraisal.* 1996 ISBN 0-7923-4028-0
185. K. Michalski: *Logic and Time.* An Essay on Husserl's Theory of Meaning. 1996 ISBN 0-7923-4082-5
186. G. Munévar (ed.): *Spanish Studies in the Philosophy of Science.* 1996 ISBN 0-7923-4147-3
187. G. Schubring (ed.): *Hermann Günther Graßmann (1809–1877): Visionary Mathematician, Scientist and Neohumanist Scholar.* Papers from a Sesquicentennial Conference. 1996 ISBN 0-7923-4261-5

Also of interest:

R.S. Cohen and M.W. Wartofsky (eds.): *A Portrait of Twenty-Five Years Boston Colloquia for the Philosophy of Science, 1960-1985.* 1985 ISBN Pb 90-277-1971-3

Previous volumes are still available.

KLUWER ACADEMIC PUBLISHERS – DORDRECHT / BOSTON / LONDON